Philosophy Begins in Wonder

Philosophy Begins in Wonder

An Introduction to Early Modern Philosophy, Theology and Science

Edited By
MICHAEL FUNK DECKARD and
PÉTER LOSONCZI

James Clarke & Co

James Clarke & Co
P.O. Box 60
Cambridge
CB1 2NT

www.jamesclarke.co
publishing@jamesclarke.co

ISBN: 978 0 227 17372 5

British Library Cataloguing in Publication Data
A Catalogue record is available from the British Library

First Published by James Clarke & Co, 2011

This edition is published by arrangement with
Pickwick Publications

To our teachers:

Bruce Ellis Benson

Herman De Dijn

Herman Parret

For the works of God are not like the tricks of jugglers,
or the pageants, that entertain princes,
where concealments is requisite to wonder;
but the knowledge of the works of God
proportions our admiration of them,
they participating and disclosing
so much of the inexhausted
perfections of their author,
that the further we contemplate them,
the more foot-steps and impressions
we discover of the perfections of their Creator;
and our utmost science can but give us
a juster veneration of his omniscience.

—Robert Boyle

Contents

List of Figures / *ix*

Preface—Michael Funk Deckard / *xi*

Abbreviations / *xv*

Introduction
—Péter Losonczi and Michael Funk Deckard / *xvii*

PART ONE: Historical, Scientific, and Religious Contexts

1 Wonder and Wondering in the Renaissance
—*Elisabeth Blum and Paul Richard Blum* / 1

2 Wonder, Magic, and Natural Philosophy: The Disenchantment Thesis Revisited—*Koen Vermeir* / 43

3 Religious Awe at the Origin of Eighteenth-Century Physico-Theology—*Miklós Vassányi* / 72

PART TWO: Wonder in Seventeenth-Century Europe

4 Descartes on the Excellent Use of *Admiration*
—*Dorottya Kaposi* / 107

5 Admiration, Fear, and Infinity in Pascal's Thinking
—*Tamás Pavlovits* / 119

6 On Thomas Hobbes's Concept of Wonder
—*Jianhong Chen* / 127

7 "Straight toward Heaven": Natural Theology and Politics in Milton's *Paradise Lost*—*Veronika Szántó* / 143

8 Malebranche on Restlessness and Curiosity
—*Roland Breeur* / 158

9 Wonder in the Age of the Saeculum: Spinoza
—*Gábor Boros* / 175

PART THREE: Wonder in Eighteenth-Century Europe

10 Berkeley's Wonderful Divine Language: Apology and Biblical Realism—*Péter Losonczi* / 190

11 "Of Curiosity, or the Love of Truth": David Hume on Wonder in *A Treatise of Human Nature* —*Michael Funk Deckard* / 211

12 A Risk of Testimony: Astonishment and the Sublime —*Baldine Saint Girons* / 234

13 Two Sources of Wonder in Early Modern Judaism —*Roberta Sabbath* / 261

14 Kant and the End of Wonder—*Patrick Frierson* / 285

15 Ways of Wondering: Beyond the Barbarism of Reflection —*William Desmond* / 310

List of Contributors / 349

Index of Authors / 353

Index of Subjects / 359

Figures

Figure 1	Medieval Bestiary, British Library, Harley MS 4751, Folio 36r / 47
Figure 2	Zoroaster, in a manuscript from 1425. Add. MS 39844, f.51 / 53
Figure 3	Diviners, in Pierre Le Brun, *Histoire Critique des Pratiques Superstitieuses* (1733), book 3, Plate A / 64
Figure 4	Metamorphosis machine, in Athanasius Kircher, *Ars Magna Lucis et Umbrae* (Amsterdam, 1671), 783 / 68
Figure 5	Georges de La Tour, *Job and his wife* (ca. 1650), Épinal, Musée Départemental des Vosges / 240
Figure 6	Michelangelo, *Pietà de Palestrina* (1553–1555), Florence, Galleria dell 'Accademia / 245
Figure 7	Michelangelo, *Pietà* (1497–1500), Saint Peter's Cathedral in Rome / 247
Figure 8	Piranesi, *Carceri d'invenzione* (1760–1761), board XV / 258

Preface

Péter Losonczi and Michael Funk Deckard

The first time I was taught David Hume's example of the billiard ball, I was baffled. How can it be that all my life I thought that causation was something so simple? When one pool ball hits another, of course it causes the other to move. This seems like just plain common sense. As an eighteen year old in my first philosophy class, I went to my dorm room thinking about causation and about how Hume explained that there is no "necessary connection" between one ball moving to another and causing the other ball to move: it is just a matter of custom and habit that we think that it does. Such causation cannot be proven or follow from logical induction. This was exasperating but amazing at the same time. Much later I learned that Hume was not the first to come up with this criticism of causation in the 1730s. Malebranche had used it earlier in his *Search after Truth* (*De la recherche de la vérité*, 1674–1675) to support the view of God as the "occasional" cause of every instance of cause and effect: For even a billiard ball to cause another to move, God had to intervene. One does not need to go into all of the details of the transformation between Malebranche's metaphysical view of the world and Hume's skeptical one, Hume could not have thought what he did without Malebranche's philosophy.

This one example from the history of philosophy invisibly motivated me to study philosophy further: what causes something else to move? For a teacher the question might be, what story or philosophical argument might impel a student to be moved to desire further study? I could tell many friends Hume's story of the billiard ball, and these friends would not necessarily be moved as I was to study philosophy, to pursue the quest for knowledge more deeply. Moreover, this story of causation led me to ask, why is there something rather than nothing? Why is it that we are here? Why is it that there is anything at all, and how do we account for this "anything"?

These questions form the background for the following examination on a silent and invisible occurrence in early modern philosophy: the move from wonder (*thaumadzein* in Greek or *admiration* in French) to curiosity. Simply speaking, every person in every society wonders about his or her existence. Even if this becomes a more scientific curiosity, the wonder need not disappear. It is only reignited in a different sense. The goal of this book is to show what this means in early modern Europe. Most of the authors included in this volume are early in their academic careers, as are the editors, who hoped this relative nearness to the initial influence of wonder might provide a novel introductory reflection upon the thinking of early modernity. In place of resorting to the antiquated method of teaching the history of philosophy by establishing Continental rationalism (Descartes, Spinoza, Leibniz) versus British empiricism (Locke, Berkeley, Hume), the editors of this volume think that looking at a broader array of thinkers is more fruitful. We encourage teachers of the history of ideas or early modern philosophy to take thinkers like Malebranche, the Cambridge Platonists, Physico-Theology, or Giambattista Vico into account—none of whom fit neatly into Frederick Copleston's (1907–1994) reading of early modernity.[1] The book in hand attempts to read the history of early modern ideas in light of the notion of wonder, which is missing from the current literature. For example, in the recent sixteen-hundred-page *Cambridge History of Seventeenth-Century Philosophy*, wonder is mentioned merely on two occasions and only in relation to Descartes.[2] We hope this work begins to fill that lacuna. Although this book is by no means exhaustive of all of the early modern philosophers, theologians, or scientists on the subject of wonder, nevertheless as the subtitle indicates, it is meant as an introduction, not only to present early modern philosophy, but ideally to

1. See F. C. Copleston, *A History of Philosophy*, 8 vols (Burns, Oates & Washbourne, 1946–1966), esp. vols. 4–6. Volumes 9–11 appeared later. This work is still the leading single-authored history of philosophy.

2. See *The Cambridge History of Seventeenth-Century Philosophy*, eds. D. Garber and M. Ayers, 2 vols. (Cambridge: Cambridge University Press, 1998), 2:933–34. Wonder is also overlooked in *The Cambridge History of Eighteenth-Century Philosophy*, ed. K. Haakonssen, 2 vols. (Cambridge: Cambridge University Press, 2006) and in *The Cambridge Companion to Early Modern Philosophy*, ed. D. Rutherford (Cambridge: Cambridge University Press, 2006).

initiate the very experience of wonder within our readers. Our aim has thus been to inspire readers with awe as well as to enable them to read historical texts in a new way.[3]

Besides following the Coplestonian reading of the seventeenth and eighteenth centuries, most scholars of early modern thought interpret this time period as one when God was considered an afterthought, or even superfluous, to the thought systems of, for example, Descartes, Spinoza, Berkeley, and the like. In place of this misreading of early modernity, we seek to show in this book that philosophy is still intimately connected to theology. Against scholars such as Jonathan Israel, who points to the secular origins of early modern thought, we think that for both Hobbes and Spinoza, theology and religion were at the very center of their work, even if theirs were fairly unorthodox theologies.[4] At the same time, a great number of thinkers were advancing and arguing for new philosophical theologies, such as "physico-theology." Many know that philosophy and theology were closely related in medieval thought, but this was also the case for most of early modernity, which is less acknowledged. During this latter period, however, another system of thinking and culture is added: science. In place of the centrality of philosophy and theology for medieval thought, philosophy, theology, *and* science become intertwined.[5] Belief in the wonder and works of God, evident in the quotation of Robert Boyle above, need not be taken as incompatible with philosophy and science. Unlike many thinkers and scientists of

3. One of the intended uses of this book is as an undergraduate textbook. For this reason, we point out that some chapters are more accessible than others for this audience. Most, especially those on single philosophers (e.g., Descartes, Pascal, Malebranche, Hobbes, Hume, Kant, and the like), are directed toward an undergraduate audience, though they may also be interesting to more-advanced scholars. The chapters considering various thinkers (e.g., "Religious Awe at the Origin of Natural Science in Eighteenth-Century Physico-Theology," "A Risk of Testimony: Astonishment and the Sublime," and "Ways of Wondering: Beyond the Barbarism of Reflection") are less accessible to undergraduates.

4. Cf. Jonathan Israel, *Radical Enlightenment: Philosophy and the Making of Modernity, 1650–1750* (Oxford: Oxford University Press, 2001); Israel, *Enlightenment Contested: Philosophy, Modernity, and the Emancipation of Man 1670–1752* (Oxford: Oxford University Press, 2006).

5. For this discussion, see David C. Lindberg, "Medieval Science and Religion," in *Science and Religion: A Historical Introduction*, edited by Gary B. Ferngren (Baltimore: Johns Hopkins University Press, 2002) 57–72. For the birth of science as a system of thinking and culture, see Stephen Gaukroger, *The Emergence of a Scientific Culture: Science and the Shaping of Modernity*, 1210–1685 (Oxford: Clarendon, 2006).

today, the early modern writers discussed in this book did not take for granted this interaction between philosophy, theology, and science.

Thanks to those whose incredible work on this volume made it much better and more aesthetically pleasing: Matt Avery, Robby Duncan, Jeremy Funk, Patrick Harrison, and to Charlie, Jim, and Raydeen.

Abbreviations

AT	*Œuvres [de Descartes]*
CSM	*The Philosophical Writings of Descartes*, vol. 1 (Stolthoff and Murdoch, eds.)

Introduction

In his *Observations*, a part of Descartes' fragmented early notebook later published under the title *Cogitationes Privatae*, he writes as follows: "In the year 1620 I began to understand the fundamental principles of a wonderful discovery."[1] This sentence can be interpreted as a symbolic expression of the essence of that attitudinal change which occurred in the European history of ideas during the period covered by the essays of the present volume. In the course of this process a radically new way of approaching reality evolved, which slowly became dominant first in natural philosophy (i.e., science) and then in other branches of philosophy, as well as in ethics and the philosophy of religion. Its principles essentially contributed to the formation of what Charles Taylor recently characterized as the pre-ontology of modern secular culture.[2] The consequences of this shift are very complex and our contemporary debates are in many respects directly related to the challenges it created. The bioethical and political disputes over the problem of human engineering and the related dilemmas concerning human self-instrumentalization are new waves of this movement that originated in that early modern paradigm shift.[3] Jürgen Habermas draws a parallel between the debates over genetic engineering and the cultural turns that evolved from the Copernican and the Darwinian paradigm shifts.[4]

Evidently the detailed analysis of the processes that constitute the historical bridge between these questions, as well as the intellectual

1. *The Philosophical Writings of Descartes*, 3 vols., eds. R. Stoothoff and D. Murdoch (Cambridge: Cambridge University Press, 1985–1991) [henceforth abbreviated CSM] 1:3; *OEuvres [de Descartes]*, 11 vols., eds. Ch. Adam and P. Tannery (Paris: Cerf 1897–1909[1], Vrin, 1964–1974[2], Vrin, 1996[3]) [henceforth abbreviated AT] 10:216.

2. Charles Taylor, *A Secular Age* (Cambridge: Harvard University Press, 2007) 3.

3. Cf. Jürgen Habermas, "The Debate on the Ethical Self-Understanding of the Species," in Habermas, *The Future of Human Nature* (Cambridge: Polity, 2003) 16–100, 53.

4. Ibid., 54.

chains that link them together, would surpass the limits of the present endeavor. However, we hope that the essays of this book may illuminate important aspects of the early history of the developments of modernity. Originally we were surprised to realize that the specific problem at the core of this volume (namely, the systematic discussion of the problem of wonder in early modern thought) is relatively underrepresented in the scholarship. However, in the last several years, discussion on the theme has grown.[5] We think that the relevance of the question and its intellectually inspiring nature can be shown not merely by its historical aspects but also by its implicit but fundamental importance for the contemporary context.

We must mention here a general characteristic of this trend: what we define as *naturalization*, or the continuously manifest tendency to rationalize reality. Simultaneously, however, we must refer to corresponding revisions of naturalization's cognitive capacities (or, in Spinoza's terminology, the improvement of human understanding) and delimitations of its consequences. The representation of the question of wonder has an intrinsic relation to these occurrences. According to Leszek Kolakowski, the early modern paradigm shift was brought about by a cultural moment that preceded scientific motifs. In his view, this shift was due to the preference for *libido dominandi* over the search for meaning in the universe and in our lives.[6] In his view, striving for dominance over nature was the consequence of this unbalanced order of preferences, and this caused a decline in religion. In a sense; then, naturalization replaces religion. Correspondingly, the dominance of a specific form of rationality and the complementary aversions toward the

5. See, for example, Lorraine Daston and Katherine Park, *Wonders and the Order of Nature 1150–1750* (New York: Zone Books, 1998); Philip Fisher, *Wonder, the Rainbow, and the Aesthetics of Rare Experiences* (Cambridge: Harvard University Press, 1998); Mary Baine Campbell, *Wonder and Science: Imagining Worlds in Early Modern Europe* (Ithaca: Cornell University Press, 1999); Laurence Renault, *Descartes ou la félicité volontaire. L'idéal aristotélicien de la sagesse et la réforme de l'admiration* (Paris: PUF, 2000); Neil Kenny, *The Uses of Curiosity in Early Modern France and Germany* (Oxford: Oxford University Press, 2004); *Curiosity and Wonder from the Renaissance and the Enlightenment*, eds. R. J. W. Evans and A. Marr (Aldershot: Ashgate, 2006); Jonathan P. A. Snell, *Rhetoric and Wonder in English Travel Writing* (Aldershot: Ashgate, 2006).

6. Leszek Kolakowski, *Modernity on Endless Trial* (Chicago: University of Chicago Press, 1990) 98. See also *Curiosité et Libido sciendi de la Renaissance aux Lumières*, 2 vols., eds. Nicole Jacques-Chaquin and Sophie Houdard (Paris: Ophrys, 1998).

symbolic aspects of human knowledge were really constituents of these disenchanting tendencies.

In his cardinal work, *The Legitimacy of the Modern Age*, Hans Blumenberg (1920–1996) elaborates another way of interpreting the situation,[7] in a thorough analysis that we may characterize as an intellectual-historical survey of the problem of the "knowing how" of approaching the world, from which follows a kind of ethics of knowing. This essentially belongs to the formation of the intellectual climate of the early Middle Ages and Scholasticism. In Augustine—not independently from his anti-Manichean strife—the ancient pagan desire for knowledge becomes theologically embroiled and subsumed into a general religio-ethical imperative. The theological themes of sin and salvation became inscribed into the discourse concerning the use of human intellectual abilities, and the topic of *curiositas* became an element of the Augustinian paradigm determined by the differentiation between *frui* and *uti*.[8] According to Hans Blumenberg's working hypothesis, the question of secularization and the accompanying debates on the legitimacy of modernity are to be addressed in light of the dilemma over the transformation of these teleologically saturated religio-ethical and theological perspectives brought about as a decline or a fulfilment of the original vision. The present volume does not aim to decide these metathematic quandaries. However, we think that in the time of "embryo politics," it is worth reconsidering those processes that resulted in the advent of a posthuman future and a cybernetic mythology.

The ambiguities about the epistemic and intellectual status of wonder were themselves generated by these processes. Part one of this book—on the historical, scientific, and religious contexts—explores what the famous Dutch historian Johan Huizinga called *Herfsttij der Middeleeuwen* (the "autumn of the Middle Ages").[9] As Elisabeth Blum and Paul Richard Blum demonstrate in chapter 1, the situation was multilayered, and the religious factor played a decisive role in the formation of the precon-

7. Hans Blumenberg, *The Legitimacy of the Modern Age*, trans. Robert M. Wallace (Cambridge: MIT Press, 1983). This book was originally published as *Die Legitimität der Neuzeit* in 1966 and was revised in 1976.

8. Ibid., 313.

9. See Johan Huizinga, *The Autumn of the Middle Ages*, trans. Rodney J. Payton and Ulrich Mammitzsch (Chicago: University of Chicago Press, 1996). Huizinga's original was published in 1921.

ditions of the modern approach.[10] In the examples of Marsilio Ficino, Giovanni Pico della Mirandola, Pietro Pomponazzi, Henry Cornelius Agrippa of Nettesheim, Giordano Bruno, Francesco Patrizi, Tommaso Campanella, and Francisco Suárez, the authors show how different, often conflicting spiritual, religious, and philosophical tendencies shaped these conditions, and this combined background makes a nuanced account necessary for understanding the problem of wonder. As Koen Vermeir also points out in chapter 2, there were different trends inasmuch as the discourse on human acquaintance with the different layers of reality is concerned, and the formulating of specific attitudes implies not merely a decline in the discussion of the question of wonder, but also in its reformulation and dissemination, so to speak. Vermeir describes this movement as one between enchantment and disenchantment, following Max Weber's "disenchantment of the world." In his essay we are presented with much more than a philosophical or scientific overview of the question, as he develops what we might call a cultural history of wonder, describing the divergent discussions and representations of wondrous phenomena. This chapter is an important piece for the theme of the book not merely because of its detailed analysis but also because of its remarks on the ambiguities of the topic. The consideration of these ambiguities is a crucial task for all, since the twenty-first century brought a clear—no less ambiguous—reenchantment after disenchantment.

Whereas the first two chapters look closely at the phenomena of magic mostly from the fifteenth through the seventeenth centuries (delving even as late as the eighteenth century), "physico-theology" is a leading theory of the universe in which a true scientific religious enchantment reigns, as the third chapter shows. Here Miklós Vassányi specifically discusses a large number of now-obscure thinkers in this regard, all of whom he interprets within the history of the development of "physico-theology": Walter Charleton, Matthew Barker, John Ray, archbishop François Fénelon, Bernard Nieuwentyt, William Derham, Abbé Pluche, Christian Wolff, Johann Albert Fabricius, Friedrich Christian Lesser, Pierre-Louis Moreau de Maupertuis, Sebastian Friedrich Trescho, Georges-Louis Leclerc, Comte de Buffon, Jacques-Henri Bernardin de Saint-Pierre, Johann Gottfried Herder, and William Paley.

10. See also Lorraine Daston, "Marvelous Facts and Miraculous Evidence in Early Modern Europe," *Critical Inquiry* 18 (1991) 93–124; Gaukroger, *The Emergence of a Scientific Culture*, and Taylor, *A Secular Age*.

Even so, the Cartesian passage we have quoted ("I began to understand the fundamental principles of a wonderful discovery"[11]) remains a paradigmatic document of the naturalizing and rationalizing tendencies of the seventeenth century. It is important to see that Descartes's discovery does not primarily concern the wonders of the universe, but the principles of science by which the universe can be understood according to a marvelous and, at the same time, disenchanting method. The rationally accessible principles of the new science—rather than the marvels of the created universe—prove to be worthy of wonder, and this implies an essential alteration not merely in the object of this specific passion but also in the understanding of the order of things. Following shortly on the death of the early modern Scholastic Francisco Suárez in 1617, many seemingly new philosophical methods arose, a number of which are shown in part 2 of this volume: "Wonder in Seventeenth-Century Europe." Descartes's three dreams are symbolic in this respect.

As Dorottya Kaposi argues in chapter 4, the Cartesian *use* of wonder is a very consciously developed endeavor in which the dominance of "disenchanted" knowledge, in the form of *mathesis universalis*, is palpable and became a paradigm for approaching the question of wonder and related topics in modernity. From the *Meditations* (1637) to the *Passions* (1649), Descartes develops wonder as a background to his thinking such that he describes it as the "first of all the passions." As Tamás Pavlovits demonstrates in the fifth chapter, Pascal takes a very specific path when he diverts from both the traditional Aristotelian and the newly invented Cartesian science. He explores the wonderful nature of reality by realizing the limitedness of human reason and by articulating the idea of a general anthropological incompleteness that constitutes the core of Pascal's philosophical attitude. All this, it is well known, generates a theological strategy that counters the dominant rationalizing and naturalizing tendencies of the epoch. Thomas Hobbes, as Jianhong Chen shows in chapter 6, also formulates wonder (or wonders) between what we might call today disenchantment and enchantment. Chen brings into the debate the revolutionary work of Hans Blumenberg, a recurring voice in the discussion of wonder, as well as Martin Heidegger's thought. Both thinkers have changed the way we see modern philosophy. Instead of neglecting being, wonder may very well reenchant it by revealing its scientific and political undersides.

11. CSM 1:3.

Veronika Szántó takes a seemingly different path when she analyzes Milton's *Paradise Lost*. Her essay convincingly reveals that Milton's masterpiece implies a specific natural theology together with its political consequences. In chapter 7 Szántó argues—and her case is embedded in the then-contemporary natural theological context—that wonder functions as a decisive element of this Miltonian (political)theological strategy. For Milton, every dimension of existence may "reveal," or point toward, its transcendent source, and wonder serves as the human capacity that can render the human being "sensitive" to this excess. However, as Szántó notes, according to Milton, wonder always touches upon idolatry; and this element of the Miltonian theological thesis carries fundamental political morals about the status of human political power and its legitimacy. We might also say that it touches upon the very essence of the early modern paradigm shift. As we have mentioned, in Descartes it is not only the proper use of *good sense* but also the proper and excellent use of wonder that counts as the neuralgic point of the early modern period. The different versions of these *dual conceptions* of wondering and knowing determine the philosophical and scientific, as well as the theological, character of the respective theories. In Milton the proper use of wonder points to the difference between idolatry and authentic worship.

In chapter 8, on Malebranche, Roland Breeur also discusses the theological adaptation of this question. In Malebranche, the aforementioned Cartesian dilemma together with the occasionalistic treatment of the body-mind relationship are introduced into a theological context through the discussion of the ambiguous character of curiosity. According to this French philosopher, curiosity has to do with the influence original sin exercises over our nature. Malebranche in fact thematizes the matter of orthodoxy in a direct relation with the "use of curiosity." Curiosity belongs to our postlapsarian character, and even its seemingly positive manifestations (in so far as it is a driving force) are the restlessness of our will and our infinite desire for novelty and knowledge. The root of all these is but our broken nature. From this moral theological ambiguity of curiosity follows its generally problematic nature for Malebranche. Thus Malebranche's diagnosis anticipates Kolakowski's above-mentioned characterization of the modern paradigm shift as the rise of *libido dominandi*. In the final chapter in part two (chapter 9), Gábor Boros provides a thorough analysis of a most influential variety of the disenchanting paradigm by revealing its Spinozist form that, despite being derived from Spinoza, can be interpreted within the general context of the "secularization" of religiously relevant forms of emotions. This Spinozist revision of the question of wonder as much as the Cartesian one represents a radical shift and demonstrates the soundness of Vermeir's thesis in chapter 2.

Whereas method dominated seventeenth-century thinking, eighteenth-century thought could be described as a resistance to method, with the possible exception of the interpretation of physico-theology mentioned above. We hope this book will dig deeper into the sources of Enlightenment wonder than any previously published source, challenging the reading of Lorraine Daston and Katherine Park: that wonder died out in the eighteenth century.[12] In the first essay of part three, Péter Losonczi asks that question of George Berkeley and finds that wonder underlies the metaphysical and theological work of the Irish thinker. Berkeley not only attributes a crucial role to wonder in his metaphysics, but he also subscribes to the "proper-use-of-wonder" discourse—whose persistence during the early modern period this book brings into focus for the related scholarship. Berkeley entertains wonder as the instrument of authentic religiosity, which does not seek *l'art pour l'art* exercises but attends to the divine language communicated through the ordinary. We can say that besides the scientific-philosophical pillar of the discourse of wonder in early modernity, an equally important theological layer can be disclosed without which neither the complexities of the issue nor the real relevance of the former aspect can be understood. Specifically pointing out a lacuna in Daston and Park's exceptional *Wonders and the Order of Nature*, Michael Funk Deckard next looks closely at David Hume to find an underlying strand of wonder in his epistemology and ethics. On his reading, offered in chapter 11, even the ability to wake up in the morning, whether for the philosopher, the scientist, or the religious believer, must be attended with some reason for caring; this reason is akin to wonder. Even as late as his *Dissertation on the Passions* (1757), a reworking of the *Treatise*, book 2, Hume repeated an earlier claim regarding the nature of custom: "When the soul applies itself to the performance of any action, or the conception of any object, to which it is not accustomed, there is a certain unpliableness in the faculties, and a difficulty of the spirits moving in their new direction. As this difficulty

12. See their *Wonders and the Order of Nature*, chap. 6.

excites the spirits, it is the source of wonder, suprize, and of all the emotions, which arise from novelty."[13]

This could have just as well come from Descartes, and yet much had changed during the interim. Providing a short history of the source of this surprise, Baldine Saint Girons uses Shaftesbury, Vico, Montesquieu, and Burke to show its power related to the sublime in eighteenth-century thought. Part of her book *Fiat Lux: une philosophie du sublime* (available only in French) this complex article (here translated as chapter 12) is embedded in her own Freudian theory of the sublime and sublimation. In this light, she ties Burke's use of Homer to Freud's *Beyond the Pleasure Principle* to show one of the central problems to be the line (if there is one) between pleasure and pain, between pleasing astonishment and terrible sublimity. It is this distinction that is a risk of testimony. Another short history comes from a Jewish perspective, which is equally inspiring and dreadful. In chapter 13, Roberta Sabbath tells a story recalling the medieval and modern trauma of the Hebraic peoples. Her story culminates in the eighteenth century, with Moses Mendelssohn's attempting to forge a particular view of wonder unlike that of his British and German peers yet still recognizable to the Enlightenment project. Following this most meaningful story is the Enlightenment thinker par excellence: Immanuel Kant. In a remarkably accessible encapsulation of the three critiques, Patrick Frierson points to Kant's wondrous system embodied by the awe-inspiring phrase: "the starry skies above and the moral law within." Readers will find that chapter 14 facilitates a better understanding of Kant's intricate system.

The disenchanted world came about through the clogging of the porosity that William Desmond examines in the concluding chapter. *Porosity* is also a central category in Charles Taylor's analysis mentioned earlier. Desmond's notion of porosity opens up an ontological-metaphysical understanding of the sources of wonder and helps us see the differences among and relations between astonishment, perplexity, and curiosity. In its ontological-metaphysical stress, Desmond's sense of porosity is not like Taylor's use of *porosity* to refer historically to forms of human life not touched by modern disenchantment. Because Desmond's sense of porosity is ontologically constitutive, not just historically rela-

13. David Hume, *A Dissertation on the Passions; The Natural History of Religion: A Critical Edition*, ed. Tom L. Beauchamp (Oxford: Oxford University Press, 2007) 28. It is worth noting that this formulation is the same as in Hume's *Treatise* 2.3.5.2.

tive, the possibility of wonder beyond the barbarism of reflection is never closed off. Desmond begins with Giambattista Vico's thought and moves quickly through many thinkers (including Aristotle, Schopenhauer, Nietzsche, Comte, and others) in a way that explores what was lost in the destruction of our capability to be astonished and perplexed. However, he shows that even the most extreme forms of scientism cannot dispense with a sort of curiosity derivable from wonder as astonishment. He provides the philosophical resources to renew proper mindfulness about wonder that at the same time introduces a postsecular (or posthumous) deepening of the problem with which we began. The beginning and end points merge so that the philosophical, the theological, and the scientific can engender a nexus without absorbing one another.

PART ONE

Historical, Scientific, and Religious Contexts

1

Wonder and Wondering in the Renaissance

Elisabeth Blum and Paul Richard Blum

WONDERING WAS IDENTIFIED—HOWEVER IRONICALLY—TO be the moving force of philosophy by Plato in his *Theaetetus* (155d). So it doesn't come as a surprise, if wondering, the possibility of an incapacity to resolve mysteries, admiration as the motivation for philosophical quest, the secrecy of the divine, and the like are treated in early modern times prevalently by Renaissance Platonists. Heirs to late ancient and medieval Christianity, they began to stress the epistemological or ontological status of miracles, thus exploring the cognitive side of amazement and the metaphysical side of any sort of spiritual intervention.

Clearly, miracles must have an ontological status that describes them as extraordinary occurrences within or without the order of nature; on the other hand, the bewilderment that necessarily accompanies the assessment of miracles, and anything else that goes beyond the normal run of things, has the epistemological status of wonder. Wonder (in Latin: *mirabilia*) is indeed the catchword used by Marsilio Ficino (1433–1499) to describe this twofold position of miracles. In his commentary on the fifth book of Plato's *Laws*, before entering the matter itself, he felt compelled to meditate about wonder. This fifth book is known to be the introduction to the legislation proper suggested by Plato, but Ficino here begins with some ethical norms concerning human perfection and imperfection. It is at this point that in his summary Ficino explains the meaning of wonders.

Miracles, he states, occur daily, for instance in magnetism, and generally in all nature as well as in artifacts. "We do not know the causes, and nonetheless for ignorance of the causes do we not deny the effects that we see."[1] And again, he insists that "stunning" achievements are made by the soul and mind that cannot be denied, although we do not know how they came about. So far he does not venture to assign any status to miracles; but immediately after this, he draws an analogy typical of his cosmology: fire, the more it likens itself to the heavens the more wonderfully it works; namely, it draws into its own form what other elements can never do. Also it never undergoes mixture, and when it seems to be divided, it actually divides other things, and it extends its own effect instantaneously as far as light reaches. Obviously the light of the sky is a case in point, which leads to the simile of a mirror: as we see the light in a mirror, so we speculate in the soul the divine things. The analogy is this: in the same way as fire is more efficient than other elements, and heaven is more efficient than fire, in the same way more efficient than the heavens are the spirits that enliven or move the heavens. Ficino concludes that superior spirits operate in our spirits, which are cognate to them, as sight does in a mirror; and in so doing we shape our spirits, i.e., our soul, so that in the end our soul operates like a celestial spirit.

We need not follow up further classifications of the workings of spirit and speculation, because Ficino's approach is obvious: the epistemic riddle of miracles needs to be transformed into a cosmological analogy between the human soul and divine spirits, so that wonder about the wonderful turns into the self-referentiality of the cognizing soul. Human souls are open to the numinous because they are equally intellectual natures, and thus the numinous is rooted in doubtless belief; by means of the will, our souls transform themselves into it, while memory contains hope. As long as the soul is embodied, she is attuned like the strings of a lyre.[2] This is, according to Ficino, the origin of prodigies, dreams, soothsaying, and oracles.

In this context Ficino refers to his *Platonic Theology* (1482). Book 13 is dedicated to the dominance of the soul over the body, of which

1. Marsilio Ficino, *Opera* (Basel: Henricpetri, 1576; reprint, Turin: Bottega d'Erasmo, 1983), 2:1501: "Mirabilia quotidie fieri videmus a magnete in ferrum . . . a tota natura vel in motu coeli, vel in artificiosa, quae gignuntur, formatione. Causas quidem ignoramus, neque tamen ob causarum ignorationem negamus [e]ffectus fieri, quos videmus." All remaining quotations of this section are from the same page. Cf. Plato, *Opera*, translated by Marsilio Ficino (Basel: Froben, 1539), 799.

2. Ibid.: "a fidibus lyrae in fides similiter temperatas"—a wordplay with *fides* (gen. *fidis*): "string" (hence the word *fiddle*) and also the constellation of Lyre, and *fides* (gen. *fidei*): "faith."

chapter 2 debates reason; and within this chapter, Ficino analyzes seven ways of "emptiness" or "release" (*vacatio*) as preconditions to open the mind for the influence of higher spirits. Wonder is one of the seven ways: The ancient priests "thought about the divinity's majesty with such overwhelming awe and veneration that their senses, phantasy, and reason instantly intermitted their activities. At that time the mind, exposed as it was to the divinity alone whether god or demon, perceived the divine commands, and especially those which concerned the affairs they had just been considering."[3] The question is, does emptying open the mind for the divine commands, or is wonder equivalent with it as a perception of the numinous? It is not excluded that Ficino is trying to describe the phenomenology of a circular movement of the soul, in so far as the divine majesty exacts awe, which presupposes its presence before its effect. However, the commanding presence cannot be perceived if the human mind is distracted by sensual, imaginative, or discursive thinking.

In Ficino's epistemology, senses and fantasy are bodily qualities of the soul that may be subject to corporeal exigencies, which therefore must and can be suppressed in favor of the higher powers of the soul. It should be noted that release and emptiness (*vacatio*; in German: *Gelassenheit*) are basic terms of mysticism. It seems, however, that Ficino is not addressing mystical experience, something that cannot be required and is therefore not apt to form a phenomenological argument for the natural predominance of the spiritual over the corporeal. He is rather referring to a power of the soul that can be enacted and performed by means of reasoning and will, as mentioned in the commentary on *Laws*. In the *Platonic Theology* we learn that the soul by referring to itself is able to vacate itself of corporeal contingencies, prompted by, or opening for, the wonderful, so that this self-opening of the soul toward the influx of pure spirits is equivalent to awe, which, as an act, can be called wonder. The commentary on Plato's *Laws* goes one step further and declares this specific capacity of the soul to be grounded in the ontological familiarity of the soul with transcendent objects.[4]

3. Ficino, *Platonic Theology*, book 13, chapter 2, number 36, volume 4, 167.

4. Pertaining to this, ignorance is not an issue in Ficino's thought regarding admiration. Rather, admiration is the cognitive side of the miracle that cannot possibly lack a metaphysical foundation. It is the distinctive approach of certain kinds of empiricism to reproach ignorance as a failure to the effect that miracles are nothing but nonmiracles in the state of suspended explanation.

Does it make sense at all to "inquire into the reasons behind miracles"? Ficino himself asked that question in the fourth chapter of book 13, which is dedicated to miracles as evidence for the governance of reason over matter.[5] Here he recapitulates his spiritual cosmology according to which the soul is "a third essence under the angel but above the whole matter of the world, an essence which is formed by the divinity but itself forms matter, and which receives spiritual forms from the divinity (*a numine*) but gives corporeal forms to matter."[6] Therefore, the soul does not admire miracles but performs them: "not only in the forming and shaping matter through the rational principle of art . . . does the human mind appropriate for itself the divine right (*ius divinum*); it also does so through its sovereignty and transmuting the species of things. The resulting work is called a miracle, not because it is the supernatural work of our soul when it becomes God's instrument, but because it induces wonder (*admirationem*), being a mighty event and one that happens rarely."[7] There seems to be nothing to wonder at about supernatural works, unless they are huge and rare, though not impossible. Therefore the essential question is much simpler: is it possible that spirits like the soul can have an impact on matter? Of course they do; this can be seen in the "complexion of the four humors and our face," which are obviously shaped by the soul. But even more, "the soul raises a heavy body upwards, contrary to its nature, and prevails in turn over the four opposing elements; and no one is amazed at this miracle."[8] If this is the case, according to Ficino, then the great, miraculous works are only a problem of the purification of the soul from its downward-weighing distractions, so that "it is possible for man's soul to be turned at times towards mind, its head, by the total concentration of its reason, just as it is turned at other times towards the phantasy . . . and towards reason."[9] We have already learned that *vacatio* is the right means to achieve this, and we need not repeat the plethora of ancient authorities ranging from the Gospel of St. John to Zoroaster and Hermes Trismegistus. It suffices to say that the cooperation of body and soul, and specifically of the mind in the human body is enough of a wonder.[10] But is it also natural?

5. See Ficino, *Platonic Theology*, book 13, ch. 4, nr. 2, 185.

6. Ibid.

7. Ibid., nr. 1, 183.

8. Ibid., nr. 6, 191.

9. Ibid., nr. 12, 201.

10. Ibid., nr. 17, 209.

This question is addressed in the final chapter of book 13 of Ficino's *Platonic Theology*. Granted that spiritual and material substances interact on all levels of the world like a "light trapped inside a heavy lantern" that nevertheless can illuminate long distances, the question remains, "By working miracles, however, isn't the rational human soul confounding the world's order?"[11] Not so in Platonic terms (*ut platonice loquar*). Divine action, angelic influences, demonic operations, and generally all kinds of interaction between spirits and bodies—that's what Platonism is about, and certainly Ficino's Platonic theology. "Miracles have been prescribed in the universal law of things and in the everlasting series of causes contributing to the same. Contributing to it is the providential care of the minds under God down to our own mind, minds singing in continual harmony."[12] We should observe at this point that Ficino is disingenuous, in a sense. On the one hand, he declares the interaction of the material and the immaterial to be natural; on the other hand, he keeps calling it a miracle. In the eyes of his readers the human soul is a miracle, and at the same time the Neoplatonist strives to explain it away as the normal run of things. Therefore for him, miracles are not constituted by lack of explanation but by overpowering cosmology.

Eventually he agrees that this kind of miracle is natural for the rational soul and at the same time nonnatural: "natural, in so far as the soul has recourse to the mind and is made the attendant and instrument of God on high; non-natural, in so far as it serves the needs of one body alone . . . and neither unfolds the full extent of its own power nor becomes a participant in the divine power."[13] Miracles, we may conclude, are nature upside down. It's natural to be an instrument of God whereas to defect from God, that's what is unnatural. What remains to admire in Ficino is the clarity, with which he forestalls empiricist skepticism.

• • •

For Giovanni Pico della Mirandola (1469–1533), in his famous *Oration on the Dignity of Man* (originally intended as an introductory speech to his public disputation of nine hundred philosophical and theological theses, published in 1486), the greatest wonder in the world and the foremost reason for amazement and admiration is man himself: "I have

11. Ibid., chapter 5, nr. 1, 209.

12. Ibid., nr. 4, 213.

13. Ibid. nr. 7, 217.

read . . . how Abdala the Saracen, when questioned as to what on this stage of the world . . . could be seen most worthy of wonder, replied that there was nothing to be seen more wonderful than man. In agreement with this opinion is the saying of Mercury [Hermes Trismegistus]: 'A great miracle, O Asclepius, is man.'"[14] However, according to Pico, this is not due, as many philosophers would have it, to man's naturally assigned place in the universe or to a central position between things temporal and eternal, between corporeal and spiritual creatures:

> I was not satisfied by the many grounds offered by many for the excellence of human nature: that man is the mediator between creatures, the intimate servant of the higher and the king of the lower beings; that by the acuteness of his senses, by the discernment of his reason, and by the light of his intelligence he is the interpreter of nature; that being the interval between fixed eternity and fleeting time, and (as the Persians say) the bond, nay, rather, the marriage song of the world, he is, on David's testimony, but little less than the angels. These are great issues, but not the principal grounds, on which man may rightfully claim the privilege of the highest admiration for himself. For why should we not admire even more the angels and the blessed choirs of heaven? But it seems to me that I have finally understood why man is the most fortunate life form, and therefore worthy of all admiration, and what is the core of this *human condition* allotted to him in the universal chain of being, to be envied not only by the beasts, but by the stars and by the minds beyond this world.[15]

The really astounding fact (*res supra fidem et mira*) about the human condition is its lack of conditioning: "We have given thee neither a secure abode, nor a proper aspect, nor any special endowment, O Adam . . ."[16] No place in the chain of being is assigned to the ultimate creature, a divine afterthought after the completion of the universe, a being meant to oversee, and thus to appreciate, the perfection of God's masterwork: "But there was neither any archetype from which to fashion a new offspring, nor any treasure to grant the newborn son as his heritage, nor a

14. Giovanni Pico della Mirandola, *Oratio de hominis dignitate*, in Pico, *De hominis dignitate, . . . Scritti*, ed. Eugenio Garin (Florence: Vallecchi, 1942), 102. All translations are ours unless otherwise noted. For reasons of terminology, we did not use the popular translation by E. L. Forbes in Ernst Cassirer et al., eds., *The Renaissance Philosophy of Man* (Chicago: University of Chicago Press, 1948).

15. Ibid.

16. Ibid., 104.

seat on the whole globe, where this beholder of the universe might be seated. Everything was already filled, all things were already placed in the highest, middle, or lowest orders." Therefore man is free to choose his position amongst all the possibilities he realizes, free to create his own rank and degree in the universe: "The definitive nature of the others is constrained by the law we have prescribed them. Thou, constrained by no straits, shalt limit thine own nature for thee, according to thy free will, in whose hands we consign thee."[17]

Who should wonder, and what should he wonder at, if it were not for the indetermination, spontaneity, and unpredictability of man? It is his ability to choose, to change, to surprise others and himself, which makes him the puzzle and awe of the world. Man, being within man's will, is beyond man's knowledge.

While exempt from binding necessity, man is of course not immune to any divine rule. It is duty, a moral obligation, that steps in for natural law; the *ought* replaces the *must*. Man is expected to choose the very best. His honor—the responsibility God bestowed upon him—requires him not to imitate the senseless plants by living a mere vegetative life and not to follow the beasts in ceding to sensuality. Rather he is to imitate at least the celestial minds, or rather even the supercelestial angels in their pursuit of knowledge (directed upwards towards the divine Truth) and in practical judgment and moral action (directed downwards towards the material world). "A certain holy ambition shall invade our mind to disdain all the mediocre and to aim at the highest, struggling with all our might to reach it, as we can, if we will it. Let us disdain the earthly and spurn the heavenly and, having finally left behind whatever belongs to the world, let us fly to the court beyond the world that is closest to the supreme Godhead. There, according to the tradition of the holy mysteries, the Seraphs, Cherubs and Thrones hold the first places . . . If only we will it, we shall be second to them in nothing."[18] And to imitate the highest orders of the angels means exactly to make perfect use of our human intellect, i.e. to philosophize:

> Great is the power of the thrones that we attain in judging; immense is the sublimity of seraphs, that we attain in loving. But how can anybody ever either judge or love the unknown? . . . Therefore it is the cherub mediating with his light to prepare us,

17. Ibid., 106.

18. Ibid., 110.

> illuminating us likewise for the seraphic ardor and the judgment of the thrones. He is the link of the first minds, the Palladian order,[19] the guardian in charge of contemplative philosophy. He is the one we must first imitate and approach, and even comprehend to such a degree that we may thence be rapt to the heights of love and thence descend well prepared and ready for the tasks of action.[20]

Thus, it turned out, the true sense of man's position in the very center of the world is that it is not a static but a dynamic stance, a starting point: "I have set thee in the middle of the world, so that thou mayest from thence look around thee most easily for whatever there is in the world. We have made thee neither heavenly nor earthbound, neither mortal nor immortal, for thee to fashion thyself as an arbitrary and honourable sculptor and inventor in whatever form thou thyself shalt prefer."[21] And it is the paradox of the human condition that his very knowledge of other things (the undeveloped seeds of every other life form installed in him for cultivation) enables him to choose his preferred model, to become a plant, a beast, or an angel—and thus, living by his own rule, to be himself unknown, unknowable: "O highest and admirable felicity of man! To whom it is given to have whatever he chooses, to be whatever he might will."[22]

His good, honorable ambition might drive man even beyond the universal chain of created beings, beyond the angelic choirs, and thus beyond all human knowledge: "and if satisfied by the lot of none of the creatures he should withdraw into the center of his unity, he will become

19. In calling the cherub *Palladian order* Pico likens him to Pallas, the divinity of wisdom, whose "order" is expressed in the third perfect number—*seven*, the sum of the first (supernatural) perfect number, *three* (standing for the immaterial world) and the second (natural) perfect number, *four* (standing for the material world with its four elements). This makes *seven*, according to Cicero, *Somnium Scipionis* 5, a universal link: "qui numerus rerum omnium fere nodus est." According to the symbolic interpretation of numbers commonly accepted in Pico's time, *seven* stands for wholeness and holiness, mystery, the human soul, and wisdom (it is also the magical number par excellence). Thus man, the cherub, and Wisdom are all identified by Pico in their function as universal link. One of his theses condemned by the church was that "the human soul is an angel" (see below, note 24).

20. Ibid., 112.

21. Ibid., 106.

22. Ibid.

one spirit with God and will abide in the solitary darkness of the Father that is above all, and will excel everything."[23]

Not a sufferer from excessive modesty, Pico had started his oration from the very top, displaying his best and most impressive findings first. This brilliant first part of the speech is far better known than the second part, where, speaking *pro domo sua*, he protests an orthodoxy that (he knew well) he was jeopardizing.[24] Two interrelated questions discussed in this second part are relevant to our topic. Even this daring and dashing young count feels an urge to justify his involvement in the so-called occult sciences, in "good" natural magic, and in Jewish Kabbalah. This leads him further to discuss the appropriate policy for dealing with different types of traditions: what to divulge and what to hold back, passing it on under the cover of enigmatic form and figurative speech only to the chosen few. This is a ticklish topic since it entails merging the sacred mystery with the secular, the superrational gift of revealed religion with the rational task of dialectic philosophy.

There seems to be just one viable way to make magic acceptable for Christian authorities: "genuine," natural, beneficent magic (that uses rare, profound knowledge of natural powers to uncommon effects while keeping strictly within divine law) must be distinguished from "false," demonic, black magic or sorcery (that breaks divine law and natural order by involving the powers of hell). The one is redeemed by condemning the other.

> I have also submitted theorems about magic, where I show that magic is twofold, the one kind consisting entirely in the work and influence of demons, a damnable and monstrous thing, so help me God! The other is, on thorough enquiry, nothing else but the absolute perfection of natural philosophy. When the Greeks speak of both, they never honor that first kind with the name of magic, but call it *goeteia*, while they call the latter by the proper and exclusive term *mageia*, in the sense of a perfect and supreme wisdom. For, as Porphyry states, in the Persian language *magus* means the same as "interpreter and worshipper of the Divine" in

23. Ibid.

24. This defense in advance did not save Pico from excommunication, nor was he allowed to hold his disputation in Rome. For details and for the thirteen theses condemned as heretical or dangerous (two of which connected with magic), see Paul Richard Blum, "Pico, Theology, and the Church," in *Pico della Mirandola*, ed. M. V. Dougherty (Cambridge: Cambridge University Press, 2008) 37–60 (condemned theses on pages 40 and 41).

> ours. For, O Fathers, between both crafts there is a great, or rather the greatest difference and contrast. That first one is condemned and banned not only by Christian religion, but by all laws and every well-constituted state, while this other one is approved and embraced by all wise men and all nations zealous for the pursuit of the heavenly and divine.[25]

To whitewash magic is a dubious enterprise, despite all testimonies quoted in its favor, not just the noblest of ancient heathens and modern Muslims, but also Christian philosophers like Roger Bacon and William of Paris (while, strangely enough, Albert the Great is omitted here). Yet, for all the asserted contrast between demonic magic and its saintly counterpart, there is but a slender, uncertain line that separates the two in the eyes of any uncommitted witness, to say nothing about a hostile judge. Who but the practitioner himself (and maybe not even he!) could definitely exclude the possibility of some demonic meddling in the seemingly most innocuous charm? The witch-hunters' excesses lie still hidden in the future, yet Pico knows well what treacherous ground he is treading.

In itself the defense of natural magic appears somewhat ambiguous and contradictory. On the one hand, learned magic is shown to be perfectly harmless since it keeps entirely within the boundaries of natural cause and effect: it is just the cream of natural philosophy, the advantage of more advanced knowledge. If so, why should it then be kept in secret? On the other hand, it claims to be mysterious and divine, with an implication of supernatural powers. This would make it a sacred art, uncomfortably close to the religious mysteries. And in that case, the only difference between the black and the white magician would be that the latter is believed to be a master of demons, while the former deliberately binds himself as their slave.

For Pico, however, the natural and the sacred are not contrary or mutually exclusive options. He describes white magic as a wholly natural and *therefore* sacred, priestly art, its occult quality deriving from the mysteries of nature herself:

> For as that first [kind of magic] makes man culpable and bound in slavery to the powers of evil, likewise this other one makes him their chief and master. For on the bottom line, that first one can claim the name neither of art nor of science; whereas, filled with the most sublime mysteries, this other one involves

25. Ibid., 148.

> deepest contemplation of the most secret things, and, eventually, the knowledge of all nature. Calling forth, so to say, from their hiding-places into the light the forces that are diffused throughout by God's bounty and disseminated throughout the world, this art, rather than working wondrous things by itself, eagerly assists nature in her wondrous workings.[26]

This is intended to go down smoothly. And yet if those spirits subjected to the good magician are the same naughty devils, the difference from the sorcerer might evaporate. If, on the other hand, the demons themselves are to be considered natural agents, there is no black magic left altogether. And, more dangerously still, in what follows, the distinction between the profession of the magician and the priest collapses—the magician appearing as the better, holier, and more efficient link to the divine:

> [As] if it were itself the master craftsman, [magic] brings forth out into the open the miracles concealed in the caves of the world, in the womb of nature, in the treasuries and strongholds of God. And like the farmer weds the elms to the vines, the magician weds earth to heaven, i.e. the inferior to the gifts and forces of the superior. Accordingly this [art] shows itself as divine and beneficent to the same extent as that one is monstrous and harmful. Mainly for this reason: that [black magic] by subjecting man to God's enemies lures him away from God, while this [true magic] incites that admiration of God's works that leads most assuredly to intimate love, faith and hope. For nothing moves us more towards religion and the worship of God than a persistent contemplation of God's wondrous works, and if we have explored those thoroughly by this natural magic we are speaking of, we will be more ardently inspired to the worship and love for the Creator.[27]

As for Kabbalah, Pico argues, it is not an exclusive property of Judaism, but the same God of the Old Testament, who is also the God of the New Testament, gave it to Moses together with the law, and therefore the Christian church legitimately inherits it. "Not only the famous Hebrew scholars, but also our Esdras, Hilarius, and Origines write that Moses had received from God on the mountain not only the Law which he left to posterity written down in five books, but also a most secret

26. Ibid., 152.

27. Ibid., 152–54.

and true interpretation of the Law."[28] It is even meant primarily not for Hebrews, but for Christians since "When I had . . . read them through with utmost diligence and unwearying application, I found in them, by God, not so much the Mosaic, than rather the Christian religion. There is the mystery of Trinity, the incarnation of the Divine Word, and the divinity of the Messiah; there I read the same about original sin, about its expiation by Christ, about the Celestial Jerusalem, the fall of the demons, the angelic orders, purgatory, and the punishment of hell, as we read every day in Paul and in Dionysius, in Hieronymus and in Augustine."[29]

But Kabbalah is a secret knowledge: not a product of human whim but rather of divine decree:

> However he [Moses] was ordered by God to proclaim the Law the people, while he was neither to write down the interpretation of the Law in books, nor to divulge it, but he himself was to reveal it only to Jesus Nave,[30] who should then reveal it under a strict obligation of silence to other succeeding high priests. It was enough to recognize by means of simple historical narration now God's power, now His wrath against the wicked, His mercy to the righteous, and His justice to all, and to be instructed through healthy divine commandments in living a good and blissful life and in true religious worship. But to lay open to the plebe most secret mysteries and arcane truths of the supreme Godhead that were hidden under the surface of the Law and under the rim of uncouth words—was this not to give the holy to the dogs and to scatter peals among the swine?[31] Therefore, it was not a human decision, but a divine commandment, to keep hidden from the vulgar what was to be shared with the perfect, among whom alone, as Paul says,[32] he speaks wisdom.[33]

Who then are the perfect, to whom the secret knowledge must be revealed, and who are the vulgar to be kept out? Again Pico offers mutually exclusive alternatives without a clear commitment. When he first introduces "the ancient mysteries of the Hebrews," he seems to include all Christians: "so that they [the mysteries of Kabbalah] may not be held

28. Ibid., 154.

29. Ibid., 160.

30. Cf. Ecclesiasticus. 46:1: "Fortis in bello Jesus Nave, successor Moysi."

31. Cf. Matt 7: 6.

32. Cf. Rom 1:17 and 1 Cor 2:13.

33. *Oratio*, 156.

for made-up stories or peddlers' tales by those to whom they are unknown, I want everybody to understand, what and of what kind they are, where they come from, by whom, and by what important authors they are supported, how sacred, divine, and necessary they are for our people to defend our religion against the impudent libel of the Hebrews."[34] But presently Christ himself is numbered among authors like Pythagoras or Plato, who revealed their esoteric teachings orally to a few selected disciples.[35]

What then is illumination? Is it an occult theology attained by the intellectual effort of exceptional wise men (heathens, Jews, and Christians alike)? Or is it rather a free gift of the divine Spirit descending upon simple fishermen and a mystical church, chosen and selected by divine grace? It is obvious that Pico tends toward the first alternative, a Platonic ascent of the god-seeking human soul, the access to the mysteries merited by human initiative and effort. But he attempts proving its compatibility, or even identity, with orthodox Christian tradition, where the initiative in choosing his people is left to God's ineffable will, and any theurgical attempts are frowned at.

For all his declared intention to reconcile opposite positions that sometimes induces Pico to embrace unresolved contradictions, he is quite clear on the essential, intrinsic secrecy of universal nature and its basic functions. The occult and mysterious is not just an accidental, relative, external feature of some truth, the expression of a contingent and easily reversible inadequacy of our present knowledge, but is rather a distinctive and defining property of the highest, most valuable truth: a nobility and holiness that, while withdrawing itself in its arcane difficulty and sacred remoteness, also needs to be protected in its splendid isolation by its keepers. The problem is, if the danger of profanation must be avoided by all means, how does one escape the opposite danger of mystification and sacred fraud? We shall encounter different solutions by theoreticians of occult philosophy in the next two centuries.

It does not come as a surprise that early modern advocates of learned magic were no less suspected of heresy than the muttering midwifes, shepherds, and blacksmiths they strove so hard to distance themselves from. While different on most counts and contrary in many respects, the one feature they shared was their infringement upon the

34. Ibid., 154.

35. Ibid., 156.

monopoly of an established religion. The administration of mystery and wonders, a privileged connection with and an exclusive understanding of the divine order, and, consequently, the office to interpret it for the people—these were issues the church could not give up; whereas the revived Neoplatonic *pious philosophy* claimed them more or less openly for itself. If an identity of Platonism and orthodox Christian theology could not be demonstrated and upheld (as it could not, in the long run), certainly the branch of that widespread family that was most awkward to integrate and most evidently heterodox was Renaissance magic. On the whole, scholarly magicians might have fared better by just claiming secular wisdom, without insisting on a superior piety.

• • •

In other words, theoretical problems can arise when competing paradigms or patterns of thought are combined. That is the case in Pietro Pomponazzi (1462–1525), who set to give natural explanations to wondrous phenomena. Therefore he endorsed Marsilio Ficino's theory of the continuity of levels of spirit pervading all things, which included, for instance, the presence of certain natural virtues in human beings on the basis of the "the general consensus that man is in the middle between the eternal beings and those that come to be and pass away."[36] As we have seen in Ficino and Pico, this can have the consequence that the miraculous is not anymore a miracle because it strictly follows the spiritual governance of the world. Therefore in Renaissance Platonic terms, wonder and miracle are limited to an attunement to that cosmic force. In accepting this line of thought, Pomponazzi managed to naturalize magic spells and other miraculous phenomena. He duly reported the objection that the Bible would be endangered because it is based on miracles.[37] The same would be true of the stigmata of St. Catherine of Siena and St. Francis.[38] The counterargument he offers is of a purely fideistic nature, namely, that "even what can be reduced to natural causes, if the

36. Pietro Pomponazzi, *De naturalium effectuum causis sive de incantationibus*, in *Opera* (Basel: Henricpetri, 1567; reprint Hildesheim: Olms, 1970), 25.

37. Ibid., 65: "quoniam leges istae potissima numina et fundamenta sumunt a miraculis. Unde secundum istos modos nulla sunt miracula, quae enim in Veteri lege et in Exodo leguntur de Moyse scilicet de stupendis per eum factis coram Pharaone, possunt satis reduci ad aliquem vel ad aliquos vel ad omnes istos tres modos."

38. Ibid., 67.

Bible declares it to be a miracle it has to be taken to be a true miracle."[39] Pomponazzi as a reformer and interpreter of Aristotle's philosophy tends to defend both naturalism and faith. By *naturalism* he understands a Neoplatonist uniformity of the cosmos—turned nontranscendent. After Renaissance Platonists had spiritualized nature, it was Pomponazzi's endeavor to naturalize spirit. In that he is consistent with his famous book, *The Immortality of the Soul*, which also gave entirely natural reasons for the interaction between soul and body leaving the belief in immortality to the teachings of the Church.[40]

• • •

In spite of his obvious affinity to Pico, his evident commitment to a Neoplatonist worldview, and his insistence on a sacred, spiritual foundation of the magical arts, the most deliberate defender of *occult philosophy* took one decisive step in the same direction. The humanist Henry Cornelius Agrippa of Nettesheim (1486–1535), best known as the author of the first comprehensive early modern work on magic,[41] is rather an encyclopedic scholar rather than an adept initiate; he was set more on debunking magic than on enthusing about it. Like Pico, Agrippa deplores the evil name magic has acquired in Christian society after having been honored in antiquity and despite such biblical testimonies as the feats of Solomon and the adoration of Christ by the magi. Like Pico, Agrippa blames the fact not on a fundamental incompatibility but on the abuse of the name by "heretical" impostors and on their forbidden practices that exploited and perverted Christian rites.[42] But to clearly distinguish the legitimate, honorable science of magic from all fakes and forgeries, he deems it necessary to first define his subject, to lay open its scientific foundation, and to make a systematic inventory of its diverse branches and disciplines. His sober, rationalistic approach distinguishes Agrippa

39. Ibid., 82.

40. Cf. Paul Richard Blum: "The immortality of the soul," in *The Cambridge Companion to Renaissance Philosophy*, ed. James Hankins (Cambridge: Cambridge University Press, 2007) 211–33 (with ample bibliography).

41. *De occulta philosophia*, was first published in Cologne in 1533, while the project and most of its composition dates back to 1510. We quote the Latin text as reprinted in Heinrich Cornelius Agrippa von Nettesheim, *Opera* I (Hildesheim, New York: Olms, 1970).

42. Cornelius Agrippa, *Opera* I, a 3 v.–a 4 r. (in the dedicatory letter to the abbot Tritemius).

from other Neoplatonist Renaissance magicians. Instead of enchanting the reader with inspired poetical rhetoric, Agrippa's presentation is extremely learned, ironic in a scholarly, dry way—at times even somewhat pedantic and boring in its exhausting exhaustiveness. It is explicitly designed to counteract the mystification and confusion of all preexisting "modern" books on magic—including the work of such venerable authorities as Roger Bacon and Albert the Great.[43] The most tangible charm of Agrippa's book lies not in its fascinating subject matter but rather in the surprising frankness and admirable intellectual integrity of its author.

By applying the Aristotelian threefold division of philosophical science (physics, mathematics, and theology)[44] to the threefold Neoplatonic cosmos (terrestrial, celestial, and supercelestial), Agrippa gains his basic division of occult philosophy in natural, mathematical, and ceremonial magic. The most advanced arts of ceremonial magic (magical rites involving the calling or banning of supernatural spirits, the evocation of the Divine Names, and the like) cannot be practiced outside the limits of true religion without the risk of damaging one's soul, and they must never be attempted without a sound knowledge of the lower grades: natural magic (physics, chemistry, psychology, and the like) and mathematical magic (astrology, the symbolic values of numbers, and the like).

There is nothing supernatural at all in the two lower ranks of magic; the spirits (or forces) involved are strictly of a natural order, like the properties of the elements and their mixtures or the attractions and repulsions between living creatures or the force of numbers and speech or the influences of celestial bodies. Such properties are termed occult whenever they are not the immediate consequence of the four basic elementary qualities of Aristotelian physics (warmth, cold, dryness, and moistness). This is a very innocuous meaning of the term *occult* since most human crafts and skills (e.g., medicine, pharmacy, metallurgy, brewing, food processing, animal taming, and the like) are based on knowledge of such occult qualities. While Agrippa's concept of science is

43. Ibid., a 4 v.: "siquidem, quos ex recentioribus ego vidi, Rogerium Bachonem, Robertum Anglicum, Petrum Apponum, Albertum Teutonicum, Arnoldum de Villanova, Anselmum Parmensem, Picatricem Hispanum, Cicchum Asculum Florentinum et plerosquam alios, sed obscuri nominis scriptores, cum se magia tradere pollicentur, non nisi aut deliramenta quaedam nulla ratione subnixa, aut superstitiones probis omnibus indignas praestiterunt."

44. Cf. Aristotle, *Metaphysics*, 6.

still premodern (based on Aristotelian qualitative physics), we nevertheless find a close parallelism between the interrelation of his two lower levels of magic and the interplay of experimental and scientific theories in modern natural sciences.

Agrippa himself was not only a humanist and a scholar but also an experimental scientist and a medical doctor. He points out how wondrous effects like *automata*, monumental buildings, changes of coastlines, etc., be achieved even without resorting to occult forces but just by applying mechanics and mathematical sciences. "Therefore," he continues,

> if a magician is well versed in natural philosophy and in the mathematical sciences and knows the combined sciences that there are between the two, as arithmetic, music, geometry, optics, astronomy, and all the sciences dealing with weights, measures, proportions, articulations, and conjunctions, and also knows the mechanical crafts that derive from these—what wonder if, excelling over other men in both art and intelligence, he will work many wondrous things which even the most wise and knowledgeable will admire highly? Are there not nowadays still traces of ancient works in existence, like the pillars of Heracles and of Alexander, the Caspian city doors, of fused bronze and locked with iron bolts[?] . . . And we read in trustworthy historians, how once, by similar arts, rocks were cut, valleys filled, mountains flattened . . . bogs dried, new islands created artificially, and others reunited with the continents.[45]

Such magnificent effects of human inventiveness and engineering, however, are prone to appear supernatural to the simpleminded individual and to a time that has lost the technical know-how:

> About suchlike the simple folk tells tales, how it is devils' work, since the art and the artist thereof have passed out of memory, and there is nobody who cares to understand and to investigate them. Therefore, whenever those intellectually blind ones happen to see any admirable sight, they bundle the making of it all upon the devils and take for miracles what is the result of either natural or mathematical sciences. Like somebody ignorant about the property of the magnet, if he sees heavy iron lifted up or hanging in the air . . . will he not keep saying that this is devils' work? But if he knew the influence of magnet upon iron and had experi-

45. Cornelius Agrippa, *Opera* 1:154.

> mented with it, then all his bewilderment would have come to an end, leaving not the least doubt that this is a natural effect.[46]

So far he is rather close to the analysis of his contemporary Pomponazzi.

Yet while his sober rationalism and skepticism seem to anticipate enlightenment notions, Agrippa remains firmly rooted in the Neoplatonist worldview, and thus he adds immediately: "But at this point you must be well aware, that in the same way as we gather natural powers from natural things, we likewise receive from the abstract, mathematical and heavenly things heavenly powers like movement, life, sense, speech, foresight, and divination."[47] Thus incantation (i.e., manipulation through the magical power of words—contained in spells or in proper names) is also to be considered a "natural" procedure.[48]

While it is wonderful in that it is admirable and in that it involves the spiritual forces that pervade and interconnect the whole material universe, achievements of the two first levels of magic are nevertheless not miraculous in the proper sense. It is simply human ignorance about the natural forces and the mutual influence of things on one another that make such effects appear stunning or suspicious. There seems to be no stringent need to keep the practices of natural magic secret even though abuse (i.e., the use of ethically neutral forces with bad intentions and to bad effect) is possible and conceded on every level. Not mystification but rather information would prevent impostors from dealing in fake magic and would establish the difference between legitimate and forbidden manipulations:

> If you find anything that displeases you, pass it over and make no use thereof. Neither do I recommend those things to you, but just report them. Yet do not, therefore, reject the rest, since it likewise happens to perusers of medical books to read, at once with antidotes and medicines, also about poisons. I further admit that magic itself teaches many most idle and ostentatiously baffling prodigies. Leave those behind as vain as well; yet do not fail to inform yourselves about their [underlying] causes. But whatever can be done without offending God or violating religion towards

46. Ibid., 155.

47. Ibid.

48. It is consequently included in the first book of *De occulta filosofia*; cf. *Opera* I: 139–46.

> the good of men, the prevention of accidents, the destruction of mischievous spells, the healing of diseases, the dispelling of spook, and the continuous prosperity of life, honor, and good luck, who would not consider that rather a necessary than merely a gain?[49]

Fake magic, like any deception, is implicitly bad, while the moral value of genuine magic, both natural and mathematical, is not contained in the means employed but depends on the good or bad intentions of the practitioner. Yet the same cannot be assumed for ceremonial magic. While the celestial spheres, and even the world-soul, might be included in a purely philosophical, secular area where unrestrained scientific inquiry is legitimate, the highest level of magic is by definition both supernatural and connected to theology. Since its significance cannot be played down, how can it be justified at all? How and why should the magician be allowed to poke his curious nose beyond the celestial spheres? Or else, why can Agrippa not limit his project of rehabilitation to the two lower levels of scientific magic, avoiding the area where a conflict with religious institutions seems inevitable?

There are three good reasons, or rather three different aspects of one good reason: personal conviction. Agrippa is a genuine Christian humanist (i.e., he is both an intellectual and a believer). Closely akin in his creed to such internal critics of the Catholic Church as his elder contemporary and role model, Erasmus of Rotterdam (1466–1536) and to Thomas More (1478–1535), but less of a politician and more outspoken in his writings than both these men, Agrippa combines skeptical rationalism with a religiosity that is neither blind nor uncritical. For him, faith and reason are not diametrically opposed but complimentary and mutually supportive faculties. Reason must be pushed as far as possible since not the least important among its tasks is to realize and to establish its own limits. Faith reaches out beyond reason, and while it confirms and purifies reason,[50] it is in its turn trimmed by reason, i.e., purged from superstition.[51] (As we will see, for Agrippa and his contemporaries

49. Ibid., a 2 v. (preface by the author).

50. Ibid., 310: "Religio enim sacra mentem purgat, redditque divinam."

51. This is the simple solution to the problem of how the "rationalistic" Agrippa, author of *De occulta filosofia*, is to be reconciled with the apparently "fideistic" author of his second important work, *De incertitudine et vanitate omnium artium et scientiarum* ("On the Uncertainty and Idleness of All Arts and Sciences"), and why, after having included magic in the number of scathingly criticized sciences, Agrippa proceeded to

superstition is a far more serious and dangerous thing than the harmless whims or amusing pastimes we might associate with the term *superstition* today.)

Both his Neoplatonist philosophy and his Christian religion made it impossible for Agrippa to deny a supercelestial realm of purely spiritual beings. And while it was philosophically feasible to reduce the inhabitants of this highest sphere to the one transcendental Godhead, theology was quite explicit about angels. Furthermore, no Christian could deny a direct influence of the caring God upon the individuals in this world (divine providence and grace), the efficacy of prayer and ritual, or the necessity of religion as such (i.e., of a personal and institutional relationship with God). Once it was granted that God was in some way accessible and that he acted not just through natural agents but also more directly through angelical messengers, some legitimate way for mortals to approach the divine being was also conceded. Was it to be limited to prayer and contemplation? And if so, how were prayer and contemplation to be defined? Agrippa would never renounce one type of contemplation that he claims at the same time as a branch of ceremonial magic: Christian Kabbalah.[52]

Then there was the problem of miracles in both the Old and the New Testaments: were they to be declared as essentially different from magical effects, or were they just the very best of magic? And what about the fallen angels turned into evil forces; and their identification with pagan polytheistic divinities; and the undeniable fact that magic predated Christianity and that many of Agrippa's sources were pagan; and the equally undeniable fact that there actually were individuals who tried to contact evil spirits in order to gain supernatural powers? Agrippa's encyclopedia of magic arts could never come close to being complete without accounting for all these issues, and this is the third reason why he simply could not ignore ceremonial magic.

---

publish his encyclopedia of magic. Like in Erasmus's *Praise of Folly* (1511), Agrippa's criticism in *De vanitate* was not directed against the use of reason but against the unrestricted confidence in reason combined with a very restricted application thereof, as demonstrated in the various disciplines. It was an outburst against the puffed-up narrow-mindedness and monomania of specialists, a cancer devouring the sciences from within.

52. Unsurprisingly Agrippa is one of the most courageous defenders of John Reuchlin's orthodoxy against the accusations of the Dominicans of Cologne.

But including ceremonial magic meant also conceding the possibility of genuine black magic—not just of deceiving and, possibly, self-deceiving tricksters and crazy hags, but of genuine magicians achieving extraordinary effects through the assistance of demonic forces. His own systematic approach deprived Agrippa of the easy way out, namely of declaring all black magic to be fake. On the other hand, he was not willing to outlaw all kinds of ceremonial magic. It follows inevitably that he conceded the reality and efficacy of both sorcery and thaumaturgy, (i.e., black and white magic in the full sense of the terms). What are, then, the criteria according to which the sheep are separated from the goats?

While ceremonial magic ought only to be based on true faith and religion, de facto it can also be founded on false belief or on superstition:

> There are two that reign over every operation of ceremonial magic, namely, religion and superstition. Religion is itself the persistent contemplation of the Divine and the connection in piety with God and the Divine *numina*[53] through pious deeds, by which in reverent service a worthy worship and a sanctified cult is paid to them and the ceremonies of the divine cult are properly performed . . . But owing to the diversity of times and regions the rites and ceremonies of religion are different. And each of the religions has something good in it, since it is directed towards the same Creator God. And granted that God approves only of the Christian religion alone, nonetheless he does not, in his grace, spurn the other forms of cult entirely, and does not leave them without all temporal (if not eternal) compensation . . . But the impious and throughout irreligious He hates as his enemies . . . for their impiety is greater than that of who followed an erroneous religion.[54]

So this is the superstition of the heathens: an ignorance of the true religion to be considered rather their bad luck than their fault, for "To the true religion, however, man cannot come by himself, but only if he is taught by God."[55]

Up to this point, Agrippa seems to give a broad-minded and generous interpretation of pagan religion and magic: it appears as a misguided and imperfect cult of the true God rather than as the cult of bad demons,

53. These *numina* are the divine emanations, identified with the divine attributes of the Kabbalah; cf. *Opera* 1: 329–34.

54. *Opera* I: 316–18.

55. Ibid., 318.

and the successes of pagan magic are the just, temporal remuneration of a good if insufficient effort. Yet this optimistic mood does not prevail, and what follows reveals a far gloomier picture:

> Thus any form of cult alien to the true religion is superstition. And any worship that renders more [additions] to the divine cult than the true religion is also superstition. Likewise also that which renders divine cult either to whom it is not due, or in an improper way. Therefore one must beware above all that no offence is done to the highest, omnipotent God and the holy *numina* below him through any perverse cult of superstition. That would be both nefarious and most unfit for a philosopher. Thus, while alien to true religion, superstition is, nonetheless, not all condemned wholesale, for in many cases it is also tolerated and observed by the religious leaders.[56]

It turns out that superstition is a serious danger after all, and is not limited to the pagans but is a trapdoor through which Christians are falling into blasphemy and heresy; it is an inner foe of the church. And while such a warning (by a layperson!) might or might not sound quite acceptable to hardline theologians, the reference to the divine *numina* and to a true philosopher's duty is, in this context, risky; and that most ominous aside about the toleration of superstition by the official church is downright provocative.

While Agrippa could diligently avoid any clash between natural magic and established religion, at this point all the intrinsic conflicts return with a vengeance. Under such auspices every magical ceremony falls under the suspicion of superstition and must be defended and exonerated singularly. And, even worse, this is not just true for ceremonies outside the official institution since, according to Agrippa, superstition does not stop at the church doors:

> But I call superstition above all that which is a certain likeness to religion, insofar as it imitates whatever there is in religion of miracles, sacraments, rites, observations, and suchlike, and therefore obtains a power far from small, and also gets no lesser force from the credulity of the practitioner. How much a persistent credulity can do has been said in the first book, and it is well known everywhere among the common folk. Thus superstition requires credulity in the same way as religion requires faith. For a persistent credulity can go so far that it even works miracles

56. Ibid.

> in false opinions and procedures. For whoever in his religion believes with the utmost assurance even false things to be true elevates his spirit by this same credulity until it is assimilated to those very spirits that are the leaders and rulers of that same religion, and performs things that are not understood by reason and nature. But a hesitation in credulity and diffidence weaken every operation, not in superstition only, but also in the true religion, and frustrate the effect, however strong the experiments aimed at it might be. But in which way superstition imitates religion I will show you by these examples: for instance, when worms and locusts are excommunicated to prevent them from damaging the harvest, when bells are baptized, and images, and more of that same kind.[57]

Superstition exists and is effective both inside and outside the church. Genuine miracles are not the criteria of true religion. And, most dangerously, the process through which the effects are achieved, namely, elevation to higher spiritual spheres, is the same in superstition as in true faith. How is a mere human being to distinguish between them?

> But since those old, venerable magicians and those who were the inventors of their art among the ancients were Chaldeans, Egyptians, Assyrians, Persians, and Arabs, whose religion was all perverse and tainted by idolatry, we must, above all, beware not to allow their errors to subdue the foundations of our catholic religion. For, this would be blasphemous and damnable, and I were a blasphemer, too, in this science, if I did not caution you about these things. Therefore, wherever you find such things written down by us, you must know that we report them out of other authors, and that I hand them down to you not as truth, but as conjecture, which has affinity to truth, and as an instruction to imitate that among them which is true. For it is up to us to gather the truth from their errors. This is a task, exclusively, for a profound intelligence, an unblemished piety, a strenuous and hardworking diligence.[58]

Thus Agrippa set out to defend magical arts and ended up questioning Christian rites. Willing to give up neither Christianity for magic nor magic for Christianity, he tried to achieve a balancing act on a tightrope that could easily have turned out to be a high-tension wire. He showed a certain tendency to demystify both magical and

57. Ibid., 319.

58. Ibid.

religious practice, but never claimed for human reason the ability to solve all mysteries. He extended the area of the purely natural, but never denied the reality of the supernatural. While quite outspoken on many dangerous issues, he still left much in his work open to different interpretations. His intention was to open, not to definitively close, a discussion: not to settle all questions and spell out every detail, but to renew a neglected and maligned field of knowledge by giving an overview as completely and as concisely as possible. He addresses a mature reader, to whom he expects will be, like himself, unprejudiced, critical, independent, sincere, and courageous: ready to leave old miracles behind and unafraid of encountering new ones.

• • •

Half a century later, when Giordano Bruno (1548–1600) addressed similar problems, the cultural climate had changed noticeably. The Christian humanists and the great Protestant Reformers were dead, and any hope of a reunited Christian church had been buried in the course of the Tridentine Council. Now conflicting truth claims battled not only in the diverse philosophical sects but also in the theologies of the different Christian denominations. Far from fulfilling the humanists' dream of a separation of the spiritual and secular powers, the Protestant Reformation had tightened the bond between single states and their local churches. These developments seemed to confirm Machiavelli's view on the merely temporal and political value of all religions.

On the other hand, while Copernicus was also dead, his *De revolutionibus* was very much alive. It had begun to replace the ancient finite, earth-centered image of the spherical cosmos with a solar system and the earth a planet among planets. It did not yet endorse the notion of an infinite number of such world-systems and a positive infinity of the material universe. This step was to be taken by Bruno himself.[59]

In a cosmos that is conceived as infinite in both space and time, there are three possible answers to the question of a divine principle: Either there is none (atheist materialism), or it is immanent to the cosmos as a whole (panentheism), or it transcends even the infinity of the cosmos (transcendental deism). Bruno's position is a combination of the latter two, and complex enough to have been often mistaken—by

59. The philosophical consequences of Copernicanism are the main topic of Giordano Bruno's dialogue *The Ash Wednesday Supper*.

friend and foe—for the first one. It is easy to see how the assumption of the positive infinity of the universe has important consequences for all questions concerning truth claims, standpoints, and the limits of the human mind.

An infinite world has no absolute center, no top and no bottom, no absolute upward or downward direction. Every individual standpoint is the relative center of its own perspective; none is more privileged than the rest; all are linked to the subject, all liable to change. Thus human knowledge acquires, from the side of the subject, a certain relativity of perspective while, from the side of the object, diverse aspects do not compromise the unity of truth. Knowledge is not merely subjective in a derogative sense but always necessarily personal (i.e., related to a standpoint). In his Latin poem *De immenso* (1591), Bruno describes this in an autobiographical passage through the changing image of two mountains: his native Cicala and Vesuvius. Both look dark, bare, and menacing when viewed from a distance, but colorful, fertile, and welcoming upon arriving there. When he first arrives in Naples, the boy speaker of the poem is amazed:

> Thus when it came to being there and to seeing Vesuvius closer,
> who is famous through Bacchus and glories in many vineyards
> abundant in grapes hanging from the vines,
> and fruit of every kind . . .
> and when I saw that he was lacking
> none of the things I had known, nay, he held a hundredfold more:
>
> Stunned by the novelty then I first accused my eyes
> of lying. Then I wondered, if that splendid brother of yours
> maybe beguiles me now with a false semblance.[60]

But observing his native mount Cicala from his new vantage point, comparing the two opposed experiences, the youngster makes a crucial discovery:

> While I enjoyed the richness of the trees, I listened with open ears
> to what the Old mountain [Vesuvius] pleasantly uttered:
>
> Stay here and spurn your Cicala's native spirits,
>
> Look how wealthy I am . . . and how full my horn of plenty.
>
> From here turn your eyes and look at Cicada.

60. Giordano Bruno, "De immenso et innumerabilibus," in: *Opera Latine conscripta*, ed. Francesco Fiorentino (Naples: Morano, 1879) I, 1, 314.

> My brother is the opposite of me. How he touches
> the sky with his black peak and hides
> his swarthy limbs in a pitch-colored cloak,
> humbly and shamefully he wraps his miserable body in blue darkness.
>
> I replied: You were like this, too, before I joined your valleys,
> and he will appear similar to you,
> when he receives me. Then the fog will stand back from there,
> and also the sky, and the world's center will return
> thither with me, as it always accompanied me
> in coming up here. And it will stay with me wherever I stay.[61]

And the more universal consequence is drawn:

> So the parents first taught the child to wonder,
> how distance changes the face of things, while the things endure,
> and how the soul remains the same throughout . . .
> so that I should not deem so easily the distant better
> or worse than what is near at hand. And so that I should find
> that we are the sky for those, who are, in turn, the sky for us.
> . . . and thus it is [vision] itself that makes the figure of the aspect:
> for it limits the space for itself, and the space does not limit vision,
> and it contains endless space in itself wherever it goes . . .
>
> Thus they even believed that earth is the inmost of all things,
> while I will find the center in the whole and in every part,
> since every part is related in the same way to all other parts.
>
> Consequently heaven is not limited by any definite margin[62]

We see how astronomy, for Bruno, is not a separate discipline but an integrated part of his epistemology. Under the conditions of an infinite universe, all knowledge is partial and relative, but while every partial glimpse of truth has its relative value, there are better and worse perspectives, more or less adequate images.

This partiality and relativity applies also to all human attempts to cognize the divine, to all religions and philosophical theologies without exception. All finite, individual things in the infinite universe are limited in time and subject to change in time. Therefore humans can not approach the absolute godhead except in its manifestations in nature. The infinity of the universe has its repercussions on the Platonic ascent too:

61. Ibid., 315.

62. Ibid., 315–16.

as upwards and downwards are no more absolute, but merely relative directions, the ascending mind, rather than arriving at the top, describes a full circle:

> Tansillo: . . . since it is neither natural nor appropriate for the infinite to be contained, it cannot render itself as finite: for thus it would not be infinite. But it is natural and appropriate that the infinite, for being infinite, is infinitely pursued (in that kind of pursuit that is not according to the law of physical movement, but by a certain metaphysical movement); and this movement is not from the imperfect towards the perfect, but rather it moves in a circle through the various grades of perfection, in order to arrive at that infinite center that is neither formed, nor form.[63]

The divine as it is reflected in nature, and immanent in the universe is the natural object of human reason while the transcendent, absolute source of being and truth is beyond its capacity. In Bruno's summary of *Eroici furori*, man is said to be naturally born blind to the absolute:

> In the fourth dialogue the nine causes of the inability, inadequacy, and deficiency of human vision and apprehension of divine things are symbolized and, to a certain degree, explained. There the first blind man, who is born blind, stands for the reason that is by nature which humiliates and debases us. The second, blind because of the poison of jealousy, stands for the irascible and concupiscent faculties that distract us and lead us astray. The third, blinded by the sudden appearance of intense light, demonstrates the reason which derives from the splendor of the object that dazzles us. The fourth, grown and nurtured a long time in view of the sun, stands for the excess in high contemplation of the unity which steals us away from the crowd. The fifth, whose eyes are always flowing with plentiful tears, demonstrates how the inadequacy of means between the faculty and the object hampers us. The sixth, who by much weeping has lost the organic humidity [necessary] for vision, symbolizes the lack of true intellectual nourishment that weakens us. The seventh, whose eyes are burned to ashes by the heat of his heart, stands for the burning affection that sometimes scatters, distresses, and devours the discerning faculty. The eighth, blinded by the wound from the point of an arrow, stands for the reason that is caused by the act of union with the aspect of the object itself, since it subdues, alters, and corrupts the appre-

63. Giordano Bruno, "De gli eroici furori" ("The Heroic Frenzies") in *Oeuvres complètes* VII (Paris: Les Belles Letters, 1999) 167.

> hensive faculty . . . The ninth, who, being mute, cannot explain the reason of his blindness, symbolizes the reason of all reasons that is the hidden divine decree which gave to man this eagerness and purpose of investigation in such a way that he can never arrive at a cognition higher than that of his own blindness and ignorance, and will deem silence more proper than speech.[64]

Yet for all the inadequacy and limits of the human mind, for all the virtue of moderation, which counsels resignation and mere contemplation of reflected light and shadows and traces of the divine as they appear in nature: it is the highest achievement and most noble aim of the elected few, the heroes among philosophers, to leave reason behind and to embark on the adventurous quest for truth unveiled. It is frenzy, because it means overreaching human power, and cannot be achieved, since human initiative annuls itself in the process. This is symbolized in the image of Actaion, the hunter who, on surprising the bathing Diana, was turned into a stag and thus became the prey of his own hounds. Truth, "caught naked," counters by baffling and frustrating reason and turning the human mind back upon itself.[65] Yet it is an heroic, virtuous, and meritorious enterprise, in spite of its madness, because it pursues the most admirable of goals, will not be satisfied by the second best, and urges human effort to its utmost limits. In the face of a limited capacity to hold and understand, it responds to the infinity of the object by an infinite capacity of admiration and desire. Even if this circuitous journey brings the philosopher back to seeking truth in nature, the result is both negative and positive at the same time: he gains a clearer insight in human nature, an instructed ignorance about the absolute, and a critical awareness of the presumption in all positive statements about God.

Of course such a pursuit of the eternal mysteries of absolute truth is the privilege of exceptional representatives of humankind. The crowd is to remain in the paradise of ignorance: "ignorance is the mother of sensual happiness which is, in turn, the garden of paradise for the animals."[66] He continues:

> There needs be artisans, mechanic workers, peasants, servants, pedestrian soldiers, ignoble people, vile and poor ones, pedants, and others of such kind, since otherwise there could not

64. Ibid., 35–37.
65. Ibid., 153–59.
66. Ibid., 99.

> be philosophers, contemplatives, cultivators of the minds, lords, captains, nobles, famous men, rich and wise ones, and others suchlike, who are heroic similar to the gods. Why should we therefore strive to disrupt the state of nature which has divided the universe into greater and lesser things, superior and inferior, famous and obscure, worthy and unworthy, and not just outside ourselves, but also inside, in our very substance, even up to that part of substance that is asserted to be immaterial? . . . Therefore I should not believe that this ought to be taken as an example, so that the order of things would be perverted, the subjects wanting to be superiors and the ignoble equal to the noble, resulting eventually in a certain neutrality and beastly equality like it is found in some wild and uncivilized states. And furthermore, do you not see what pitiful condition the sciences have come to, because of the pedants who had wanted to be philosophers, to deal with natural things, and to intrude and meddle with defining things divine?[67]

To the masses, contact with the divine is not given in a personal philosophical quest, but mediated through the creeds and cults of an established religion. Religions are bound to the people and country of their origin (this is why Christianity, as an "imported" religion, compares unfavorable to the autochthonous religions of ancient Egypt and ancient Rome[68]), and while their alleged aim is to serve God, their real purpose is to establish the order and identity of a state and to encourage social virtues in the people as a whole and in the individual person:

> no law must be accepted that is not directed towards the practice of human community life . . . Whether it comes down from heaven, or out of the earth, a constitution or law must not be approved and accepted, if it does not bring about the utility and profit that leads us to the best goal. And we cannot attain a greater goal than that which directs the souls and reforms the minds in such a way that they bear profitable and necessary fruits for the life of the human community. For that must certainly be a divine thing, and the art of arts and discipline of disciplines, by which men are to be ruled and restrained, who are amongst animals the ones with the most different complexions, the most various habits, the most diverse inclinations, the most instable impulses.[69]

67. Ibid., 373.

68. This is one of the central topics of the dialogue *Spaccio della bestia trionfante* (*Expulsion of the Triumphant Beast*).

69. Giordano Bruno, "Spaccio della bestia trionfante," in *Oeuvres complètes* (Paris: Les Belles Letters, 1999) V, 2, 197.

"For of all the laws some have been given by us [gods], and others invented by men, primarily for the advantage of human life. And as some do not see the fruit of their merits in this life, the good and evil, the price and punishment of the next life, according to their deeds is promised them and placed before their eyes."[70]

"Because the gods care only for those things for which humans can care, and are not moved or angered by anything that is done, or said, or thought for them, except if it caused the loss of that respect by which the republics are maintained. For the gods would not be gods, if they had pleasure or displeasure, sadness or joy in what men do or think . . ."[71]

Thus established religions are always necessarily pious fraud, and not only because their pretended aim (to honor God) does not correspond with the true aim (the good of the people in this life), but also because they are always fashioned for the ignorant and profane by those who know better, the founders of religion and the initiate priests. It is the task of the religious leaders to translate the truth they know into rites and fables for the people, and the criteria that make a good or bad religion are not to be found in its sincerity (since it is not desirable for the people and the priests to believe the same things in the same way), but in the appropriateness of the fable for the people and state it serves as a surrogate or disguise for truth. Comparing the traditions of the great flood in various religions Bruno suggests ironically:

> he should make an end of the controversy, whether he had been up to now in heaven as the father of the Greek, of the Hebrew, or of the Egyptians, or others. And if his name is Deucaleon, Noah, or Otrius, or Osiris . . . And he should teach regarding these two different ways of telling (as they cannot be both history) which one is fable, and which one history. And if they are both fables, which one is the mother and which the daughter. And he should see to it, if he can reduce them to a metaphor of some truth worthy to be concealed.[72]

Wonders produced by magic (natural magic and demonic magic being a difference in degree, not in kind) are a legitimate means employed by the founders of religion to make the vulgar believe what they do not believe themselves: that it is an *immediate* divine intervention.

70. Ibid., V, 1, 147.

71. Ibid., V, 2, 201.

72. Ibid., 453–55.

But while such actions are not morally reproachable in themselves (since they are a necessity in the life of a people), they are bad if they lead to a bad and perverse religion (i.e., one that does not promote virtue). Far from despising the role of a religious leader for a philosopher, Bruno seems to have at least considered that possibility for himself.[73] But this does not, of course, justify any kind of fraud and abuse of the gullibility of people in religious matters by unworthy priests of useless religions: "Therefore a scoundrel will never be capable of gaining honor by serving, with the assistance of fiends, as an ape and mocker of blind mortals . . . to prevent future scandal . . . I [Jupiter] command that he lose all faculty to make tricks, impostures, legerdemains, elegant turns, and other marvels that are good for nothing . . . I see how easily deceived this world is, and accordingly inclining to craziness and ready for any corruption and indignity. Therefore I do not want that our reputation should be at the discretion of himself or others of his kind."[74]

Since for Bruno philosophy measures higher than any established religion, appreciation of the divine and numinous is reserved to the few—and consequently factual religions are the field of operation of leaders for good or for ill.

• • •

Though Francesco Patrizi (1529–1597) was Bruno's senior by nearly a whole generation, his reflections on wonder, on the admirable, and on the epistemological function of admiration were composed at a later date than Bruno's Italian dialogues. They belong to that part of Patrizi's monumental poetics project which remained unpublished during the author's lifetime. In fact, while the first two parts (*La deca istoriale* and *La deca disputata*) appeared in 1586, parts 3 through 7, though almost completed inside the following two years, were apparently abandoned by their author, who then chose to concentrate exclusively on another even more ambitious project: the *Nova de universis philosophia* (completed by 1591). The manuscript of the posthumous parts of the *Poetica* were

73. This is apparent from the denunciations by Giovanni Mocenigo and by Francesco Graziano, a fellow prisoner of the inquisition in Venice, who are both lacking the imagination for inventing such an accusation. See Luigi Firpo, *Il processo di Giordano Bruno* (Roma: Salerno Editrice, 1993) 158 and 251.

74. Giordano Bruno, "Spaccio della bestia trionfante," in *Oeuvres complètes* V, 2, 467.

discovered by Paul Oskar Kristeller in 1949, and were first published in the critical edition in 1969.[75] This makes it, of course, impossible for the text to have exercised any direct influence on succeeding generations, though its main ideas might have circulated in Ferrara, where Patrizi lectured on Platonic philosophy. One of the main theses touching the wondrous actually appears on a list of *Paradoxa poetica*, which were probably discussed at the *studio* of Ferrara.[76]

After having (in the second part of his poetics, titled *La deca disputata*) energetically rejected the Aristotelian notion of poetry as mimesis (imitation of reality), Patrizi explores the nature, purpose, and effect of poetry in the third part (the first and most complete portion of the posthumous manuscripts) under the title *La deca ammirabile*.[77] This is arguably the most innovative and original part of his poetics if not of his philosophy as a whole. Interpreting the myth of Orpheus' taming of the wild beasts, Patrizi declares that "The working of miracles . . . through symbolizing in poetry and by deeds beyond it" belongs to the offices the gods entrusted to the poets.[78] All specific properties and devices of poetry (such as verse, enigmatic and unusual speech, allegory, exaggeration, fiction, variety, ornament, paradoxes, presenting the divine as well as the monstrous) have the one common trait: they aim at producing wonder and admiration.[79] The wondrous (*mirabile*) is both the formal and the final cause of poetry:

> It seems to us that the wondrous must certainly be such in the poem as the mind or reason is in man. And if this reason is to be called his form, the wondrous will be the form of poetry; and if, as it is, reason is not [the form] of this or that other man, but of all men altogether, thus the wondrous must be [the form] not of one or the other poem, but of all poetry there is. And owing to that wondrous [the poet] surpasses all other writers of whatever kind in the same way as man, owing to reason, surpasses all other animals.[80]

75. Francesco Patrizi da Cherso, *Della Poetica*, ed. Danilo Aguzzi Barbagli (Firenze: Istituto Palazzo Strozzi, 1969) vols. 2–3.

76. Cf. Francesco Patrizi da Cherso, *Lettere ed opuscoli inediti*, ed. Danilo Aguzzi Barbagli (Firenze: Istituto Palazzo Strozzi, 1975) 168.

77. Patrizi, *Della Poetica*, 2:331–68.

78. Ibid., 240.

79. Ibid., 257–69 and 305.

80. Ibid., 329.

> Since the poet is the producer of the wondrous, and that, in its turn, the producer of wondering, the wondrous will be the scope of the poet's production, and wondering the scope of that production itself, and, consequently, wondering will be the scope of both the poet and the poem ... And these are the two final causes and universal properties of all poetry.[81]

And, according to Platonic tradition, poetry is the cradle of all human culture, of law, philosophy, and science:

> Plato called them *fathers and leaders of science*. And he was right in saying so, since we see that the poets have been the first ones to deal with every kind of wisdom. And therefore Strabo claimed that poetry had been the first philosophy. And this was true. That is, for many centuries before the usage of prose (as we have shown) all sciences of things, divine as well as natural, or human, were written in poetry.[82]

The second half of the work (chapters 6–10) deals with the central questions: what is the wondrous (*mirabile*), and what is wondering (*maraviglia*)? The wondrous, Patrizi contends against contemporary Aristotelians, cannot be produced inside the limits of the mere credible and probable.[83] It is rather achieved by a union or mixture of the opposites credible/incredible, probable/improbable, common/uncommon, true/false, but always with an emphasis on the second element of these couples, since the credible and ordinary never produce the essential effects of novelty and suddenness.[84]

> [T]o conclude, we say that the wondrous is, by its nature, produced neither through the order of the credible alone, nor through the order of the incredible alone, but when one order is mixed with the other, and when one credible thing appears to have assumed the face of an incredible, or an incredible one appears to have assumed the face of a credible. Thus it is the mixing of both, credible and incredible, that will produce wondering, and the wondrous will be nothing else but such a conjunction, by which any incredibles become credible, or any credibles become incredible.[85]

81. Ibid., 343.
82. Ibid., 237–38.
83. Ibid., 297–300.
84. Ibid., 305–7.
85. Ibid., 310.

As the detailed specification of presumably all such possible conjunctions (close to infinite in number, according to Patrizi himself), which is given in the next chapter, is both tedious and confusing, let us just draw the consequence concerning the subject of poetry: "That the matter of poetry is all being and nonbeing."[86] Thus the poet, leaving all other human artists far behind, constitutes the third order of genuine creators (besides God and Nature), he creates a second world out of words, and at times Patrizi seems to concede to him even the ability to create *ex nihilo*.[87]

But the most daring and original thesis of Patrizi's aesthetics as a whole is contained in the last chapter, dedicated to the question what this sense of wonder or marveling is in itself.[88] For here he states that *maraviglia* is neither a part or product of the rational faculty of the soul, nor an affect, but rather a commotion produced by a special faculty of the human soul, a *potenza ammirativa*, intermediate between the intellectual and the affective, and communicating with both—indeed, connecting them with one another:

> Therefore cognition proceeds from it (wondering) in the same way, as itself proceeds from the affects. Therefore apparently one can say that it has its place also in the cognizing faculty, or, rather, that it has an intermediate place between the cognizing and affective faculties.[89]

> Thus it will be an intermediate faculty between those two, and its motion a third motion in between those, running through all of them, from the middle position up to the high and down to the low. This motion being called admiration, it will have to give its faculty the name *potenza ammirativa*, since the other two have likewise been named after cognition and affect . . . the now so-called faculty of admiration would be neither cognizing, nor affective, but separate from both and communicating with both.[90]

86. Patrizi, *Lettere ed opuscoli inedite*, 168: "Materiam poesos esse omne ens et non ens."; cf. also *Della poetica*, 2:248–49.

87. Patrizi, *Della Poetica*, 2:285–87.

88. Ibid., 355–68: "Che cosa sia la maraviglia."

89. Ibid., 358.

90. Ibid., 361.

The function of this faculty is to transfer the force of reason down to the affects, and, vice versa, the force of the affects up to reason, thus triggering the mutual influence of these opposed faculties on one another.[91]

These are exciting ideas, though it must be confessed that the extremely condensed presentation makes them sound much more striking than they appear in the original. Unfortunately the erudite humanist Franciscus Patritius has a knack of getting in the way of the independent thinker Patrizi, urging him to seek and find corroborating quotations from antique authors even for his most original theses. This cannot but lead to repetitiveness, inconsistencies, and some contradictions. Consequently the most promising starts in Patrizi's philosophy end up as partly unfulfilled promises. This also applies to the fascinating theory of the *potenza admirativa*. Given that Patrizi defines wondering as the motion accompanying all doubtful and incomplete notions, and, on the other hand, given that virtually all human knowledge is incomplete,[92] we might expect the newly found faculty to play a leading role in the *Nova de universis philosophia*. If so, to our complete disappointment the only mention of admiration there appears to be a comparatively uninspired passage at the very beginning of the first book, where perceiving leads to admiration, admiration to contemplation, and contemplation to philosophizing.[93] This is never taken up or enlarged upon throughout the whole work. A faculty of admiration simply does not figure in Patrizi's more conventionally Neoplatonic opus magnum.

A possible reason for this is the unfinished state of this central theory of the poetics. Much remains undefined in the actual gnoseological function of the power of admiration, and, above all, its own epistemological value. (What does a connection between the true and false entail?) Consequently the relationship between poetry and philosophy as well as between divine and human creativity remains dubious. On the one hand, the poet is presented as the original, divine "better" philosopher, endowed with magical power and the last traces of an original

91. Ibid., 361–62.

92. Ibid., 290 and 363–67.

93. Francesco Patrizi, *Nova de universis philosophia* (Ferrara: Apud Benedictum Mammarellum, 1591) 1r.–1v: "Per haec, prisci homines, sublimia, & media, & ima conspexerunt. Conspecta sunt admirati. Admirando sunt contemplati. Contemplando, sunt philosophati. Philosophia ergo, lucis, luminis, admirationis, contemplationis proles est verissima."

language able to express the essence of things;[94] on the other hand, as in Giordano Bruno's *Spaccio*, the poet is primarily a producer of useful lies, that keep the ruder folks (children, women, and stubborn villains), who could not be reached by true science, inside the rules of the community.[95] It is not improbable that in spite of the formal completeness of *La deca ammirabile*, Patrizi himself felt that there were too many open questions and loose ends, and that it was not yet quite ready for publication.

• • •

Tommaso Campanella (1568–1639) has elaborated an appreciation of miracles in the framework of a philosophy of religion. Although he also refers to miracles based on cosmological links between various ontological levels when he speaks about magic and astrology, the religious meaning of miracles is separate from their natural meaning, because the religious meaning emphasizes only the supernatural as a constitutive feature of religion. Out of his many works it is sufficient to focus on his *Atheism Overcome* (*Ateismo trionfato*/*Atheismus triumphatus*, 1605–1607). The first chapters aim at establishing what is necessary in a religion: Divine reason pervades all things so that religion is, in a sense, natural to all (chapter 3). But given the variety of religions, a distinction is to be made between natural religion and regional customs (chapter 9). Consequently Campanella examines a variety of cults and founders of religions (chapter 13) and establishes criteria to assess their sincerity and veracity (chapter 14), which further requests scrutiny of the testimonies and of miracles (chapters 15–16).

Both to defend Christianity and to understand philosophically what constitutes religion, Campanella raised the question, what are the criteria to distinguish true from false religions, or, in other words, what makes religion authoritative? In chapter 14, Campanella lists six such criteria,[96] which, however, may be reduced to consistency and sincerity. His criteria are

94. Francesco Patrizi, *Della Poetica*, 2:237–40, 257–62, and passim.

95. Ibid., 291 and 348–54.

96. Tommaso Campanella, *L'Ateismo trionfato overo Riconoscimento filosofico della religione universale contra l'antichristianesmo macchiavellesco*, ed. Germana Ernst (Pisa: Scuola Normale Superiore, 2004) ch. 14, 186 sq.; cf. *Dio e la predestinazione*, 2 vols (Florence: Vallecchi, 1949–1951) lib. 1, vol. 1, 24–28, with ten criteria, and *Cristologia*, 2 vols. (Florence: Vallecchi 1958) vol. 1, 38 sqq., where the criteria are applied to the incarnation.

1. Miracles and martyrdom.
2. Prediction of the testimonies through prophets.
3. Prophecy by the believer himself.
4. A fellowship of miracle workers and martyrs.
5. Righteousness, firmness in speaking, and knowledge of divine issues.
6. Continuity over time.

The first and the fourth criteria are the most interesting for us: "Whoever is sent by God should have his patent with him as every King gives it to his ambassador. God's patent is divine and of effects and proofs that are not natural like those sent by nature but supernatural, namely, (1) miracles, (2) the capability to work great supernatural things, and then (3) staunchness against any adversity without fearing to die . . . Those who gave blood without miracles have a seal without patent like many heretics."[97] Note that these two, miracles and martyrdom, are inseparable in that they are both supernatural "patents" and mutually confirm each other. In religious terms, miracles and martyrdom testify to the authenticity of religious leaders. But Campanella is aware that they are also potential means of deceit. This raises the question concerning their epistemic status. In the framework of religion given here, mutual confirmation of testimony is the only solution Campanella has to offer. The best scenario he can think of is a fellowship of miraculous martyrs (criterion no. 4) that as a whole reiterates and confirms the message: "If there are more people who follow that doctrine and do the same thing with miracles, blood, and prophecy, then one can clearly conclude that the master of all is the true one, since there are not so many people capable of undetected fraudulence and accepting death in support of other persons' fraudulence."[98]

97. Campanella *Ateismo*, 186: "Chiunque è mandato da Dio, si conviene c'habbia la sua patente, come da ogni Re al suo ambasciatore. La patente di Dio è divina, e di effitti e dimostrationi non naturali, come quelli che da la natura son mandati, ma sopra naturali, e questi son li miracoli, far gran cose sopra naturali, e poi l'intrepidezza contra ogni avverstità, e non temer di morire . . . e chi ha sangue senza miracoli, ha il sigillo senza la patente, come molti heretici." Unless stated otherwise, all translations are ours.

98. Ibid., 187: "E se altra gente, sequitando quella dottrina, fa il medesimo con miracoli, sangue e profetia, si vede chiaro che, non potendo tanti esser capaci di furberia che non si scopra, né di consentire alla morte propria per fondar la furberia altrui, ne sequita che sia vero il mastro di tutti."

In reviewing a number of religious leaders, among whom are Mohammed, Pythagoras, and Socrates, all of them lack at least one of these criteria. This is especially true for the ancient Greeks, who have "no mandate from God" precisely for want of miracles, although they are "worthy to be taken as Christians in terms of natural law."[99] Here we need to observe that miracles constitute authority, and not vice versa. Skeptics might think that imposters set up miracles in order to impress the masses. To them (and to Bruno) Campanella must respond by saying that miracles cannot be set up or feigned, and therefore miracles (those that deserve that name) constitute authority. To wonder, then, must be a virtue of recognition. Consequently, sincerity is Campanella's major criterion to assess religions and their founders: lies and deceit are his accusations against Mohammed, Numa Pompilius, Genghis Khan, Romulus, and others throughout chapters 13 and 14. That is to say, religious lawgivers are either authorized by God (testified to according to the criteria mentioned), or they are mean spirited (if not devilish), or they follow natural reason.[100]

At one point he admits that the belief in the Eucharist is the most difficult for Christians to gain and is only warranted by the gospel, but then he associates it with natural magic, which receives its power from belief, so that the eucharistic mystery amounts to the "supernatural magic of Christ."[101] Therefore Campanella believes in Christ "philosophically"[102] because Christ is a reasonable consequence of the philosophical concept of God. "Once we believe in this incarnation, everything else proceeds by way of reason: it is appropriate for God to be born from a virgin, to be resurrected and ascend to heaven and to work miracles; also that he dies and suffers equally conforms to humanity, and all problems are easily solved."[103] For an understanding of the premodern interpretation of miracles, it is critical that Campanella makes miracles dependent upon either belief or authority or both; and vice versa: authority may not necessarily depend upon miracles, but miracles do constitute, or confirm,

99. Ibid., 189; Justin Martyr is the source here.

100. Ibid., 180.

101. Ibid., 170–73.

102. Ibid., 203: "Io di Christo sto molto ben persuaso filosoficamente."

103. Ibid., 134: "Creduta poi questa incarnatione, tutto il resto va per via di ragione: che nasca di vergine a Dio conviene; che resusciti, vada in cielo, faccia miracoli, e che moia e patisca tanto convenne all'humanità, e tutto si resolve de facili."

authority. This is notably neither an epistemology nor a metaphysics of wonder. There is no experiential tinkering with the validity of miracles; cause and effect are not on the horizon. Although divinity is invoked in this refutation of atheism, Campanella's affirmation can be best described as sociology of religion in the sense that it is a description of the factors that play a role in religious attitudes (in this case, in attitudes toward miracles). Epistemology and ontology have become separate from religion as such.[104]

Campanella is best known as the author of *The City of the Sun* (1602), clearly a theocracy in its purest form, in which every detail is based on divine rules and astromagical practice. Now it becomes clear why there is no talk about wonder in that perfect state: the religion is entirely natural, the authorities are uncontested, and therefore they do not depend on miracles, bloodshed, or prophets. On the other hand, Campanella's philosophy of nature entirely depended upon some kind of animism and magic. One of his earliest works was dedicated to the "sense of things and magic," in which he argued that everything at any ontological level is endowed with some kind of sense that partakes in the general guidance of the world. In its concluding formula Campanella takes on the motive of admiration in such a way that it merges again with the universal sense of the world: "every thing shows to every other thing the beauties of the eternal Idea. Whoever admires them knows them, who knows them carries them out, who carries them out becomes a friend of God, partaker of the great universal ever blissful glorious wisdom."[105]

104. The indeterminacy between the naturalness and the supernaturalness of events that may be termed miracles led Marsilio Ficino to play down the importance of miracles worked by Christ. They are meant to convince believers of his divinity but there is no urge to explain them philosophically. This goes along, obviously, with why he is underemphasizing of the two natures in Christ, given the contiguity between all hypostases. In Tommaso Campanella it also becomes evident that under the circumstances of heightened rationality in sciences, miracles tend to drift toward the social and psychological meaning of religion. This will come to its fulfillment in an Enlightenment philosophy of religion. For instance, Johann Friedrich Wilhelm Jerusalem (1709–1789) endorsed the basically mechanistic worldview that does not allow for exceptions, and consequently he declared miracles to be not unnatural but to be serving educational purposes that suit divine wisdom. See Andreas Urs Sommer, "Neologische Geschichtsphilosophie. Johann Friedrich Wilhelm Jerusalems Betrachtungen über die vornehmsten Wahrheiten der Religion," *Zeitschrift für neuere Theologiegeschichte / Journal for the History of Modern Theology* 9 (2002) 169–217; 185.

105. Tommaso Campanella, *Del senso delle cose e della magia*, ed. Germana Ernst

• • •

Looking forward and backward into Scholasticism may solve some of the riddles of wonder. It is well known that Aristotle, in the second chapter of the first book of his *Metaphysics*, had explained that philosophy takes its origin from wonder. Francisco Suárez (1548–1617), the best known among the early Jesuit philosophers, commenting on this passage, differentiated it in the following way: "Aristotle said only: He who doubts and wonders, clearly thinks that he does not know. Therefore, he joined two things, namely, doubting and wonder. Thus, it does not seem necessary that everyone who wonders is ignorant, but only one who while he wonders is also doubting."[106] Suárez found this passage relevant for theology because it was an important issue discussed by Thomas Aquinas concerning the humanity of Christ, where wonder or admiration features between fear and fury. Aquinas had maintained that although Christ could not be surprised from his divine point of view, and not even in terms of human knowledge, still he could make new experiences on the level of daily occurrences.[107] What Aquinas calls experimental knowledge (*scientia experimentalis*) attracted Suárez's interest, and it is his commentary on Aquinas to which he refers in his commentary on Aristotle. The question is for him, can there be true and genuine wonder in Christ?[108] The answer can be given by determining

---

(Bari: Laterza, 2007) 236: "ogni cosa, e mostrarà ad ogni altra cosa le bellezze dell'eterna Idea. Or chi l'ammira le conosce, chi le conosce l'opera, chi opera fia amico di Dio, partecipe della gran sapienza universale sempre beata e gloriosa."

106. Francisco Suárez, *A Commentary on Aristotle's Metaphysics* (*Index locupletissimus in Metaphysicam Aristotelis*), trans. John P. Doyle (Milwaukee: Marquette University Press, 2004) book 1, qu. 18, p. 28; 255: "Oportet autem advertere Aristotelem tantum dixisse: *Qui dubitat et admiratur, plane se ignorare existimat.* Duo igitur conjunxit, dubitationem scilicet et admirationem; non ergo necessarium videtur ut omnis qui admiratur, ignoret, sed solum is qui dum admiratur dubitat." Suárez used Pedro Fonseca's text and commentary: Petrus Fonseca, *Commentariorum [. . . .] In Libros Metaphysicorum Aristotelis Stagiritae, Tomvs primus* (Frankfurt: Schönwetter, 1599). (This edition was accessible to me.) Fonseca does explain this passage in Aristotle at lib. 1, cap. 2, explanatio, col. 125/126, but only by quoting Plato's *Theaetetus* 155d from Marsilio Ficino's translation.

107. Thomas Aquinas, *Summa Theologiae* III, qu. 15, art. 8, c: "In Christo autem non poterat esse aliquid novum et insolitum, quantum ad scientiam divinam, qua cognoscebat res in Verbo: neque etiam quantum ad scientiam humanam, qua cognoscebat res per species inditas; potuit tamen aliquid esse sibi novum et insolitum secundum scientiam experimentalem, secundum quam sibi poterat quotidie aliqua nova occurere."

108. Francisco Suárez, *Commentariorum et disputationum in tertiam partem Divi*

the possible modes of admiration. First of all, according to the Jesuit, admiration entails a cognitive state and presupposes some kind of desire (*in appetitu*). There are three kinds of wonder:[109]

1. Admiration arises occasionally from the ignorance of the cause, when the effect is seen, and consists in the desire to know.
2. Admiration arises from the observation of something great and unique, even if it is otherwise known. (In this case it is important that the thing is seen in a somewhat new way, so that an appropriate desire ensues, which may be called wonder.)
3. It also can happen that, although effect and cause are known, they are not understood, and the emanation of the effect from the cause is unclear, and that from this knowledge arises appreciation of the thing known (if pleasant), or fear, or the interest to inquire further into the thing. (This kind of wonder may have taken place in Christ's soul in some sense.)

Suárez explains for us that Ficino was talking about this second kind of wonder, namely, the apprehension of something magnificent that leads us to worship and to desire to know it in an appropriate way. He might also have had Ficino in mind when shaping definition number 3, for Ficino is certain about the origin of the event, and invites his readers to scrutinize and appreciate it within the human soul. While Suárez's first definition of wonder seems to be closest to what Aristotle had in mind when he was advocating the natural inclination of men toward philosophy, it is also the definition closest to any empiricist interpretation of knowledge. The fact that this kind of knowledge/wonder does not apply to Christ forced the Jesuit to redefine admiration. He explains that the first kind of wonder cannot apply to Christ because "causes are not known through an experiment as distinct from the experience of the effects that follow from the experiment; rather, one proceeds from the experiment to science."[110] As to the Bible story (Christ wondering about a believer's faith), Christ was not astonished about something within sensual experience, but about the person's faith that lies beyond experi-

---

Thomae tomus primus (Venice: Zenarius, 1593), commentary in qu. 15, art. 8, p. 525.

109. Ibid., 525–26.

110. Ibid., 525: "causae non cognoscuntur experimento distincto ab experientia effectuum, qui ab illis procedunt, sed ex hoc experimento ad scientiam proceditur."

ence. As to science, Suárez clearly describes the basic tenet of empiricism. Wonder number 2 certainly covers something lofty, like religious belief, but even more clearly it is defined by its object. Wonder number 3, finally, covers both experimental science and curiosity on all levels and rises to a phenomenology of wonder.

Since Suárez was compelled by Aquinas's solution of the puzzle to think of wonder in terms of cause and effect, he not only differentiated the conditions of amazement, but also delineated the scientific setting of causal knowledge. The latter being basically the paradigm of empirical (experiential) epistemology, the epistemic status of experience becomes questionable: Is it true that knowledge through experience yields scientific knowledge? Skeptical objections are to be expected. However, miracles—regardless of their origin—are no longer qualified by their explanation (or lack thereof), but by the magnitude of either the event or the impact on the observer's mind. It takes a miracle to believe in miracles—certainly nothing trivial. Miracles are not ruled out, and they are not necessarily to be identified with divine intervention, although that would certainly qualify as magnificent. Renaissance miracles are natural throughout, and the challenge is to understand nature.

2

Wonder, Magic, and Natural Philosophy

The Disenchantment Thesis Revisited

Koen Vermeir

> They say miracles are past, and we have our philosophical persons, to make modern and familiar, things supernatural and causeless. Hence is it that we make trifles of terrors, ensconcing ourselves into seeming knowledge, when we should submit ourselves to an unknown fear.
>
> —William Shakespeare, *All's Well That Ends Well*, II, iii, 1–6.

INTRODUCTION

EARLY MODERN MAGIC MIGHT be characterized in a paradoxical way as *the science of the wondrous*. This article will be an exploration of this phrase and of its consequences. The central terms in it, such as "magic," "science," and "wonder," have changed considerably and in the early modern period they were understood very differently from today. We will have to look into what exactly these concepts meant and what place they took in the broader intellectual culture of the time.

Wonder is a very complex category, with an intricate history. There was a long, classical tradition of interest in the marvelous, and an increasingly rich vocabulary was developed to classify and describe it.[1] But

1. In Greek, the word for "wonder" was *thauma* and *to thaumaston*, from the word for "to see." The Latin used variations of *mirare* (such as *admiratio*, *admirabile*, *mirabilia*, *miracula*, or *ammiranda*), which had its roots also in the verb "to see." The German *Wunder* is from unknown origin. All of these linguistic roots have left their traces in the English language, and from the fifteenth century onwards, a renewed interest in wonder could be expressed in a subtle taxonomy, including words such as *wonder*, *marvel*,

it was especially the role of the wondrous in society and in intellectual culture that changed radically over time. In most of history, wonders were despised and dismissed as trivialities. In the early modern period, however, wonders suddenly took centre stage in various domains of culture, and they played a crucial role in epistemology and "science."

Until the early modern period, "science" [*scientia*] was about certain, contemplative, and abstract knowledge, and it meant something very different from the laboratory practices we associate it with today. Theology and mathematics were examples of "sciences," because their statements were derived from certain first principles. In natural inquiry, certainty was much more difficult to attain. Aristotelian natural philosophy claimed to be a science and aspired to demonstrate its conclusions deductively from certain premises. As a consequence, however, many more empirical and practical aspects of the study of the natural world could not be part of natural philosophy. It was only in the course of the early modern period that natural particulars, experimentation, and hands on knowledge became accepted as part of natural philosophy.

Before the seventeenth century, "science" was only concerned with universals and the general order of nature. Wonders and marvels were irrelevant in natural philosophy because they were anomalous, contingent phenomena of unknown cause. Therefore, the characterization of magic as the "science" of the wonderful is paradoxical from an Aristotelian perspective. First, a science of marvels is impossible because marvels are unique particulars while science deals with universals and regularities. Second, wonder stops where science and knowledge begin. Nevertheless, there was a general consensus in the early modern period that magic was the 'discipline' that studied and manipulated wonders.

At the time that magic started to disappear as a credible pursuit, the notion of "science" was revalued and it began to include wondrous and particular phenomena. This can make us wonder about the relation between magic, science, and the wondrous. Did science take the wondrous out of the world? Did science reject and replace magic, or did science revalue and incorporate aspects of magic? In this article, we will enquire whether the world was "disenchanted" by science, or whether science was "enchanted" by magic. During a relatively short but crucial time span in European history, roughly between 1600 and 1700, it seems

miracle, *admiration*, and *thaumaturgy* ("wonder making"), as well as *astonishment* ("to turn into a stone") and *bewilderment* ("to become wild").

that it would not be so preposterous or paradoxical to describe magic as "the science of the wondrous." This is the period when the notions of magic, science, and wonder were revalued, when they interacted and started to play new roles in intellectual culture. It is this period with which I will be dealing here.

WONDROUS PHENOMENA

Wonders are extraordinary phenomena. Throughout history, there was some ambiguity however as to whether their unusualness was objective or was only in the experience of the beholder. In the early thirteenth century, the English nobleman Gervase of Tilbury (c.1150–c.1228) outlined three categories of wonderful things: "We embrace things we consider unheard of, first on account of the variation in the course of nature, at which we marvel; then on account of our ignorance of the cause, which is inscrutable to us; and finally on account of our customary experience, which we know differs from others."[2] The first category describes really extraordinary phenomena, in the literal sense that they are outside the ordinary course of nature, such as monstrous births or strange apparitions in the sky. The second category may include more common phenomena that were incomprehensible at the time, because their cause was unknown or unknowable, i.e., the causes are "occult" or "hidden." The best example here is magnetic attraction, which did not really fit into the conceptual framework of Aristotelian natural philosophy and one did not know its cause. The third category is only uncommon in the eye of the beholder, and it signals ignorance or lack of experience. Examples might be exotic species that are unknown in the West, or better known objects that are only new to the ignorant beholder.[3]

2. Gervase of Tilbury, *Otia imperialia*, in *Scriptores rerum Brunsvicensium*, ed. Gottfried Wilhelm Leibniz (Hanover: Nicolaus Foerster, 1707) 960; quoted in Lorraine Daston and Katherine Park, *Wonders and the Order of Nature, 1150–1750* (New York: Zone, 1998) 23.

3. On the history of wonder, see especially Daston and Park, *Wonders and the Order of Nature*; Stuart Clark, *Thinking with Demons: The Idea of Witchcraft in Early Modern Europe* (Oxford: Clarendon, 1997); William E. Burns, *An Age of Wonders: Prodigies, Politics and Providence in England 1657–1727* (Manchester: Manchester University Press, 2002); Mary Baine Campbell, *Wonder & Science: Imagining Worlds in Early Modern Europe* (Ithaca: Cornell University Press, 1999); Charles T. Wolfe, , ed., *Monsters and Philosophy* (Texts in Philosophy 3; London: College Publications, 2005); Zakiva Hanafi, *The Monster in the Machine: Magic, Medicine, and the Marvelous in the Time of the Scientific Revolution* (Durham: Duke University Press, 2000).

There are many kinds of wondrous phenomena, and what was considered wondrous changed throughout the centuries. There was a canon of wonders, however, which remained relatively stable. One example is monstrous births, which had always been marvels and took pride of place. Especially human monstrous births were poignant and unsettling, because they put into question the nature and limits of humanity. They are uncommon, the causes that generated them are unknown, and they violate the ordinary course of natural generation and reproduction. A second example is the loadstone, which was not yet well known. Although regular and orderly, its mysterious attractions were a continuous cause for wonder, because of their incomprehensible nature.

It was in general not difficult to identify curious phenomena, but the debate evolved around the causes of these phenomena, and the same possibilities were always restated.[4] Of course, many wonders were considered to be of divine origin in one way or another. In the early modern period, marvels such as monsters, comets and strange meteorological phenomena were often still interpreted as portents, for instance, as divine signs that had to be interpreted. Portents were distinct from miracles, however. Portents were uncommon but natural phenomena, which had nevertheless a sign function. They could signify that God was unhappy with the current events, such as with political changes or with the moral degeneration of the people. In contrast, miracles were not natural phenomena but direct supernatural interventions from God. They were not related to worldly events but had an exclusively religious and theological meaning.

Not all wonders came from God. In the Middle Ages and early modern period, nature was not understood as acting according to inviolable laws. Nature, therefore, was not completely regular. Nature was active and ingenious and there was place for plays and sports of nature. Fossils, for instance, were interpreted as just such jokes.[5] It was thought that nature imitated animals in rocks, sometimes changing them into unrecognizable species, just for the sport of it. Occasionally, nature even

4. Koen Vermeir, "The 'Physical Prophet' and the Powers of the Imagination. Part I. A Case-Study on Prophecy, Vapours and the Imagination (1685–1710)," *Studies in History and Philosophy of Science* 35C (2004) 561–91.

5. Paula Findlen, "Jokes of Nature and Jokes of Knowledge: The Playfulness of Scientific Discourse in Early Modern Europe," *Renaissance Quarterly* 43 (1990) 292–331.

MEDIEVAL BESTIARY—According TO medieval bestiaries, barnacle geese, a marvelous species of geese, come from trees that grow over water, straddling the boundary between plant and animal. British Library, Harley MS 4751, Folio 36r.

---

depicted a religious scene in a stone or in a curious landscape. Nature created her own wonders.

More often, however, wondrous phenomena were the result of demonic action. Demons and devils could not act beyond the laws of nature. Only God could do supernatural deeds. Demons were considered to be like powerful natural magicians who knew all about nature and could manipulate it. Therefore, demons were often held responsible when nature did not act according to its ordinary course. They could change the circumstances, bring in obstacles and manipulate nature in many other ways in order to bring about monsters, strange apparitions and other marvels. If cause and effect were held to be disproportionate—something that could almost serve as a definition of a wondrous phenomenon—people suspected that a demon was involved. People who could do wonderful things were often identified as witches and evil magicians, who were accused for dealing with demons and making contracts with the devil.

The fourth possible origin of wonders was human. Nature herself sometimes acted as a master artist when she played with and transgressed her own rules. On the other hand, artists and artisans tried explicitly to imitate and enhance nature in their work. This already indicates the flexibility between the natural and the artificial in early modern reflection about wonders. Mankind had always manipulated nature, sometimes to wondrous effect. Artists and craftsmen began to construct objects that went beyond what could be imagined before. Sixteenth- and seventeenth-century cabinets of curiosity or *wunderkammern* were filled with marvels and wonders both natural and artificial. Early modern pleasure gardens were equipped with elaborate fountains in which moving statues spouted water at unsuspecting spectators. Speaking statues gave passersby a fright and, totally confounded, they sometimes interpreted the message as if it were an oracle. Often, artificial wonders would trick the spectator in an illusion, confounding the credible and the incredible.[6]

An excellent example is Athanasius Kircher's (1602–1680) museum at the *Collegio Romano* in Rome, which was filled with such artificial wonders.[7] Kircher describes, for instance, how the ingenious setup of mirrors in his museum can transform spectators in the most unexpected ways. The unwary visitor sees the color of his face changed in that of a liver patient. He suddenly sees a horn growing on his head, or sees his neck craning, making him similar to a real crane. In one of Kircher's instruments, the body of the spectator was reflected with his head substituted by an animal head. In this way, Kircher held a mirror with a moral message to the courtier, who could not discern his mask from his real face anymore. Kircher's museum was a real catoptrical theatre, in which the spectator doubts his own identity and is disguised for himself.[8]

6. On collecting and cabinets of curiosity, see Paula Findlen, *Possessing Nature: Museums, Collecting, and Scientific Culture in Early Modern Italy* (Studies in the History of Society and Culture 20; Berkeley: University of California Press, 1994); Barbara M. Benedict, *Curiosity: A Cultural History of Early Modern Inquiry* (Chicago: University of Chicago Press, 2001); Krzysztof Pomian, *Der Ursprung des Museums: Vom Sammeln* (Wagenbach Taschenbuc 227; Berlin: Wagenbach, 1988).

7. On Kircher's museum, see Giorgio de Sepibus, *Romanii Collegii Musaeum Celeberrimum* (Amsterdam, 1678); Gaspar Schott, *Magia Universalis* (Würzburg, 1657); Johannes S. Kestlerus, *Physiologia Kircheriana experimentalis* (Amsterdam, 1680); Filippo Buonanni, *Musæum Kircherianum* (Rome, 1709); see also Michael J. Gorman, "Between the Demonic and the Miraculous: Athanasius Kircher and the Baroque Culture of Machines," in *The Great Art of Knowing: The Baroque Encyclopedia of Athanasius Kircher*, ed. Daniel Stolzenberg (Stanford: Stanford University Libraries, 2001) 59–70; and Koen Vermeir, "Athanasius Kircher's Magical Instruments: An Essay on Science, Religion and Applied Metaphysics," *Studies in History and Philosophy of Science* 38 (2007) 363–400.

8. Kircher, *Ars Magna Lucis et Umbrae* (Amsterdam, 1671), 775, 783, 786; Koen

This short discussion of the four kinds of wonders already suggests that the reception of an uncommon phenomenon was always seen from a specific framework. A wondrous phenomenon could be interpreted as a divine intervention, as the result of a playful nature, as the effect of demonic involvement or as human artifice. Sometimes, extra conditions and circumstances could clarify the situation, but usually the same phenomena stood open for widely divergent interpretations. Many wondrous phenomena were therefore highly contested and the subject of fierce controversies. For instance, when on the third of February 1688 a young woman suddenly started to prophesy, convulsed and spoke various languages as if in a trance, she could be interpreted as being a holy prophet, a deluded hypochondriac or the victim of demonic possession. A fourth possibility was also open, that her prophecies and convulsions were artificial and brought about by a certain trick. Which of those four possibilities was the right one was open to vigorous debate, often with strong moral, political or religious consequences. In this case, these prophecies sparked a Huguenot revolution that would rage in the South of France for years.[9]

Contemporaries used categories which had informed similar debates about unusual events for centuries. Godly miracles, the cunning of the devil, hidden natural principles and human artifice were persistently evoked as causes of wondrous phenomena. One could make a further distinction between the *effects*, as real and imaginary. Each of these two could be caused by God, nature, demons (good or evil) or man. If the perceived effect was only *imaginary*, this could be caused by God or his messengers (mystical visions, real prophecy, when the imagination is pure and perfected), nature (the disturbed imagination of an ill person) or demons (which delude the imagination). The last possibility was the involvement of human art (mostly seen as fraud, or as artificial magic), which did not really delude the imagination, but which deceived the spectators with an *apparent* effect.

One kind of imaginary wonders was created in the literary imagination. A work of Francesco Patrizi (1529–1597), *La deca ammirabile* (1587), was an important one for Renaissance literary theory, breaking

---

Vermeir, "Mirror, Mirror, on the wall . . . On the Aesthetics and Metaphysics of 17th Century Scientific/Artistic Spectacles," in *Kritische Berichte, Zeitschrift für Kunst- und Kulturwissenschaften* 32:2 (2004b) 27–38.

9. Vermeir, "The 'Physical Prophet' and the Powers of the Imagination. Part I."

with Aristotelian tradition. Patrizi argued strongly in favor of creating wonder as the task of literature. In an expanded taxonomy of wonder, he introduced twelve sources of wonder available to the poet: ignorance, fable, novelty, paradox, augmentation, departure from the usual, the extranatural, the divine, great utility, the very precise, the unexpected, and the sudden. In particular, Patrizi argued against verisimilitude and the imitation of nature, and stated that the maker of the marvelous had a duty to refashion reality. This was in contrast to the Aristotelian stress on credibility, verisimilitude, possibility, necessity and truth.[10]

Patrizi's literary theory attests to the new prominence of wonder in early modernity. Similar ideas were advanced in the other arts, but here not only imagined literary wonders were constructed but also *real* marvels were built. The best known manifestation of early modern wonder is probably the vogue of *wunderkammern*, in which these real artificial wonders were collected. The *wunderkammer* is a sign of the times, and attests to the prominence of wonder, but it also reflected distinctions present in the broader cultural setting. Kircher's museum, for instance, was like a big show, similar to illusionist paintings or baroque theatres excelling in stage machinery, devised to baffle courtiers and dignitaries. I have already related that Athanasius Kircher's artificial marvels sometimes had a moral significance, but more often it represented religious and political messages. It was a museum of wonders for the elite, where the latest techniques and inventions found a place in a Catholic framework, and Kircher played with hidden messages that served to reinforce his religious position. The common man, it was supposed, was prone to superstition. Only the educated gentry were able to enjoy the play and read the hidden messages. Marvellous devices referred to the incapacity of the vulgar and helped constitute a distinct elite epistemic culture.

The third category of wonders discussed by Gervase of Tilbury referred to the ignorance or lack of experience of the beholder. This category of wonder reflects some intellectual elitism, because knowledge of wonders distinguished cognoscenti from the vulgar. Knowledge and wonder went hand in hand. Therefore, collecting marvels and wonders became a sign of social elevation and prestige. This tradition of collecting exploded in the early modern period into a craze of wonders and *wunderkammern*, providing some social standards and distinctions in a

10. Peter Platt, "'Not Before Either Known or Dreamt of': Francesco Patrizi and the Power of Wonder in Renaissance Poetics," *Review of English Studies* 43 (1992) 387–94.

time of social, political and religious instability. The rare and extraordinary was valuable, because it fascinated, but also because it was scarce. Marvels therefore played an important role in a gift culture. Scholars and courtiers aspired for the patronage of the mighty and tried to lure them with gifts: the attribution of a book, a marvelous natural rarity or an exceptional work of art. Galileo even named his new discovery which was one of the greatest wonders of the time, the four moons revolving around Jupiter, after Cosimo II De Medici, as a gift in order to win the Grand Duke's patronage. In order to smoothen international relations, diplomats often exchanged wonders, which attests to the value attributed to them.

The excited activities of collecting and creating wonders had many goals. Besides aiming at the emotion of wonder, combined with pleasure, curiosity or fear, they had important social functions in an early modern court society. Wonders, once presented as gifts to a King or Prince, often ended up in a cathedral or cloister, to evoke awe at the wondrous variety of God's creation. They had also an epistemic significance in that they could hint at new understandings of nature. It was not always clear, however, whether a wondrous effect was natural or whether it was caused by God, the devil, or human artifice.

MAGIC AND WONDER

Magic is a term almost impossible to define, but among early modern writers, there was a broad consensus that magic was the enquiry into the wonderful. The Protestant philosopher and encyclopaedist Johann Heinrich Alsted (1588–1638) wrote, "magic is the art which is concerned with wondrous effects [*apotelesmas*], commonly known as incredible."[11] The Jesuit scholar Gaspar Schott (1608–1666) for his part defined magic as "whatever is marvellous and goes beyond the sense and comprehension of the common man."[12] Of course, such a 'definition' cannot be taken as comprising necessary and sufficient conditions. Some of the alternative descriptions of magic were more specific, or singled out one

11. Johann H. Alsted, *Encyclopaedia*, 7 vols. in 4 bks. (Stuttgart: Frommann-Holzboog, 1989; reprint of original edition of 1630) 4:2267: "Magia est ars occupata circa apotelesmata mirabilia, & vulgo incredibilia."

12. Gaspar Schott, *Magia universalis naturae et artis* (Würzburg: 1657–1659) 1:18. See also Norbert Henrichs, "Scientia Magica," in *Der Wissenshaftsbegriff: Historische und Systematische Untersuchungen*, edited by Alwin Diemer (Meisenheim: Hain, 1970) 30–46, for various early modern and Enlightenment definitions of *magic*.

specific aspect. Francis Bacon (1561–1626), the English statesman and philosopher, favors the operative character of magic, for instance, when he writes that he understands magic "as the science which applies the knowledge of hidden forms to the production of wonderful operations; and by uniting (as they say) actives with passives, displays the wonderful works of nature."[13] This operativeness is not an essential characteristic, however, and the occult philosopher Agrippa von Nettesheim (1486–1535) also stresses the contemplative potentials of magic, making it akin to mysticism: "Magick is a faculty of wonderfull vertue, full of most high mysteries, containing the most profound Contemplation of most secret things."[14] What binds these different definitions of magic together is the stress on the wonderful, and it has preserved these connotations until the present day, together with related words like magical, fascination, enchanted, and bewitched.

In early modern Europe, scholars and intellectuals relegated magic to the domain outside science and outside religion, which dealt with the ordinary course of nature and with supernatural events respectively. Magic belonged to the realm of the *preternatural*, i.e., of the wonderful but not supernatural. This was one of the early modern epistemic categories that guided the understanding of phenomena. According to traditional scholasticism, magic comprised what was excluded by supernatural and "regular" phenomena. The supernatural was a strictly defined category that described God and God's actions in the world (like miracles), but excluded the actions of demons and spirits, since these could not do anything beyond the course of nature. The sciences, on the other hand, conceptualized the regular workings of nature along Aristotelian lines by means of manifest qualities. Everything that was inexplicable in this way,

13. See Francis Bacon, *De augmentis scientiarum*, in *The Works*, 14 vols., edited by James Spedding et al. (London: Longman, 1857–1874), 4:366–67.

14. See Heinrich C. Agrippa, *Three Books of Occult Philosophy*, edited by Donald Tyson, translated by James Freake (St. Paul, MN: Llewellyn, 1998/15311) I, 2. The quotation continues: "together with the nature, power, quality, substance, and vertues thereof, as also the knowledge of whole nature, and it doth instruct us concerning the differing, and agreement of things amongst themselves, whence it produceth its wonderfull effects, by uniting the vertues of things through the application of them one to the other, and to their inferior sutable subjects, joyning and knitting them together thoroughly by the powers, and vertues of the superior Bodies." Operativeness is often part of magic, but is not a necessary condition. Other traditions stressing the contemplative and mystical nature of magic are Hermeticism and Lullism. See also chapter 1 above, by Elisabeth Blum and Paul Richard Blum.

The magician Zoroaster performing demonic magic, in a manuscript from 1425. Add. MS 39844, f.51.

like magnetism and all kinds of particular virtues of plants and metals, was explained by reference to "*occult*" or "hidden" qualities. Such qualities could not be generalized, and this is why, according to Aristotle and the scholastic philosophy, they could not be part of a real *scientia*. These occult virtues, often specific to a particular substance, were studied in "magic" instead. (This professed lack of generalization already explains the strong empirical tendencies in magic.) The traditional scholarly canon thus "defined" magic in opposition to science and religion, which proved to be crucial for the history of magic in the West.[15]

15. On the preternatural, see Daston and Park, *Wonders and the Order of Nature*; Clark, *Thinking with Demons*; Daston, "Preternatural Philosophy," in *Biographies of Scientific Objects*, edited by Lorraine Daston (Chicago: University of Chicago Press. 2000) 15–41.

The four different ways of explaining wondrous phenomena were reflected in a common classification of magic into four kinds. It was widely agreed that God (directly or by means of the angels), demons, nature or humans were the possible causes of any marvel.[16] Different groups of people, theologians, natural philosophers, legal experts, physicians, gentlemen, and others, had specific preferences as regards these explanatory categories. Nevertheless, this distinction in four categories was a standard cognitive tool for all of them, and taken together these categories exhausted the possible causes that contemporaries could conceive of. Except for a direct intervention by God, which would be a supernatural event, contemporaries considered and negotiated the possible distinct kinds of magic related to these causal categories, i.e., angelic magic, demonic magic, artificial magic, and natural magic respectively.

I should add the caveat that the four explanatory categories that could be invoked, i.e., good spiritual entities (God and angels), demons, nature or humans, were usually underdetermined by the specific phenomena. The example of the young woman prophet already showed that the difference between divine, demonic, natural and artificial marvelous phenomena was often very hard to make in practice. Most characteristics of such phenomena could be interpreted in various ways. Different factions (physicians, theologians, magicians) had other ideas about what criteria were relevant for interpreting them. Marvelous phenomena assumed a different shape—e.g., as prophecy, madness, possession or politically inspired fraud—depending on who is looking. Therefore, the different kinds of magic are impossible to relate to specific practices or phenomena.

Angelic magic or 'theurgy,' propounded and practiced by Giovanni Pico della Mirandola, Agrippa von Nettesheim, and John Dee amongst others,[17] verged on religion, and often involved an ascetic spirituality that aimed at a mystical union with God. This kind of magic was very contentious indeed, especially for orthodox religion, which feared new heresies or even an alternative religion. Yet even the magicians admitted

16. See also, e.g., Alsted, *Encyclopaedia*, 4:2267, who classifies miracles "in the broad sense" in these four categories. (Miracles in the broad sense are wonders. In the strict sense, miracles are strictly divine interventions.)

17. For Pico and Agrippa, see the previous chapter by Elisabeth Blum and Paul Richard Blum; for John Dee, see Stephen Clucas, "Enthusiasm and 'Damnable Curiosity': Meric Casaubon and John Dee," in R. J. W. Evans and Alexander Marr, eds., *Curiosity and Wonder from the Renaissance to the Enlightenment* (Aldershot: Ashgate, 2006) 131–48.

that it was very difficult to make sure that only angels instead of demons were involved, and some contemporaries rejected the possibility of angelic magic. That is why the Jesuit demonologist Del Rio (1551–1608), when he classifies magic "under the heading *efficient cause*," only distinguishes between "Natural, Artificial, and Diabolic [magic] because all its effects are to be ascribed to the innate nature of things, or to human agency, or to the malice of an evil spirit."[18]

Artificial magic includes all techniques used to create wondrous effects, and comprises the more spectacular aspects of technology (as opposed to the usual crafts). Also, instruments that created illusions like the magic lantern or stage machinery were part of this tradition. Jugglery and legerdemain (and "fraud") are part of magic caused by human skill, but scholars usually did not include them in the more prestigious category of "artificial magic." Some contemporaries used "mathematical magic" as a synonym for artificial magic, but the former is a more ambiguous term and was sometimes taken to include any kind of magic that involved numbers, such as parts of astrology, parlor games and the Kabbalah. Since human ingenuity works by means of natural objects, many practitioners considered artificial magic to be part of natural magic. Therefore, when we do not consider angelic magic and when artificial magic is seen as part of natural magic, the basic distinction in magic can be reduced to the opposition between the natural and the demonic.

Giambattista Della Porta (1535–1615) writes: "There are two sorts of Magick; the one is infamous, and unhappy, because it has to do with foul Spirits, and consists of incantations and wicked curiosity; and this is called Sorcery; an art which all learned and good men detest ... The other Magick is natural; which all excellent wise men do admit and embrace, and worship with great applause; neither is there any thing more highly esteemed, or better thought of, by men of learning."[19] Demonic magic works by means of demons. These demons were not "supernatural," but

18. Martin A. Del Rio, *Disquisitionum magicarum libri sex* (Louvain, 1599–1600) I, 2. Apart from his distinction as regards efficient causes, Del Rio also classifies magic in terms of final causes: "It may be divided into (a) 'good magic', provided this be done with good intention and uses lawful methods, something which applies only to Artificial and Natural Magic; and (b) 'evil magic', whose methods and ultimate aim are both depraved. This refers particularly to forbidden magic which is tacit idolatry and a type of superstition." We see here the typical divide between natural and demonic magic as good versus bad.

19. Giambattista Della Porta, *Natural Magick* (London, 1658) bk. 1, ch. 2.

they were considered to know how to manipulate the subtlest aspects of nature. Magicians and witches communicated with the demons by means of strange symbols, incantations or magical objects such as talismans. This kind of magic was known as the black art and was often called necromancy.

In contrast, natural magicians only worked with natural means. No demons, angels or other spiritual beings were involved. They studied nature and made experiments in order to be able to manipulate nature and to create wonderful effects. The different chapters in Porta's book give a nice sample of the diversity of subjects in which natural magic was involved. Porta discusses the generation of animals and plants (showing how living creatures of diverse kinds may be mingled and coupled together, that from them, new and yet profitable kinds of living creatures may be generated), the changing of metals, counterfeiting gems, the wonders of the loadstone, ways to beautify women, distillation, perfuming, artificial fires, cookery, fishing, fowling, hunting, invisible writing, optical experiments, static experiments and pneumatic experiments. Porta himself was the most famous proponent of natural magic, and after rejecting other definitions of magic (such as "the practical part of natural Philosophy" or the "science whereby inferior things are made subject to superiors, and earthly subdued to heavenly"), he concludes: "I think Magick is nothing else but the survey of the whole course of Nature."

ENCHANTMENT OR DISENCHANTMENT?

There is a general perception that our culture has changed from the magical worldview of the Middle Ages into a modern world characterized by rationalism and empiricism. The driving force behind this transformation is usually identified as the scientific revolution, which took place in Europe in the seventeenth century. Somewhere on the way, because of the instauration of a scientific worldview, magic and wonder were eliminated from the world. One strong expression of this view is the catchphrase 'the disenchantment of the world,' made famous by Max Weber (1864–1920) in his text, *Wissenschaft als Beruf* (1918).

As it turns out, Weber did not write so much about disenchantment, and he focused especially on the disenchanting effects of the protestant reformation and the increasing importance of calculability. But together with his thesis that the rise of Protestantism was crucial for the origin of science, a thesis later adapted by Robert Merton (1910–2003), the disenchantment thesis came to represent the idea that the rejection of magic interlocked with the origin of science. The impact of this idea has been enormous, and later commentators have cashed in on this expression in different ways. The oldest versions of the disenchantment thesis coincide with triumphant accounts of the scientific revolution. They recount a story of secularization and naturalization. Nature became divested of spiritual qualities and science and religion were separated.[20]

The historical record, however, shows us a different picture. Important natural philosophers tried to find experimental evidence for spirits, for instance. Robert Boyle (1627–1691), founding father of the Royal Society and one of the most prominent natural philosophers of the seventeenth century was very interested in expanding his natural philosophy to the spirit world. Furthermore, for almost all early modern natural philosophers, natural inquiry and religion went together. For some of them, natural philosophy was even a form of religion. In the recent literature on science and religion, it is persistently stressed that religion often had a very positive influence on the development of science, and even Catholic religious orders such as the Jesuits made many significant scientific contributions. Even the Galileo affair, often used as a case to illustrate the antagonism between religion and science, has been reinterpreted and Galileo's condemnation has been explained as the result of local power politics and personal relations between the main historical actors.[21]

20. Max Weber, "Wissenschaft als Beruf," in *Gesammelte Aufsätze zur Wissenschaftslehre* (Tübingen: Mohr, 1922) 524–55. Originally, the phrase was coined by Schiller. On Weber, see Lawrence Scaff, *Fleeing the Iron Cage* (Berkeley: University of California Press, 1989) esp. 223–41. On disenchantment, see e.g., Ralph Schroeder, "Disenchantment and its discontents," *The Sociological Review* 43(2) (1995) 227–50; Charles Webster, "Paracelsus, Paracelsianism, and the Secularization of the Worldview," *Science in Context* 15(1) (2002) 9–27; Keith Thomas, *Religion and the Decline of Magic: Studies in Popular Beliefs in Sixteenth- and Seventeenth-century England.* (Harmondsworth: Penguin, 1973). For the impact of Weber's disenchantment thesis, see e.g., John McCormick, "Transcending Weber's categories of Modernity?" *New German Critique* 75 (1998) 133–77; Richard Jenkins, "Disenchantment, Enchantment and Re-Enchantment," *Max Weber Studies* 1 (2000) 11–32. For Merton, see Robert K. Merton, *Science, Technology and Society in Seventeenth-Century England* (New York: Howard Fertig, 1938); Merton, "Puritanism, Pietism and Science," in *Social Theory and Social Structure* (New York: Free Press, 1957) 574–606.

21. Maurice A. Finocchiaro, *The Gallileo Affair: A Documentary History* (Berkeley: University of California Press, 1989); Mario Biagioli, *Galileo Courtier: The Practice of Science in the Culture of Absolutism* (Chicago: University of Chicago Press, 1993).

Stories of disenchantment usually stress that a magical and mystical worldview became rationalized. Pre-modern societies are dismissed as irrational. After some reflection it will become clear, however, that "rationality" might be an ambiguous term in this respect. It is questionable, for instance, to characterize whole societies as "irrational." As an answer to such accusations of irrationality, historians have argued for a specific rationality of magic.[22] Furthermore, advocates of a rationalization often ignore that early modern natural philosophers were involved in different pursuits, such as alchemy and other kinds of magic, and saw them as integral parts of their scientific work. If we accept the dominant disenchantment stories, we have to accept that rationality and irrationality were mixed in the early modern period. It becomes clear that judgments of rationality too often relate to our own presuppositions, and have little value to come to an understanding of early modern people, on which our presuppositions of rationality do not fit. Grand narratives of the scientific revolution as a triumphant movement of progress and rationality distort historical reality. Furthermore, they are too self-centered and conceited from our side, because they implicitly characterize us as the inevitable endpoint of the march of progress.

Recent historians have told very different stories, that are much more subtle and nuanced, but they can still be identified as variations on the disenchantment thesis. Peter Harrison has argued that the early modern period saw a move away from allegorical readings of the Bible towards a focus on the historical and literal, mainly under impulse of humanist scholars and Protestants. This new attitude towards the Bible also led to a new approach to the book of nature, which opened up a way to a new "scientific" worldview based on mathematics and taxonomy.[23] Lauren Kassell and others have proposed that the rise of collecting and of antiquarianism led to a kind of disenchantment. Studying ancient magical traditions as a historical phenomenon, and collecting wonders in order to order and classify them, brings with it an objectifying attitude, which might have taken the wonder out of their subjects.[24] William Burns gave

22. Richard Kieckhefer, "The Specific Rationality of Medieval Magic," *American Historical Review* 99 (1994) 813–36.

23. Peter Harrison, "The Bible and the Emergence of Modern Science," *Science & Christian Belief* 18 (2006) 115–32.

24. Lauren Kassell, "The Economy of Magic in Early Modern England," in *The Practice of Reform in Health, Medicine, and Science, 1500–2000: Essays for Charles Webster*, edited by Margaret Pelling and Scott Mandelbrote (Aldershot: Ashgate, 2005) 43–57.

political reasons for the demise of wonders in England. Radicals and republicans had used wonders to question the legitimacy of the Royal house. The conservative Tories tried to dismiss these wonders by ridiculing them. When the progressive party finally gained power, they had no use anymore for wonders, and the Tories could not use them in a credible way, because they had always dismissed them as superstition and enthusiasm.[25]

All these positions argue for new understandings of how wonders disappeared from the modern world. They should not be read as sweeping claims that aim at explaining the origin of modernity. These are particular arguments that aim at elucidating particular historical transformations, in religion, in antiquarian and historical study, and in politics, and their relevance for the history of wonders. Nevertheless, they are forms of a disenchantment thesis, and they bear witness of the continuing influence of the notion of disenchantment.

The early versions of the disenchantment thesis exemplify that the rejection of magic is commonly seen as one of the victories of modern science. When one looks at the broader context of early modern culture, however, it turns out that "magic" and "science" often coexisted. Porta's pronouncement that "Magick is nothing else but the survey of the whole course of Nature" already indicates that there might be a similarity between magic and what we today would call "science." Lynn Thorndike and Frances Yates have studied some of the interactions between what we would call "magic" and "science."[26] Charles Webster nicely traced magical ideas and influences, from Paracelsus and Porta via Bacon to the Royal Society with Boyle and Newton.[27] Famous examples of the interaction of magic and science interaction include Newton, who is the prime hero of advocates of modern science, but was termed the "last of the magicians" by Lord Keynes because of his alchemical interests. Also luminaries like Bruno, Bacon, Tycho, Kepler, and Boyle were involved in practices that scientists now would regard with suspicion indeed. Some of these observations led Frances Yates to make the vigorous claim that

25. Burns, *An Age of Wonders.*

26. Lynn Thorndike, *A History of Magic and Experimental Science*, 8 vols. (New York: Macmillan & Columbia University Press, 1958); Frances A. Yates, *Giordano Bruno and the Hermetic Tradition* (London: Routledge Classics, 2002/1964);

27. Charles Webster, *From Paracelsus to Newton: Magic and the Making of Modern Science* (Cambridge: Cambridge University Press, 1982).

science *originated* in magical theories and practices—a claim that became known as the *Yates thesis*.

In *Giordano Bruno and the Hermetic Tradition*, Yates argues that Giordano Bruno was not a martyr in the cause of modern science, but—like Ficino or Pico—an heir of the Hermetic magical tradition. Bruno was one of the first to take Copernicanism as a physical theory, extending it to an infinite universe filled with an infinite number of worlds. Yates argues that Bruno's reason for adopting heliocentrism was not scientific but magical, and was part of his sun-centered religious ideas. At the end of the book, she suggests that Hermetism caused an increasing interest in the material world and so fostered the development of science. Yet she still keeps Renaissance magic and Modern science strictly separate: "The procedures with which the Magus attempted to operate have nothing to do with genuine science. The question is, did they stimulate the will towards genuine science and its operations?"[28]

In "The Hermetic Tradition in Renaissance Science," Yates develops a more radical view in which science becomes *mixed* with magic. She claims that "the Renaissance magus was the *immediate ancestor* of the seventeenth-century scientist," that "[Mechanics and conjuring] belonged to the *same worldview*" and that there was a basic confusion "between mechanics as magic and magic as mechanics." There is a slow development from magic to science with many intermediate stages in which both are mixed and interdependent.[29]

Yates's work stirred scholars from a range of disciplines and stimulated an interest in the history of magic. It was taken as a new account of the origin of science, and she figured in philosophical debates about relativism, pre-scientific thought, other cultures and irrationalism. Critics took issue with both the historical rigor and the philosophical consequences of her work. The result of all this, however, has been a lasting influence, and she has been the source of inspiration of major scholars in the field. Some of her historical claims had to be revised, and Brian Copenhaver has done a great deal to sketch a better and more modest picture of Renaissance magic. Although some of the specifics of Yates' claims had to be reassessed, theses similar to those of Yates are

28. Yates, *Giordano Bruno and the Hermetic Tradition*, 488.

29. Frances A. Yates, "The Hermetic Tradition in Renaissance Science," in *Art, Science, and History in the Renaissance*, edited by Charles S. Singleton (Baltimore: Johns Hopkins University Press, 1968) 258–60.

still attractive. Today, theses about the "origins of science" have become unpopular, and it is recognized that modern science came about by complex processes, which are not to be pinned down to one factor and a limited time span.

One of the revisions important to the magic and science thesis is the recognition that "science" and "magic" mean very different things today than what they meant in the seventeenth century. There was not something like early modern "natural science" as we know it today, which came only into existence in the nineteenth century. Instead of one "natural science," there were various disciplines like natural philosophy, mathematics, and mixed mathematics. Natural philosophy, the contemplative study of nature, was strongly interwoven with religion. Some considered natural magic the practical and operative part of natural philosophy, which dealt with particulars instead of universals. Such views on the relationship between natural magic and natural philosophy were contested, but they became more and more accepted in the course of the early modern period. The gist of what is usually called the "scientific revolution" might be reconsidered as the integration of aspects of natural magic (empirical attitude, focus on particulars and experiments, interest in number and mathematics) into natural philosophy. John Henry writes: "The reason why natural magic has disappeared from our conception of magic is precisely because the most fundamental aspects of the tradition have now been absorbed into the scientific worldview. Or, to put it in another way, the scientific worldview developed, at least in part, out of the wedding of natural philosophy with the pragmatic and empirical tradition of natural magic."[30]

Claims about the entanglement of science and magic still speak to the imagination and variations to the Yates thesis continue to be proposed. The end of the seventeenth century, Keith Thomas has claimed, coincided with the "decline of magic." Keith Hutchison reassessed Thomas' thesis by arguing that, despite the rhetoric against magic, occult qualities became incorporated into the mechanical philosophy. Recently, Stuart Clark provocatively but cogently argued that *demonology* played a vital role during the Scientific Revolution and made a significant contribution to the development of science. Lauren Kassell has claimed that

30. John Henry, *The Scientific Revolution and the Origins of Modern Science* (New York: Palgrave, 2002) 55.

the study of astrology contributed to the development of case-based reasoning and statistics.[31]

The enchantment thesis argues that "science" partially developed out of early modern magic. In some versions of the thesis, it follows that essential parts of the magical tradition are absorbed by natural philosophy, and in extension, by the natural sciences. In its strongest form, which might be called the strong Yates thesis, as *the* origin story of the modern sciences, the thesis is very controversial. But more limited, subtle and nuanced versions of the thesis are much more promising and they enrich our understanding of the early modern interaction between various forms of natural enquiry and manipulation of nature, including natural magic, natural philosophy, medicine, alchemy, mathematical magic, mathematics, mixed mathematics and artificial magic. Wonder and knowledge turn out to be closely connected in a dynamic relationship.

THE WONDER OF WONDER

In the previous section, I have exposed two major theses about the relation between "magic" and "science." As sweeping and exclusive theses about the development of modernity, neither the disenchantment thesis nor the enchantment thesis is plausible. But there might be a kernel of truth in both of them, and it is a challenge for contemporary historiography to find out what exactly can be made of them. Many variations and nuanced versions have been proposed, and many of these claims are not incompatible. The idea that a new way of reading the Bible sparked off a new attitude to the book of nature is compatible with the idea that advanced studies in demonology refined early modern thinking about the natural world. Partial enchantments and disenchantments can be going on at the same time, and such a complex and dynamic picture will be an antidote to oversimplistic stories about the origin of modernity.

Wonders and the Order of Nature, a magisterial study of the history of wonders by Katharine Park and Lorraine Daston, can also be read as the combination of an enchantment and disenchantment thesis.[32] They

31. Thomas, *Religion and the Decline of Magic*; Keith Hutchison, "What Happened to Occult Qualities in the Scientific Revolution?" *Isis* 73 (1982) 233–53; Clark, *Thinking with Demons*; Lauren Kassell, "Casebooks in Early Modern England: Astrology, Medicine and Written Records," revised talk, delivered at 'The Cultural History of Medicine Seminar,' Goldsmiths, University of London, 2/21/2007.

32. In a previous article ("Unnatural conceptions"), Daston and Park described three successive notions of the monstrous: in the Middle Ages, prodigies with disturb-

solve the apparent conflict between the theses by positing a temporal distance between both developments. In their account, wonder is brought into "science" (or better: natural philosophy) at the end of the sixteenth century and beginning of the seventeenth century. From then onwards, it was increasingly recognized that natural particulars that violated the ordinary course of nature, usually studied in practical medicine and natural magic, could be important for natural philosophy. The seventeenth century saw the apogee of wonders in natural philosophy. Bacon had encouraged the gathering of strange facts as a collective enterprise in order to increase the factual basis of natural philosophy. This supported an epistemology that had its basis in "pure" facts, established by observation and experiments and attested to by credible witnesses, which stood apart from any theory. This approach suited the new scientific societies as well as a political system plagued by civil war, religious strife and political revolutions, because well-attested particular facts were less likely than theory to provoke controversies and unsettling disputes.

At the end of the seventeenth century, a new dynamic took place. Wonders were more and more expelled from natural philosophy and elite culture in general. This had little to do with the rationalization and naturalization of wonders. The elite rejected marvels as destabilizing, vulgar, and indecorous. In the course of the seventeenth century, the general course of nature had been replaced by invariant natural laws, and exceptional occurrences were either irrelevant or anomalies, a vulgar sensationalism or a destabilizing nuisance. Park and Daston read in this transformation also a reaction to the political, social, and religious upheavals of the seventeenth century. As a reaction to these disturbances, absolutism instated an absolute law that could not be violated.

This periodization is an interesting solution for mitigating the apparent conflict between the disenchantment and enchantment theses. This strategy is not necessary, however, if one recognizes that the disenchantment and enchantment theses are only in conflict when we take them as exclusive and all explaining grand theses about the origin of modernity. They do not necessarily conflict in the more particular variations on their themes. This also comes to the fore when one studies the

ing religious implications; starting in the sixteenth century, natural wonders that evoked delighted appreciation; and finally, objects of scientific inquiry, wholly explicable in terms of natural causes. In their book, however, they repudiate this teleological master narrative of the triumph of naturalization.

DIVINERS showing how to use the dowsing rod, in Pierre Le Brun, *Histoire Critique des Pratiques Superstitieuses*, book 3, Plate A.

---

complexity of wonder and disenchantment in particular historical case studies. I will give a brief example of this by elaborating a wondrous event at the end of the seventeenth century.

In 1693, a diviner found out that he could find criminals with his divining rod.[33] He received wide acclaim for his skills, but was also subjected to experiments and tests by his critics. It turned out in this case

33. Koen Vermeir, "The 'Physical Prophet' and the Powers of the Imagination. Part II. A case-study on dowsing and the naturalisation of the moral (1685–1710)," *Studies in History and Philosophy of Science* 36C(1) (2005) 1–24.

that many natural philosophers easily took in this wondrous and magical phenomenon and devised Cartesian and other theories to explain it. Theologians, however, reacted dismissively. They argued that there must be some kind of demonic involvement. They accused the natural philosophers of "credulity," for too easily believing the claims of the diviner. They accused them of "superstition," because the natural philosophers were lured by the tricks of the devil into too eagerly naturalizing an essentially religious phenomenon. We can see here that the natural philosophers who "naturalized" this wondrous phenomenon gave credibility to a marvelous and magical practice. They "enchanted" the world by giving it the status of a natural fact, and they "enchanted" their theories by devising extraordinary theories to explain the phenomenon.

On the other hand, the skeptical theologians, who wanted to discredit the phenomenon, resorted to a spiritual explanation. A third party in the debates, one of the Royal Princes, had staged his own experiments. He did this in order to unmask the diviner as a cheat. We see here that the elite refused to accept the phenomenon as genuine, representing the most strongly "disenchanting" position, but at the same time it shows that the views of most of the natural philosophers and physicians cannot be identified with the views of the elite on this matter. This episode shows the complexity of the dynamic relation between wonder, knowledge, and social structures. In other cases, this relation might have different characteristics, and it will be difficult to draw sweeping generalizations.

The difficulty of making generalizations is confirmed if one notices the ambiguities in attitudes towards wonder in the work of one author only. There are so many different kinds of wonder, and one can look at them from so many perspectives, that it is no wonder that philosophers seem to contradict themselves when we try to interpret their pronouncements in the light of an enchantment thesis or a disenchantment thesis. Daston and Park treat Francis Bacon as the lynchpin of their story, because they consider him responsible for bringing wonders and strange facts into natural philosophy. On the other hand, Bacon famously dismissed wonder as "broken knowledge." This difference can partly be explained by recognizing that wonder was not only a phenomenon, but also an emotion or even a cognitive state. Even if wonders as singular phenomena can be allowed in natural philosophy, this does not imply that the cognitive emotion of wonder is acceptable. For early modern philosophers, wonder was an emotion closely associated with the acquisition of knowledge.

Daston and Park already explained that "wonder" could denote an emotion as well as a phenomenon. Both are sometimes interlocked, but I will explain that this is not necessarily the case. On the one hand, the emotion of wonder can be inspired by marvelous singular phenomena that do not fit into our epistemic categories or are even beyond our ordinary experience. For early modern philosophers, wonder at curious phenomena that are beyond current comprehension could spur natural inquiry. They posed a challenge to give them a place in their conceptual framework by rethinking their categories of thought and their natural philosophical theories. Reducing this kind of wonder was necessary in the early modern aspiration to overcome their ignorance. They followed the Aristotelian view that "a man who is puzzled and wonders thinks himself ignorant . . . they philosophized in order to escape from ignorance."[34] More knowledge dispels wonder.

On the other hand, and this is important, the passion of wonder could also be directed towards the ultimate order and design of nature. Given our experience of the disorder and chaos in our everyday world, the regularity and order of celestial movements can strike us as surprising and wondrous. By extension, the discovery of the design behind the disorder in the sublunary world, coming to see the unity behind the variety of natural phenomena, goes beyond our ordinary experience and can therefore cause wonder. In this case, more knowledge increases wonder. In early modern natural philosophy, the wonder for curious objects that defied explanation was usually supposed to disappear with the advancement of knowledge. In natural theology, however, the wonder at the order that natural philosophy laid bare induced reflections about a divine being. More knowledge heightened this wonder, even if there was a recognition that the divine plan ultimately would transcend comprehension.

Many early modern philosophers expressed the idea that wonder dissipates with the advance of knowledge. The mathematical practitioner Simon Stevin (1548–1620) famously said: "wonder is gheen wonder" ["Wonder is not wonder"], when he argued that what comes across as strange will nevertheless have a natural explanation. Following Stevin,

34. Aristotle, *Metaphysics*, I.2: 982b12–22, in *Complete Works of Aristotle*, trans. W. D. Ross, ed. Jonathan Barnes (Princeton: Princeton University Press, 1984).

his compatriot and teacher, Isaac Beeckman (1588–1637) wrote: "In philosophy one should always proceed from wonder to not wonder, i.e., one should research as long as what looks strange does not seem strange anymore." His student and friend, René Descartes (1596–1650), also wrote that the explanation of a marvelous thing expels the wonder.[35] Beeckman, however, added that in theology, another attitude was called for: "but in theology one should proceed from not wonder to wonder."[36] It is clear that what is marvelous, and how it relates to knowledge, crucially depends on our experience and on the perspective we take.

Wonder is not a simple emotion. Descartes characterized wonder as "the first of all the passions," because it can happen before we really know the object. He described wonder as a "sudden surprise of the soul," when encountering something for the first time. But wonder is often more than just surprise, there are different varieties, and it intertwines with many other emotions. Seeing a curious phenomenon could inspire wonder with a tinge of pleasure. Sometimes, this could lead to curiosity and an urge to come to know more about it. On the other hand, a marvel was sometimes interpreted as a portent, which signified catastrophic events or divine wrath and retribution. In such a case, wonder was mingled with fear and awe. Wonders were sometimes also signs of God in a positive sense. Exotic plants and animals, for instance, signaled the divine bounty and evoked wonder, mixed with awe and admiration, at the diversity of God's creations.

Descartes's attitude towards wonder is even more ambiguous than Bacon's. For Descartes, wonder was serviceable, because it excited curiosity, which could lead to knowledge. Indeed, he wrote that "those who have no natural inclination to this passion are commonly very ignorant."[37] But Descartes also recognized the danger of excessive wonder, that would make wonder antagonistic to knowledge: "rather than wondering too little, more often we fall into excess wonder and are astonished by things that are little worthy of our attention, if at all, a state of mind

35. See the chapter below by Dorottya Kaposi.

36. See Isaac Beeckman's diary, *Journal tenu par Isaac Beeckman de 1604 à 1634, vol II (1619–1627),* edited by Cornelis de Waard (Amsterdam/The Hague: Martinus Nijhoff, 1953) 375: "In de philosophie moet men altyt procederen van wonder tot gheen wonder, dat is te segghen, men moet so langhe ondersoecken totdat hetgene ons vrempt dunket, ons niet meer vrempt en schyndt, maer in de theologie moet men procederen van gheen wonder tot wonder."

37. See René Descartes, *Les Passions de l'Ame*, art. 75–78.

METAMORPHOSIS MACHINE, with perspectival cones added on the wall and the roof, in Athanasius Kircher, *Ars Magna Lucis et Umbrae* (Amsterdam, 1671) 783.

---

which may completely eradicate or pervert the use of reason. This is why, although it is good to be born with some natural inclination towards this passion since it disposes us to the acquisition of science, we must strive to extract ourselves from its sway as quickly as possible" (ibid.). Excessive wonder was akin to enthusiasm and a strong imagination. The best way to drive away wonder was "to acquire knowledge of most things and to apply oneself to the consideration of all those which might seem the most rare and strange" (ibid.). Here, similar to other early modern philosophers, Descartes states that knowledge dispels wonder. Wonder could lead to knowledge, but it could also pervert the use of reason and make knowledge impossible.

In *La recherche de la vérité,* Descartes explains that he will display the works of man on nature, "and having struck wonder into you with the most powerful machines, the rarest automata, the most specious of visions, and the subtlest impostures that artifice can invent, I shall reveal to you their secrets, which are so simple and so innocent that you will henceforth wonder at nothing in the works of our hands."[38] The artifices of artificial magic are wondrous, but Descartes promises to demystify the artifice and expel the wonder by explaining their principles. Again, knowledge dispels wonder. Remember that for Descartes, the world is like a great artifice, and that his mechanical philosophy lays bare the principles of the world. His philosophy in general can be seen as bearing the promise to take wonder out of the world.

We find a similar but slightly different dynamics in actual performances of artificial magic. Kircher's demonstrations of his instruments, for instance, belong to the tradition of natural and artificial magic, and he knew well the different ways of creating wonder. In 1625 at the Jesuit College in Heiligenstadt, Kircher had devoted a great deal of care to the mise-en-scène of a theatrical play. An astounding display of moving scenery and fireworks was put on to the delight of the audience. The spectacle was so astonishing and the effects were so successful that it was rumored black magic was involved. Here, the public marveled at the extraordinary phenomenon that seemed inexplicable. The apparent causes seemed to be incongruous to the effects, and this evoked suspicions of demonic involvement.

To a select group of visitors, however, Kircher demonstrated how he had created these special effects, uncovering all kinds of strings and pulleys, because he anticipated that this uncovering would cause a different kind of delight to those who could appreciate it.[39] Part of natural and artificial magic was about provoking wonder in the cognoscenti by showing the hidden contrivances behind marvelous phenomena. Similarly, showing the hidden principles at work behind the wondrous effects of nature could evoke wonder. A complicated rhetoric of veiling and unveiling is at work here, and Kircher employed these two kinds of wonder where appropriate.

Here, we can see intrinsic ambiguities within the dynamics of magic itself and the way it plays with wonder. There is a fundamental paradox

38. René Descartes, *Oeuvres Philosophiques*, edited by Ferdinand Alquié (Paris: Garnier, 1963–1973) I.576.

39. Athanasius Kircher's autobiography, *Vita admodum Reverendi P. A. Kircheri*, retold in Joscelyn Godwin, *Athanasius Kircher: A Renaissance Man and the Quest for Lost Knowledge* (London: Thames and Hudson, 1979).

at work here. In our last example, disenchantment or the unveiling of the wonder itself provokes wonder! Artificial magic's complicated play with wonder lays bare the limits of categories such as "enchantment" and "disenchantment." It also allows us to understand a tradition that connects science and wonder also long after Daston and Park's demise of wonder at the end of the seventeenth century. Think of the spectacular demonstrations of science, in the vein of Abbé Pluche's scientific shows and his *Spectacle de la nature* (1732–1742), and think of the awe evoked in natural philosophy and natural theology for the order and immensity of the cosmos.[40] Or consider the wonder that the greatest scientific minds today feel for symmetries and mathematical order in science.

Instead of taking Daston and Park's path in arguing that wonder disappeared from science in the eighteenth century, we could speculate that wonder was actually integrated and formalized in the modern sciences. Strange phenomena were considered anomalies, and even if these are neglected during the first stages of the development of a paradigm, Thomas Kuhn has described their crucial role in starting a scientific revolution and in the shaping of new paradigms. Anomalies, the preternatural phenomena of old, have become the locus of scientific fascination, and the centerpiece of modern philosophy of science. Given the competitiveness and aim for novelty in current scientific research, instead of being neglected, anomalies are probably more than ever the focal point of scientific research. Another kind of wonder, the wonder for the order of nature, is embedded in aesthetic considerations (such as simplicity and symmetry) that play a fundamental theoretical role in many modern sciences. Theories or phenomena need to exhibit certain symmetries and simplicities in order to become accepted as promising or true. It is these aspects of science that often still evoke the greatest wonder in scientific practitioners. According to many modern physicists, if a theory is not wonderful, it cannot be true.

This is to show that there are many kinds of wonder, and this ensures me that there will remain a certain ambiguity at the core of our theorizing about and our histories of wonder. Wonder can evoke fear or pleasure. Wonder can induce or inhibit knowledge. Wonder can occur in the veiling or in the unveiling of hidden principles. Nevertheless, the two theses that I have discussed in this paper, the disenchantment thesis and the enchantment thesis, can still serve as guiding threads, because

40. This is the starting point for the next chapter by Miklos Vassanyi.

they represent two extremes in the spectrum of possibilities. If one keeps both in mind when studying wonder, they will serve as good heuristic principles, because they will pull the mind in opposite directions, precluding short-sightedness or prejudice. Of course, we should not take any of these theses literally as grand claims about the origin of modernity. Instead, they should inspire us to do more detailed research into wonders and their history. The other chapters in this book will provide us with such a detailed analysis of the function of wonder in particular early modern philosophers.

3

Religious Awe at the Origin of Eighteenth-Century Physico-Theology

Miklós Vassányi

INTRODUCTION

"WHAT IS MAN?"—THE WORDS of Psalm 8:5 incite physico-theologist Abbé Pluche (1688–1761) to begin a devout train of thought on account of David, who is portrayed as a simple herdsman in the title engraving of part 5 of Pluche's *The View of Nature* (*Schau-Platz der Natur*, 1750).[1] As Pluche says, David, traditionally considered the author of the Psalms, "wonders at the great might given to man, and in the quiet night-time, sings of the unspeakable grace of Him who gave so many good things to man. The pale moonlight allows him to behold a part of them . . . In their pen, his sheep warm up the land he wants to till, and the Pleiades show him, by virtue of their position, the hour in which he has to let the sheep into another pen. The entire Earth is ready to satisfy his will, and even the sky revolves at the service of man."[2]

This pious introduction to a volume of one of the most popular French physico-theologists of the eighteenth century may provide us with a preliminary idea of what physico-theology is, and what role religious awe might play in it. Classical eighteenth-century physico-theology is a science which intends to prove the existence and the chief attributes (omniscience, omnipotence, bounty) of God by systematically

1. German translation (Vienna & Nuremberg, 1748–1750) of Pluche's *The View of Nature* (*Le spectacle de la nature*), on which see section 2 below.

2. "Was ist der Mensch? Ps. 8, 5 etc." (ibid., p. X 2 verso.) All translations are by the author of the present chapter.

pointing out with natural scientific means that the physical universe displays manifest signs of deliberate purposeful planning. On this account, the physico-theological argument is also called the "argument from design." As the universe, had it been produced by mere chance, would not be bearing evident signs of intelligent design, it must have been made and ordered by virtue of divine pre-meditation, i.e., by the considered choice of the most perfect being, God. In its classical early modern form, physico-theology thus posits the operation of supernaturally determined final causes in order to explain natural teleology, whereby it makes a philosophically argued transition from nature to the supernatural condition of nature. While our citation from Abbé Pluche does not yet reveal the strictly natural scientific character of the fundaments of the physico-theological argument (as we shall see in sections 3 to 5), it does show in outline how the authors belonging in this tradition argued from the purposeful arrangement of all created things (in favor of man) to the existence of an all-good and almighty God, and how the science of physico-theology begins in religious awe or *wonder* at the pervasive order of Creation.

As we shall see, other distinctive features of eighteenth-century physico-theology are the outspokenly edifying tenor of its (moral) theological conclusions, and its tendency poetically-spiritually, sometimes even homiletically to elaborate on the metaphysical final result: the existence of God, of its natural scientific enquiry. Eighteenth-century physico-theology is hence, generally speaking, a natural science in the service of (systematic and even pastoral) theology and religion, spirituality and devotion; a natural science the ultimate aim of which is the scientific-argumentative reinforcement of the believer in his or her faith, and the noncoercive conversion of the atheist or non-believer to a broadly conceived Christian creed.

In the contribution that follows, we shall first try to give a systematic historical introduction to the sources of eighteenth-century physico-theology. Then we shall give an idea of the natural scientific valor of this ideological movement by adumbrating, as an example, the new cosmological picture of the world physico-theology advocated and relied upon. This is to reveal physico-theology as a natural science, and to show that despite its explicit ideological pledge, it was no retrograde or subprime strand in respect to scientific value in the palette of the then existing natural scientific attitudes. Next, we present and analyze the philosophical core of the physico-theological argument, examining three counter-arguments (the problem of evil, the hylozoistic alternative, and Kant's criticism), which may be set against it. The chief query throughout our investigation will be, however, the following classical question: *is the physico-theological argument for the existence and attributes of God cogent?*

MAJOR SOURCES OF EIGHTEENTH-CENTURY PHYSICO-THEOLOGY

It seems likely that the term 'physico-theology' had been coined by Walter Charleton (1619–1707), whose book *The Darkness of Atheism Dispelled by the Light of Nature* (1652) carried the subtitle *A physico-theologicall treatise.*[3] The origins of early modern physico-theology thus go back to the middle and late seventeenth century. The first two authors our enquiry covers still belong to the century of Locke, who also tried his hand at physico-theology in one of his early *Essays on the Law of Nature* (*Quaestiones de lege naturae*, eight scholastic questions composed in Latin around 1660).[4] We will briefly characterize the other, chief physico-theological sources in chronological order.

Matthew Barker's (1619–1698) *Natural Theology, or, the Knowledge of God, from the Works of Creation, Accomodated, and Improved, to the Service of Christianity* (1674) tells us about the ideological objectives of its author already in its title. In accordance with the title, Barker's chief scriptural reference is to Romans 1:20, where St. Paul sets down as it were the very principle of physico-theological research as he professes the fundamental theological thesis that "the invisible things of Him from the creation of the world are clearly seen, being understood by the things that are made, even His eternal power and Godhead."[5] Though in terms of natural scientific quality, Barker's 220-page-long treatise does not

3. Cf. *Historisches Wörterbuch der Philosophie*, ed. J. Ritter and K. Gründer (Basel: Schwabe, 1971–2005) vol. 7, s.v. "Physikotheologie" (S. Lorenz) 948 and 952. Among the precursors of the eighteenth-century physico-theological movement, the author cites S. Parker's *Physico-theological Attempts concerning God* (*Tentamina Physico-Theologica De Deo*, London, 1665) as well, which we could not consult.

4. See Essay 2 of John Locke, *Essays on the Law of Nature*, ed. W. von Leyden (Oxford: Clarendon, 1:1954, 2:1988) 122–32, esp. 132.

5. Matthew Barker, *Natural Theology, or, the Knowledge of God, from the Works of Creation; Accomodated, and Improved, to the Service of Christianity* (London, 1674) 1; see also ch. 1.

compare to the physico-theological achievement of later, professional naturalists, it clearly follows the inferential pattern of the argument from design as it asserts that "when we see all things in Nature in *motion*, and every wheel of Nature moving to a *rational* end, we may conclude, this was done by some *Infinite Intelligence*."[6]

John Ray (1627–1705), whose *The Wisdom of God Manifested in the Works of the Creation* (1691), lays great stress on the religious awe the scholar feels for "the admirable Art and Wisdom that discovers itself in the make and constitution, the order and disposition, the ends and uses of all the parts and members of this stately fabrick of Heaven and Earth."[7] An educated and cultivated philosopher, Ray defends the operation of final causes in creation, and in harmony with these, rejects the respectively Epicurean and Cartesian, mechanical cosmogonies.[8]

Archbishop François Fénelon (1651–1715) composed his *Treatise Concerning the Existence of God* toward the end of his life, in 1712. The first part, a *Demonstration of the existence of God drawn from the view of nature and from the knowledge of man*,[9] of this work is a systematic physico-theology. Since we notice, says Fénelon, admirable, superior art (*art*) and skill (*industrie*) in the internal structure and external concatenation of natural substances, it is reasonable to conclude that (an inconceivably) great wisdom (*sagesse*) and power (*puissance*) has constructed and ordered them. His presentation of the physico-theological argument describes God as the source of proportion, measure and moderation in, and as the First Mover of, the natural world, but only accidentally as the Creator. Fénelon is further perhaps the only physico-theologist who explicitly deals with, and offers a philosophical solution to, the problem of evil (cf. part I, ch. 88).

The Dutch Doctor in Mathematics, Bernard Nieuwentyt (1654–1718), is perhaps the first truly systematic physico-theologist, whose work *The Right Way of Using the Contemplations of the World* (1714) may be considered the ideal type of a physico-theological treatise.

6. Ibid., 22 (Barker's emphases).

7. John Ray, *The Wisdom of God Manifested in the Works of the Creation* (London: S. Smith, 1691) 12.

8. Cf. ibid., 13–40.

9. See Fénelon, *Traité de l'existence de Dieu: Démonstration de l'existence de Dieu tirée du spectacle de la nature et de la connaissance de l'homme*, ed. J.-L. Dumas (Editions Universitaires, *sine loco*, 1990).

As its full title reveals,[10] the ponderous volume was intended to convince two categories of people: the *ongodisten* or atheists, of the existence, wisdom and goodness of God; and the *ongelovigen* or, in Nieuwentyt's own definition, the deists, of the authority of Scripture (and therefore, of the truth of the Christian religion).[11] This natural, science-based apology of Christian faith relies on the demonstration of the purposive functioning of the human body, the natural elements, natural laws, chemical and astronomical objects. In each case, the author points out the impossibility of the supposition that the natural phenomenon at stake could have been produced or teleologically ordered by any other agent than an omnipotent and omniscient God.

The distinguished William Derham (1657–1735), holder of the chair established by the equally pious natural philosopher Robert Boyle (1627–1691) for the defense of the Christian religion, authored the eponymous book of the movement: *Physico-theology Or, a Demonstration of the Being and Attributes of God from his Works of Creation* (1713).[12] Derham, essentially an astronomer, published a continuation to this work

10. *The right way of using the contemplations of the world, proposed in order that the atheists and non-believers may be convinced* (*Het regt gebruik der werelt beschouwingen, ter overtuiginge van ongodisten en ongelovigen angetoont, door Bernard Nieuwentyt*; we have used the fifth edition of 1730, Amsterdam, ca. 900 pages). As the title indicates, the book is a series of contemplations of nature, each of them closely followed by its proper physico-theological interpretation. Kant and Rousseau referred to the French version. In this version, Nieuwentyt totally rearranged the material, and changed the title of the book, which now became *The existence of God demonstrated by the marvels of nature, in three parts; where the structure of the human body, the elements, the stars and the several different effects of these are discussed* (*L'existence de Dieu, démontrée par les merveilles de la nature, en trois parties; Ou l'on Traite de la Structure du Corps de l'Homme, des Elemens, des Astres, et de leurs divers effets* 1725; we have used the 3rd edition of 1760). The name of the translator (probably Nieuwentyt himself) is not indicated.

11. We read the following in the dedication *To the reader* (*Aan den leser*, XX): "The objective with which these contemplations have been written is to convince atheists of the wisdom, power and bounty of their God, the wonderful Maker and Governor of the universe; and to convince non-believers, who do acknowledge a God but by no means the authority of the Holy Scriptures, of the suprahuman origin of the Scriptures; and so to show both these kinds of people the right way of using the contemplations of the world."

12. William Derham, *Physico-theology or, a Demonstration of the Being and Attributes of God, from his Works of Creation.* Being the Substance of XVI Sermons Preached in . . . London, . . . in the Years 1711 and 1712 (London: W. Innys, 1713; reprint Hildesheim–New York: Georg Olms Verlag, 1976). There were many editions and reprints of this work throughout the eighteenth century.

two years later, under the title *Astro-theology: Or a Demonstration of the Being and Attributes of God From a Survey of the Heavens.*[13] Derham was a professional natural scientist at the cutting edge of the astronomical research of the time, who in the *Astro-theology* floated the idea that our solar system (let alone planet Earth) is not in the center of the universe, and consequently rejected the "old vulgar Opinion, that *all things were made for Man*."[14] Though this contradicts, e.g., Abbé Pluche's more archaic conviction, Derham is none the less a true physico-theologist as he argues, admiring the works of creation, that "this glorious Scene of *God's works*, the Heavens, plainly demonstrate the Workman's infinite Wisdom to contrive, his *Omnipotency* to make, and his infinite *Goodness* in being so indulgent to all the Creatures, as to contrive and order all his Works for their good. For what less than *Infinite* could effect all those grand things, which I have . . . shewn to be manifest in the Heavens?"[15]

The great systematic Leibnizian scholar, Christian Wolff (1679–1754), contributed to the physico-theological movement with his *Rational Thoughts Concerning the Purposes of Natural Things* (1724),[16] which stands out from the crowd by virtue of its philosophical quality. In this respect, only the early Kant compares to Wolff, who continued these thoughts with a second volume about the teleology of the bodily organs of humans, animals and plants (*Rational Thoughts Concerning the Use of Parts in Humans, Animals and Plants*, 1725).[17] In the work on

13. William Derham, *Astro-Theology: Or a demonstration of the Being and Attributes of God, From a Survey of the Heavens* (London: W. Innys, 1715); the French translation, *Théologie astronomique ou démonstration de l'existence et des attributs de Dieu, par l'examen et la description des cieux* (some 200 pages), does not contain the translator's name, was based on the fifth English edition (1726), and came out in Paris 1729.

14. *Astro-Theology*, 39 (Derham's emphasis); cf. *Physico-Theology*, 55, footnote 3.

15. *Astro-Theology*, 210. The apologetic character of the *Astro-theology* is made also clear by Derham's statement at the end of the *Preliminary Discourse* (p. lviii): "And now for a Conclusion, I shall only intreat all my Readers to joyn with me in their earnest Prayers, that as this Work is designed for the good of Mankind, particularly for the Conviction of Infidels, for the Promotion of the Fear and Honour of God, and the cultivating of true Religion, so it may have its desired Effect."

16. Christian Wolff, *Vernünfftige Gedancken von den Absichten der natürlichen Dinge* (Halle: Rengerische Buchhandlung, 1724).

17. Christian Wolff, *Vernünfftige Gedancken von dem Gebrauche der Theile in Menschen, Thieren und Pflanzen* (Frankfurt & Leipzig, 1725). Wolff further composed the physico-theological treatise, *On the Method of Demonstrating the Existence of God From the Order of Nature* (*De methodo existentiam Dei ex ordine naturae demonstrandi*) in 1731.

"the purposes of natural things," a book concerned with the final causes operating in nature, he affirms that the overall objective God wanted to achieve with the creation of the physical universe is His self-glorification in the eyes of the creatures. This chief objective (*Haupt-Absicht*) entails that the world is as it were a mirror of the divine perfections: "the main objective of the world is to reveal the magnificence of God, i.e., that God decided to bring forth the world . . . in order that one may come to know His perfections from it."[18]

The physico-theological bestseller of the time was written by Abbé Noël-Antoine Pluche (1688–1761) under the title, *The View of Nature, or Dialogues Concerning the Peculiarities of Natural History which seemed the most adapted to raise the curiosity of young people, and to form their mind.*[19] Formally a dialogue composed for the education of the well-to-do youth, the nine little volumes of this book survey the generation and organization of plants and animals, the skies, the particular constitution of man, and they include a separate volume on "man in company with God" (*l'homme en société avec Dieu*), which is intended to prove the necessity of the Christian revelation. The author constructs his physico-theological theory on the operation of final causes in nature, as he argues that, "all can please and instruct us in nature, because all is full of design, proportion and precaution . . . The particular structure of the bodies around us, their tendency towards an end designate the intention of the maker."[20]

The work by Johann Albert Fabricius (1668–1736), *Hydro-Theology, or an Essay Concerning the Bounty, Wisdom and Power of God, manifested in the creation of water* (1734),[21] examines one by one the attributes and properties of water, and points out the advantage each of

18. *Vernünfftige Gedancken von den Absichten der natürlichen Dinge*, part I, chapter I, §2; p. 2.

19. Pluche, *Le spectacle de la nature, ou entretiens sur les particularités de l'histoire naturelle, qui ont paru les plus propres à rendre les Jeuns-Gens curieux, & à leur former l'esprit* (Paris: 1:1732–1742, 2:1735–1752, published anonymously).

20. Preface of the author, reproduced in a nineteenth-century abridgement entitled *Beauties of the View of Nature* (*Beautés de Spectacle de la nature*, ed. F. L. Jéhan, Tours, 1844).

21. We consulted the French translation of this text, *Théologie de l'eau, ou essai sur la bonté, la sagesse et la puissance de Dieu, manifestées dans la création de l'eau* (The Hague, 1741), which arranged the original text into longer chapters, shortened some chapter titles, transferred less important material into the footnotes, and added some footnotes.

them assures for creatures. On account of the teleological character of a natural phenomenon like, for example, evaporation, he draws the following theological conclusion about the power of God: "The benefit we draw from this evaporation of the waters, the uninterrupted circulation it maintains with the aim of nourishing the creatures, keeping them alive and rendering them fertile, offers us a sensible proof of the wise power of the Creator." The learned Fabricius, who was also a famed classical philologist, the German translator of Derham's *Astro-theology* (1728), the author of a *Pyrotheologia* (1732) and an outline of *Aerotheologie* (published as a part of Lesser's *Litho-Theologia*, 1735, all in Hamburg), belongs among the intellectually less demanding philosophers of the physico-theological tradition.

Furthermore, Friedrich Christian Lesser (1692–1754), in his *Insecto-Theologia* (1738),[22] widens the boundaries of physico-theology toward the micro-world with the help of Leeuwenhoeck's invention, a simplified and stronger microscope. Lesser, a Leibnizian in metaphysics and author of several other physico-theological studies,[23] bases the argument from design on the teleological constitution of insects, which can reveal, he says, the attributes of God just as well as the constitution of any other animal genus can: "so there is reason to honour with sacred admiration the great Creator of these small animals . . . who vested with the necessary articulations, skin, muscles, parts, sinews and nerves even those insects which can hardly be seen without magnifying glasses, so that one has to acknowledge this as an admirable piece of work of His endless power and wisdom."[24]

22. Lesser, *Insecto-Theologia, oder: Vernunfft- und Schrifftmäßiger Versuch, wie ein Mensch durch aufmercksame Betrachtung derer sonst wenig geachteten Insecten zu lebendiger Erkänntniß und Bewunderung der Allmacht, Weißheit, der Güte und Gerechtigkeit des grossen Gottes gelangen könne*, Franckfurt und Leipzig, Michael Blochberger, 11738, 21740. Visit http://edoc.hu-berlin.de/ebind/mfn/lesser-2006-Mn016L2272/XML/index.xml/ for a full online copy of the second edition. The French version, based on the second edition of the original and published in The Hague (1742), is more a paraphrase than a translation. Moreover, it drops the paragraph numbering of the original.

23. Cf. *A Short Sketch of a Theology of the Stones* (*Kurzer Entwurf einer Theologie der Steine*, Nordhausen, 1732); *Litho-theology, i.e., The Natural History and Spiritual Consideration of the Stones* (*Litho-Theologia, das ist, Natürliche Historie und geistliche Betrachtung der Steine*, 1735); and *Testaceo-theology, or a Consideration of the Snails and Mussels* (*Testaceo-Theologia, oder Betrachtung der Schnecken und Muscheln*, 1744).

24. Introduction, p. 11 of the second edition.

A tone of critical reflection on physico-theology set in with a work by Pierre-Louis Moreau de Maupertuis (1698–1759), *Essay in cosmology* (*Essay de cosmologie*, 1750). The first president of the re-founded Royal Prussian Academy of Sciences in Berlin, Maupertuis was a physician (cf. his vitalist theory of conception and development of the embryo),[25] mathematician, astronomer, and philosopher. In his *Essay in cosmology*, he established the principle of the least quantity of action (*principe de la moindre quantité d'action*) in mechanics as the universal law of motion,[26] and used it as an upgraded physico-theological proof of the existence of God. It is not enough to show the skill (*habileté*) of the Eternal Architect in the construction and arrangement of the "marvels of nature," says Maupertuis, critically of previous physico-theology. The conviction an argument for the existence of God carries depends on whether the objective (*but, motif*) of the intentions (*desseins*) the argument attributes to God is more *reasonable* than other possible objectives (final causes) or not: "That on a thousand occasions, this universe presents us series of effects tending toward some end, is only a token of intelligence and intention; but wisdom must be sought for in the objective of these intentions. The skilfulness of execution is not enough; it is necessary that the motive be reasonable . . . What does it avail to wonder that every planet moves regularly, in the same direction . . . if we do not recognize it was better to move them in this rather than another way?"[27]

The 'pre-critical' Immanuel Kant (1724–1804) carried on the constructive criticism of physico-theology in *The Only Possible Premise for a Demonstration of the Existence of God* (*Der einzig mögliche Beweisgrund zu einer Demonstration des Daseyns Gottes*, 2nd part/5–6; 1763). Kant here asserts that the physico-theological argument in its traditional form can only prove the existence of a divine *arranger* of pre-existent matter,

25. Cf. the *Essay concerning the formation of organic bodies* (*Essai sur la formation des corps organisés*, also called the *System of nature* [*Système de la nature*]), first (lost) Latin version, 1751.

26. Cf. *Essay de cosmologie*, in Tonelli ed., tome I, p. 42.

27. *Essay de cosmologie*, part one: *Where the proofs of the existence of God are examined that are drawn from the marvels of nature* [*Où l'on examine les preuves de l'existence de Dieu, tirées des merveilles de la Nature*], ibid., 19–20)—Our experience that certain series of effects in the natural world concur to realize specific objectives can only prove that some intelligent design, not that divine *wisdom* produces natural teleology. The general objective (final cause) of the operations of nature must be more reasonable than any other objective, if we are to draw a valid conclusion that the universal efficient cause of natural teleology is entirely rational, i.e., divinely wise.

not that of a *creator* God.[28] In order to extend the scope of the argument, Kant, reaching back to Maupertuis's reform of it,[29] proposes an improved version in II.6 (*Improved Method of Physico-theology*). The existence of a *creator* God may now be proved on the ground that the order of nature is regulated by general and universal rules useful and productive in their effects to man, and coordinated with each other under a principle of unity. Nature is thus in its very conception and structure an organic whole, which it could not have been, had its individual parts been differently conceived by their omniscient cause. In this sense, the unity of the world ontologically depends on the pre-determined possibilities of existence of natural things, while the essence and existence of these depend on God as an ultimate ground: "An entirely different judgment is formed if we notice that not all natural perfection is achieved in an artificial-arbitrary manner, but highly useful rules go together with necessary unity as well, and this agreement lies in the possibilities of the things themselves . . . Since . . . this unity . . . is grounded in the possibilities of things, there must be a wise Being, without which all these natural things are not even possible, and in which as a grand ground the essences of so many natural things are united in so regular relations."[30] We shall discuss the critical Kant's alleged destruction of the physico-theological argument in the transcendental theology of the *Critique of Pure Reason* (1781) below.

Next, Sebastian Friedrich Trescho (1733–1804), poet, pastor, and philosopher, author of the *Effusions on Account of Nature during some Summer Hours* is an 'archaizing reformer' of physico-theology by virtue of his sentimental-contemplative approach to nature, and his criticism of professional, academic natural science. For this pen-pal of Herder's, the traditional disciplines of physico-theology (biology, astronomy, etc.)[31]—when conducted with an exaggerated rigor—do not immediately show the Creator by means of the instrumentality of creation. They do not

28. Cf. II.6.ii; AK (Akademie-Ausgabe) II.125: "In the constitution of an animal, the body parts related to sensation are joined with those of arbitrary locomotion and life with such an art, that one must be wicked (since it is impossible for a man to be so unintelligent) . . . not to recognize a wise Author who has brought into such an excellent order the matter out of which an animal body is composed. Nothing more follows from this. It is here by no means determined whether this matter is in itself eternal and independent, or produced by the very same Author."

29. Cf. II.1.i; ibid., 98–99.

30. II.6.ii; ibid., 125.

31. Trescho reprimands the barrenness or futility of, in particular, Buffon's highly academic research (on which see below); cf. *Effusions*, 26.

bring us religious conviction, though that is what they should do in the first place, as they will invariably remain unable to fulfill their alleged scientific mission: to discover the real intrinsic constitution of natural substances. Although the natural sciences have to renounce themselves of an insight into the true internal structure of things, they may yield enough knowledge to teach the *heart* (by way of theological conclusions drawn from empirical evidence) about the existence of God and the moral obligations of man: "Everywhere, he [the natural philosopher] will only want to find God, Jehovah, the Father of nature. To discover Him for the world by showing His wisdom, His eternal order, and His manifold bounty—to feel Him— . . . this will be his duty. If I take a stone in the hand, I will . . . ask . . . which is the feature of it by which my Creator manifests Himself? I will find that feature admirable . . ."[32]

It has been questioned whether Georges-Louis Leclerc, Comte de Buffon (1707—1788), author of an immense *Natural history* (*L'histoire naturelle générale et particulière,* 1749–1789, co-written with two others), really belongs to the physico-theological movement. In our interpretation, he does. In theology, Buffon seems a Scriptural monotheist who, by an allegorical interpretation of the Biblical text, maintains that the six days of Genesis are historically identifiable with great epochs of nature (*les époques de la nature*).[33] For him, the ultimate accomplishment of science is the transition from the systematic observation and description of nature to a spiritual contemplation of the immaterial God beyond the bounds of the physical world, a natural science-based elevation of the soul to the vision of God: "The Earth is a delectable residence, where all is alive and directed with a power and intelligence which fill us with awe and exalt us towards the Creator."[34] In the sciences, especially biology, Buffon is unequalled in the eighteenth century. The grandness of the un-

32. *Effusions, First summer hour* (*Erste Sommerstunde*), 24–25.

33. Cf. *Of the epochs of nature* (*Des époques de la nature*), in Buffon, *Histoire naturelle. Oeuvres complètes de Buffon,* 80 vols, Nouvelle Édition (Paris, 1830) IX, 37–40.

34. *The History and Theory of the Earth* (*Histoire et théorie de la terre*) vol. II of *L'histoire naturelle,* 6. Since the ultimate aim of Buffon's natural science is theological, there may be no contradiction between theology and natural science. Cf. what Buffon says in this respect about his interpretation of the cosmogony of Genesis: "I have allowed myself this interpretation of the first verses of Genesis only in order to do the great good of reconciling the science of nature with the theological science forever; to my mind, they can only be in ostensible contradiction, as my explanation seems to demonstrate" (*Des époques de la nature*, in Buffon, vol. IX, 49).

dertaking of the no less than 80 volumes of the *Natural history* is unparalleled by any other scientific project of the eighteenth century,[35] except for Diderot's and D'Alembert's *Encyclopedia* (on which Buffon also collaborated with some significant articles, e.g., "Animal"). The connection between natural science and the theology of divine power is ensured by the sentiment of wonder the philosopher experiences at the sight of the organic, living, antagonistically-harmonious universe: "it looks that the Creator's hand has thrown . . . an infinity of harmonious and conflicting combinations, and a perpetuity of destructions and regenerations. What an idea of power does this sight offer us! What sentiment of respect do they inspire in us for their author!"[36]

Jacques-Henri Bernardin de Saint-Pierre (1737–1814), whose three-volume natural-scientific treatise, the *Studies of Nature* (*Études de la nature*, 1784), is the last text we cover here. Of this author, only one book is still read today, his at least sentimental if not *kitsch* novel *Paul et Virginie* (published in 1788 as part of the third edition of the *Études de la nature*). Among other things, Saint-Pierre was a friend (so long as this was possible) and a biographer of Rousseau's (cf. his *Essai sur Jean-Jacques Rousseau*). Inspired perhaps by Rousseau's attitude to nature (cf. the *Creed of the Savoyan vicar*, *Profession de foi du vicaire savoyard* in book 4 of *Émile*),[37] he introduced a relatively new point of view into

35. The series originally comprised only thirty-six volumes. In later editions, however, the title was extended to cover all the works of Buffon, so it came to include even the books of his that had first been published apart from the original series *L'histoire naturelle*. In particular, the book *Des époques de la nature* had first been published, in 1778, independently of the series, but in nineteenth century editions, it appears as a part of *L'histoire naturelle*. A further source of bibliographical confusion is that the first three volumes of the series, published in 1749, carry exactly the same title as the whole series itself.

36. *L'Histoire naturelle*, vol. I, *On the manner of studying, and treating of, natural history*, 101.

37. Claiming that Nieuwentyt's demonstration of the existence of God is methodologically mistaken, Rousseau here rejects Nieuwentyt's painstaking full induction and systematic physico-theology. He proposes a more contemplative-meditative approach, which concentrates on the harmony of nature as a whole: "I have read Nieuwentyt with astonishment, and almost with indignation . . . Were his book as big as the world, he could not have exhausted his subject; and as soon as one wants to enter into the details, the greatest wonder is missed, which is the harmony and concord of all . . . I meditate on the order of the universe, not in order to explain it by empty systems, but constantly to admire it, to adore the wise author, who is felt in it" (Jean-Jacques Rousseau, *Oeuvres complètes*, eds. B. Gagnebin and M. Raymond [Paris: Gallimard, 1959–1995, Bibliothèque de la Pléiade] vol. 4, 580 & 605).

physico-theological methodology. Instead of examining separately the particular purposive constitutions of the several different natural agents, he shifted the emphasis of argument to showing how they harmoniously cooperate with each other to achieve their common good in a local (and universal) *ecosystem*. While he developed this fecund physico-theological approach toward the perfect coordination of nature as a whole, this holistic scientific methodology, natural science still remained for him ultimately an affair of the religious sentiments of the *heart*, which instinctively drew man towards the Divinity.[38]

To a certain extent, Saint-Pierre's fundamental idea concerning the holistic understanding of natural teleology as an argumentative basis of the physico-theological theory is in concord with what Johann Gottfried Herder (1744–1803) suggested in *God: Some Conversations* (*Gott. Einige Gespräche*, 1787). Herder stated that "traditional physico-theology had come to an end," and that the time had come for this kind of theology to acknowledge and to philosophically exploit the global regularity and teleology of nature as a unified system.[39] In the nineteenth and twentieth centuries, then, physico-theology continued to exist (a major source of the early nineteenth century is e.g., William Paley's *Natural theology; or Evidences of the Existence and Attributes of the Deity, Collected from the Appearances of Nature*,1802),[40] but it remains a historical fact that its classical epoch, the time when it was a true intellectual movement and

38. As an apology for his "intrusion" into the domain of the natural sciences, Saint-Pierre, an engineer by profession, argued that, "I have written about the plants and the animals, though I am no naturalist. As natural history is not enclosed in libraries, it seemed to me that it is a book in which everyone might read. I believed I saw the meaningful characters of a Providence in it, and I talked about it not as about a system which absorbs my attention but as about a feeling which fills my heart" (*Études de la nature, Preface of the first edition*, ed. Aimé-Martin, vol. I, 2).

39. Cf. the *Third Conversation* of *Gott*: "Philolaus: Methinks that the ordinary physico-theologies are drawing to a close. Theophron: In their time, they were very useful and actually nothing but childlike-beautiful, popular applications of a new, solid doctrine of nature. Therefore, their basis will always remain: nay, the truth in them will be incomparably more glorified, when people no longer snatch at particular, minor ends on account of every minor, individual circumstance, but more and more acquire an understanding of the Whole, which is one single system even in its minutest ramifications, and in which the wisest bounty reveals itself according to unchangeable intrinsic rules" (in J. G. Herder, *Sämtliche Werke*, 32 vols., edited by B. Suphan [Berlin, 1877–1913] vol. XVI, 492–93).

40. On the nineteenth- and twentieth-century history of physico-theological theory, see *Historisches Wörterbuch der Philosophie*, 951–55.

an important, even keynote rational theological strand, had been the eighteenth century.

PHYSICO-THEOLOGY AS A NATURAL SCIENCE: COSMOLOGY

In respect to natural scientific quality (quantification, precision, methodology, etc.), the best of classical physico-theology met (or even set) the highest standards of unprejudiced, professional natural philosophy. The natural scientific fundament of their theological conclusions was a systematic, objective, and quantified analysis of empirical data. We shall try to show this, in brief, on the example of their cosmological doctrine.

Acknowledged already by Ray,[41] called "the new System" by Derham,[42] and generally received by 1750, the cosmological model physico-theology propounded depicted an essentially Copernican universe, with the difference that the spatial extension of the world was thought to be at least indeterminate or perhaps infinite.[43] This development of Copernicus was achieved through the application of better telescopes. The indefinitely large universe was thought to be filled with an infinite amount of 'Solar Systemes' (as Derham put it), probably with our immobile Sun at its centre.[44] Physico-theologists postulated that all such solar systems consist of a sun and planets that are, by virtue of their position

41. Ray, *The Wisdom of God*, 45.

42. Derham, *Astro-Theology*, xl–xlvi.

43. In the *Preliminary Discourse* of *Astro-theology*, Derham says about the new astronomical system that it "extends the Universe to a far more immense compass, than any of the other Systemes do, even to an indefinite space; and replenishes it with a far more grand Retinue than ever was before ascribed unto it" (xl). He adds that "as Myriads of Systemes are more for the Glory of God, and more demonstrate his Attributes than one, so it is no less probable than possible, there may be many besides this which we have the priviledge of living in" (pp. xliv–xlv; the French translation of 1729 exaggerates here).

44. In the *Astro-theology*, Derham says that "[in Figure No. 3,] the Solar Systeme is set in the Center of the Universe . . . And so it may be looked upon by us . . . But whether it be really so, whether it be in the Center of the Universe . . . is a difficulty . . . above our ability to fathom, although not at all improbable" (pp. xlii-xliii). Maupertuis's statement (*Essay de cosmologie*) is only to the effect that the Sun is immobile (but this probably implies that the Sun is at the centre of the entire Universe): "the immobile or almost immobile Sun in the part of the Heavens where it has been placed, had a revolving movement round its axis" (vol. I, 52). Nieuwentyt, on the contrary, argues that it is just possible that not the Sun but the Earth is at the centre of the Universe for, he says, experts disagree on this point (cf. *30th Contemplation: On the unknown*). To judge by his argumentation, he would have preferred the geocentric model.

around their respective suns, inhabitable and possibly inhabited, but they found no scientific evidence concerning the physical constitution of the postulated extraterrestrials.[45] Telescopic observation suggested that the basins in the Moon may be seas,[46] and the hypothetical conclusion was proposed that there is an atmosphere and life in general in the Moon.[47] The resulting general hypothesis of the constitution of the universe was a kind of many-world theory, based on strict scientific observation and inference.

Physico-theology could avail itself of very exact measurement data about the respective sizes of the planets in our solar system and about the length of their orbits, as well as about their velocities, albeit only after the first six planets, together with some of their moons, had been discovered. It was also precisely known that the shape of these planets is not a perfect globe, and that their orbits are elliptical rather than circular.[48] As we have anticipated, it was possible to observe the geography of the Moon, to distinguish the lunar mountains together with their vast shadows; but it was also possible to perceive Saturn's rings, the *maculae* in the Sun and some distant stellar *nebulae*, as well as to measure, astonishing as it may seem, the velocity of the light traveling from the Sun to the Earth.[49]

45. Derham, *Astro-Theology*, lvi: "the Maintainers of the new Systeme conclude those Planets, yea all the Planets of the Sun and of the Fixt Stars also, to be *habitable Worlds*; places as accomodated for Habitation, so stocked with proper Inhabitants. But now the next Question commonly put is what Creatures are they inhabited with? But this is a difficulty not to be resolved without a Revelation, or far better Instruments than the World hath hitherto been acquainted with." Maupertuis is equally cautious: "these enormous bodies of the planets, since they have so many things in common with the Earth, may have even this in common with her, that they are inhabited" (*Essay de cosmologie*, vol. I, 55).

46. Cf. Derham, *Astro-Theology*, li-lii: "that there are Seas, or great Collections of Waters, and consequently Rivers, Clouds, Air and Vapours in the Moon, I shall make out from some of my own Views and Observations" (apparently, the names of the "lunar seas," such as e.g., Pontus Euxinus were imposed at this time).

47. Cf. Maupertuis, *Essay de cosmologie*, vol. 1, 55–56: "The star which illuminates our nights, the Moon, is one of these secondary planets . . . The bodies of the secondary planets, opaque as those of the planets of the first order, may make us surmise that they are inhabited like the others are."

48. The angle of the tilt on the plane of the ecliptic, of the axis of the Earth, was also exactly measured. On the form of the planets, and on that of their orbits, see Maupertuis's *Essay de cosmologie*, vol. I, 66. The whole third part of the *Essai de cosmologie* is a fact-filled *status quaestionis* of cosmology. Part of Buffon's text *De la nature* agrees with these descriptions of our solar system (*Histoire naturelle*, vol. 1, especially 173–75).

49. Cf. Maupertuis, vol. I, 65: "they [the rays of light] arrive to us in seven or eight minutes." This is approximately the same as what is affirmed by present-day astronomy.

In a final analysis, then, eighteenth-century physico-theology contended that the universe is in spatial terms half-open or virtually open, and suggested (but no longer fully warranted) that our solar system occupies the central position in the universe, in which other kinds of creatures might concur with the human race for the cares of Providence.

The cosmological and astronomical doctrine of eighteenth-century physico-theology was hence all but rudimentary. Possessed of mighty instruments of research, technical and mathematical, it had all the necessary methodological rigor. But was this natural scientific competence utilized in a logically valid manner in the inferential pattern of the physico-theological argument for the existence and attributes of God?

PHYSICO-THEOLOGY AS A PHILOSOPHICAL SCIENCE

Nieuwentyt's *The Right Way of Using the Contemplations of the World* outlines more technically the logical structure of the physico-theological argument as it was conceived in the eighteenth century. His scheme reveals the logical pattern according to which classical early modern physico-theology generally reached its theological conclusions based on the observation of natural phenomena. Nieuwentyt marks out the logical skeleton of the argument with a series of questions.[50]

The physico-theological argument (the argument from the direction of things, *ex gubernatione rerum*) does not rely on the ontological contingence of the natural universe (as the cosmological argument, the argument *a contingentia mundi* does). In its Nieuwentytian formulation, it first points out that the physical (proximate or remote) efficient causes producing natural change do not have the intelligence necessary for the degree of coordination without which it is impossible to reach the infinitely complex outcome, the purposeful (and universally useful) operation of the physical world. This leads to the preliminary conclusion that the virtually perfect coordination of nature is due to the operation of final causes. This first phase of the argument is formally a (natural philosophical) full induction.

In the second phase, a topological move is made from natural science to theology, through a (formally not developed) syllogism. Essentially, this contends that the concept of final cause (reached in the first phase of the argument) is logically incompatible with the operation

50. See Nieuwentyt, *Het regt gebruik der werelt beschouwingen*, 23, §29.

of chance or unconscious natural laws. The notion of intentionality is only compatible with reason, more precisely—if the objectives set are morally perfect—wisdom. Thus the physico-theological argument does not concern the ontology, but only the teleology of nature; it considers the natural agents only *qua* co-ordination and substantial form or essence, but not *qua* existence.

A LOGICALLY FORMALIZED EXPOSITION OF THE PHYSICO-THEOLOGICAL ARGUMENT

We may now try to reconstruct a little more formally the general or synoptic version of the argument. It has just been said that the first momentum of the argument is a natural philosophical full induction. Physico-theology examines empirically in a theoretically endless series of case studies whether the characteristic substantial forms and the thereby determined natural operations of indefinitely many natural agents tend toward an objective or are entropic (disorderly); and if they do, whether they tend to the individual, special, general, and universal good. In this context, the individual, special, general, and universal good may be conceived as the realization of the essence, and the conservation, of an individual, its species, genus and the entire natural universe. In our authors, this systematic examination concerning substantial form and natural operation seems more specifically to involve the six following, objectively observable characteristics of the phenomenal universe: 1) convenient arrangement of the parts of a natural whole, both *a)* in the macro-world and *b)* in the micro-world; 2) regularity of motion of inanimate bodies, which breaks down into *a)* orderly character and b) periodicity of motion; 3) coordination of the motion of parts in a whole; 4) uniformity of the organic species; 5) regular generation of organic natures; 6) control of potentially catastrophical elements or substances like, e.g., fire. These characteristics of the natural world, examined in a (theoretically) full induction, lead us to the conclusion that in every case, the specific intrinsic constitution and the dependent operation of natural agents is purposive, that is, that there is a universal teleology or design (*oogh-merk*, Nieuwentyt; *noble Ends*, Derham; *but*, Maupertuis) in the constitution and functioning of nature.

We have said above that in the next stage, the argumentation is reducible to syllogistical form. The following two syllogisms may be set up here:

maior$_1$: the operation of final causes (universal natural teleology) is attributable either to chance, unconscious natural laws, or intelligence;

minor$_1$: but the concept of teleology analytically implies premeditated intelligent design;

conclusion$_1$ *and maior*$_2$: (therefore) an intelligence is responsible for the operation of final causes in the functioning of the phenomenal universe.

minor$_2$: but this intelligence, by reason of its quasi-infinite effect, cannot be finite or natural;

conclusion$_2$: therefore, it must be infinite and supernatural (transcendent), i.e., a God.

Hence we are logically entitled to draw the conclusion that there is a (supernatural) intelligence, which is responsible for the teleological operation of nature. As we have seen, the physico-theological argument does not first conclude that there is a God and then, that this existing God is omniscient, omnipotent, and all-good, but that there is an intelligence, which, by reason of its virtually infinite coordinating capacity, may reasonably be called a transcendent being, 'God.' It intends to prove an *existing divine intelligence* from the very beginning.

This (logically) first inference concerning virtually infinite, existing intelligence is accompanied by two complementary conclusions about virtually infinite, existing power and bounty.[51] Logically speaking, the first of these conclusions seems a corollary to the thesis concerning existing divine intelligence. Besides intelligence, virtually infinite power is also a *sine qua non* of the actual purposive order of the universe. The attribution of bounty to God on the basis of the physico-theological argument demands a separate judgment of value, insofar as the objective

51. Cf. Derham, *Astro-Theology*, 209–10: "As God's *Works* have been shewn to be manifest Demonstrations of his *Existence*; so they are no less of his *Perfections*, particularly of his infinite *Power*, *Wisdom* and *Goodness*; inasmuch as every Workman is known by his Work. A Palace that should have nothing defective in Situation, Beauty, or Convenience, would argue the Architect to have been a man of sagacity" (emphasis Derham's).

of universal natural teleology must be acknowledged as good or perfect (and implying as little evil as possible).

We deliberately say "*virtually* infinite intelligence, etc." since the argument is a full induction in theoretical but not practical terms. It would be impossible to effectively carry out an induction which implies that every substance, system, and subsystem of the entire world is checked for purposeful operation, and cooperation with the rest, in absolutely every conceivable aspect. If the first logical movement of the physico-theological argument is an induction, then it is one carrying no *categorically* demonstrative conviction anyway, according to Aristotelian logic. It is therefore philosophically justified to say "*virtually* infinite, etc." regarding the divine attributes as they are determined by this argument. The *actual* infinity of a cause cannot be proved with a practically incomplete (because interminable) induction, which moreover regards the world of contingency (experience). Ideally, however, the physico-theological argument carries an indefinitely great conviction, insofar as the degree of probability it reaches may approximate infinity. As the induction covers more and more individual cases in its theoretically endless series of particular investigations, so it can approach categorically demonstrative certainty in an infinite approximation. But in the spiritual-devotional elaborations of classical eighteenth-century physico-theological sources, the qualification "virtually" (a mark of the probabilistic character of the argument) was dropped, and the argument was closed within the ethical-pastoral-religious application of the theological final result.

At this point, there is still a long way to go before we can deliberate the cogency of the argument. For it seems that classical physico-theologists hardly ever considered the problem of evil; that there may be a hylozoistic alternative to the above interpretation of the observed natural phenomena; and third, the physico-theological argument was confronted by the "critical" Kant with an allegedly devastating criticism. A considered position taken on the physico-theological argument must first face these difficulties.

THE PROBLEM OF EVIL AND THE PHYSICO-THEOLOGICAL ARGUMENT

"Si Deus est, unde malum?"—"If there is a God, where is evil from?"[52] In the case of physico-theology, an argument from the perfect co-ordination of *nature*, this question will concern first and foremost natural (and not moral) evil. The problem of natural evil affects the second phase of the argument from design, in which not only the intelligence and power, but also the bounty of God is proved. That there are ravaging earthquakes, typhoons, and volcanic eruptions does not in itself sap the conclusion concerning the intelligence and power of the Maker. But it may evidently query the (perfect) bounty of God. As is known, the great Lisbon earthquake of 1755 made Voltaire think that the amount of physical suffering in nature is compatible with the existence but not the goodness of the Creator, and that without religious belief in a better future state it is impossible to reconcile natural evil with divine bounty (cf. *Poème sur le désastre de Lisbonne*, 1755). But natural evil consists not only in catastrophic "acts of God"—it includes the spontaneous corruption of the intrinsic natures or functioning of the different forms of life in individuals (disease) or entire species (epidemic) as well. It may be proposed that an even greater source of natural evil is the ruthless and endless struggle going on in the vegetative and sensitive kingdoms and deriving from the difference between the respectively individual, special, general, and universal final causes. All this implies the physical evil of suffering. Hence it may seem that the coordination of the several different ends of an indefinitely great amount of natural agents is anything but successful or even possible, let alone perfect, in the actual condition of nature. Is the argument from design philosophically able to resolve this knotty problem?

A POSSIBLE RESOLUTION TO THE PROBLEM OF EVIL

In historical terms, we may say that most classical physico-theologists do not appear to have identified natural evil as a problem in the way of the argument, though at least Fénelon offered ingenious counter-arguments against it, based on the logical-conceptual connection between finitude and imperfection. Fénelon, in the *Treatise Concerning the Existence of*

52. This classical question is put, e.g., by Leibniz in the *Essays on Theodicy*, part one, §20.

God, produces no less than five counter-arguments to resolve the problem of (moral and) natural evil: 1) generally, man is responsible for moral evil; 2) otherwise, evil is God's instrument of punishment or 3) temptation and perfection; 4) or human beings simply do not see the wider context of what they believe to be evil but what is good within the whole; 5) and last, whatever is defective in the work of God is a sign that nature has come to be by creation out of nothing, so it still bears the mark of its essential nothingness.[53] Of these, 2), 4), and 5) concern natural evil. Number 4 is a case that, according to Fénelon, does not imply the presence of any real evil in the world, while 2) is actually a mark of divine providence, for though it does imply suffering, it only seems, but is not essentially, evil. Finally, 5) is a complex case, as Fénelon says that God wanted creation to keep reminders of its origin from nothingness, while at the same time he adds that evil (an effect of imperfection) to some extent necessarily follows also from the finitude of creatures: "All that which is not God can have but a limited perfection; and that which has but a limited perfection remains forever imperfect . . . The creature would be the Creator Himself, if it did not lack anything; because it would be vested with the fullness of perfection, which is the divinity itself."[54] This account may be called a metaphysical rationale for the problem of evil.[55]

Although this is a good philosophical answer in its early eighteenth-century intellectual context,[56] it still presupposes that the opponent accepts at least the necessity of Creation. Now in a more systematic approach, one may maintain that despite this, the argument from design still conserves the overwhelming part of its convincing power, even when

53. Cf. *Traité de l'existence de Dieu*, I.88: "I do not mean to criticize this grand work (*scil.* the universe). The flaws one finds in it come from (1) the free and dissolute will of the man who produces them by his dissoluteness; or from the ever saint and equitable will of God, who now (2) wants to punish the unfaithful, now (3) wants to exercise, by the instrumentality of the mean, the good, whom He wants to render perfect. Frequently, (4) that which in an isolated spot of the whole seems a flaw to our narrow minds, is an ornament in respect of the general pattern, which we are unable to behold with a view angle wide and simple enough in order to come to know the perfection of the whole . . . But after all, (5) the real flaws of this work are but the imperfections God has left in it in order to let us know that He has drawn it from nothingness" (86–87).

54. Ibid., 87.

55. A more down-to-earth rationale of evil is proffered in Lesser's *Theology of the Insects*, vol. II, book II, part II, chapter IV.

56. Leibniz puts forward a very similar argument in the *Theodicy* (1710), part I, §§20 and 30–31.

the problem of evil has detracted from its likelihood. The first thing we have to call to mind is that the physico-theological argument for the existence and attributes of God has been a calculus of probabilities from the very beginning, so it is not destroyed at once by some evidence to the contrary. To our mind, then, the effective philosophical intention of the argument is not to prove the existence of God *on the ground that* absolutely every individual substance of the world may actualize all the potentialities inherent in their respective constitutions, *but* on the ground that each single one of an indeterminably great amount of substances receives the maximum possibility of existence that is just allowed by the equally maximal realization of the existential possibilities (inherent potentialities) of all other individual substances of the world, in a manner that the whole—as a system coordinated on all levels as perfectly as possible—constantly and teleologically operates, in the condition of a dynamic balance, for the preservation of itself.

Thus the philosophical accent of the physico-theological argument, systematically speaking, is on that each individual being or substance receives the maximum existential possibility which is possible at all within the frame of the whole (the phenomenal universe), since outside that frame, the existential possibilities of the individual are by all means null and void. If the aggregate of all finite substances is to be a world, then each individual substance contributes to the universal overall good by its own particular struggle round truth with the rest of the substances in the natural (as well as in the moral) world.

This interpretation may remedy the alleged insufficiency of the universal coordination of the several different ends of the indefinitely great amount of natural agents, but does it resolve the problem presented by the "acts of God" and the (individual and special) corruption of the internal natures? A separate answer seems necessary here. In our opinion, it is impossible to disprove the physico-theological argument with reference to the problem of natural evil ("acts of God" and morbidity), insofar as there could be absolutely no question about evil and disorder taking place in the (natural and moral) world, if there were no *order* in the first place. Only the existence of a universal order and teleology makes it possible that natural disorder *can* occur at all. Order is the ground and condition of existence, while evil is the locally and temporarily limited lack of order. The pervasively ordered character of existence is thus our referential frame whenever we speak about disorder—so much so, that if the measure or proportion of disorder were to reach a critical threshold value, all existence would become utterly impossible at once. *We would all die a sudden death if the principle of existence were not order.* Hence the world *is* fundamentally ordered (κόσμος, *mundus*), and the argument from this order to the bounty of the ordering principle is philosophically justified. The dialectical momentum of the problem posed by the "acts of God" and morbidity, even together with the argument from the relative insufficiency of universal coordination, do not destroy the thesis concerning the divine attribute of goodness.

THE ATHEISTIC HYLOZOISTIC ALTERNATIVE TO PHYSICO-THEOLOGY

Hylozoism is the natural philosophical position that all matter is alive and active by itself (so there is no part of the material universe that could be considered lifeless and merely passive). A degree of hylozoism may be philosophically compatible with theism (cf. Maupertuis, Buffon) or deism (cf. Robinet), but it is its atheistic strand which offers a full alternative to physico-theology (cf. the mature Diderot) and which we consequently have to adumbrate here. In historical terms, Diderot's full-fledged, materialistic hylozoism apparently relies on Locke's hypothesis concerning thinking matter,[57] on Toland's theory of essentially active matter,[58] on Maupertuis's theory of the spontaneous creative degeneration of the embryo,[59] and probably on Robinet's biological transformationalism as well.[60] The complex, atheistic vitalistic (hylozoistic) theory Diderot

57. See John Locke, *An Essay Concerning Human Understanding* (1690), Book IV, Ch. iii, §6.

58. See John Toland, *Motion Essential to Matter*, in the *Letters to Serena* (London, 1704; reprint Stuttgart–Bad Cannstatt: Frommann, 1964), Epistle V.

59. See his *Essay Concerning the Formation of Organic Bodies* (*Essai sur la formation des corps organizés*, 1751). Maupertuis published a *Dissertatio inauguralis metaphysica de universali naturae systemate* (a misleading title) under the pseudonym Dr. Baumann in Erlangen in 1751, which was thus usually referred to as the "*dissertation d'Erlangen*." Very soon, it became known that the author was Maupertuis, who then issued a second, bilingual (Latin and French, *sine anno* and *sine loco*), and a third, monolingual (French, 1754) edition, allegedly in Berlin but really in Paris. The third edition carried the non-misleading title *Essai sur la formation des corps organizés*. But because the original Latin title was also translated into French as *Système de la nature* and later used as a designation of the work, the very same text has since been referred to under no less than four different titles.

60. This theory, expounded in his anonymous *Of Nature* (*De la nature*, 1761),

innovatively conceived departing in part from these authors contends that all natural phenomena identified by physico-theology as the result of preliminary divine design exclusively derive from the spontaneously operating, inherent powers of universal matter. All motion of matter (whether regular or not) is thus self-generated, and the emergence of life is due to the self-organization of matter. Hence, in a biological respect, Diderot's hylozoism may be defined as the theory of evolution without the Darwinian component of natural selection. As expounded in *The Dream of d'Alembert* (*Le rêve de d'Alembert*, 1769), in the *Philosophical Principles Concerning Matter and Motion (Principes philosophiques sur la matière et le mouvement*, 1770), and in the unfinished but systematic manuscript *Physiology* (on which he worked until his death in 1784), his materialistic rationale for the motion of matter and the phenomena of life is a complete philosophical alternative to—or better, a total negation of—physico-theology. When confronted with the syllogistical scheme of the theistic argument from design (cf. pages 88–89 above), Diderot would reject minor$_1$ (that "the concept of teleology analytically implies pre-meditated intelligent design"). He does not seem to directly deny that the phenomenal world is teleologically ordered (non-entropic) or coordinated, but he certainly does not accept divinely instituted final causes. By attributing the order of nature to the spontaneity of matter, he jettisons both theological conclusions. On his hypothesis it will not be necessary to posit a transcendent intelligence to explain natural teleology. But is it reasonable to drop minor$_1$ of the physico-theological syllogism, that is, to derive order from spontaneity? How does the argument from design relate to this idea?

PHYSICO-THEOLOGY AND THE ATHEISTIC HYLOZOISTIC THEORY

In chronological terms, atheistic-materialistic vitalism is an intellectual phenomenon of the last phase of the Enlightenment, while physico-theology was a characteristic mode of thought in the first half or first two thirds of the eighteenth century (Diderot himself had accepted physico-theology in his early *Philosophical Thoughts*, 1745).[61] It is hence

states that there has been only a single, created prototype for all species of plants and animals.

61. *Pensées philosophiques*, thoughts 18–20.

difficult to say exactly how the great physico-theological classics of the earlier eighteenth century would have reacted to Diderot's radical theory concerning the spontaneous motion, sensibility, and life of matter. But it is possible to show how they related in general to the anti-Newtonian thesis that "Action is essential to Matter," since this had been proposed at least by Toland at the beginning of the eighteenth century.[62] At least two classical physico-theologists, Nieuwentyt and Maupertuis, explicitly set down the metaphysical principle that corporeal substance is in itself inactive (inert) so bodies cannot intrinsically move themselves—which would impair only one component of Diderot's complex radical theory. Nieuwentyt argues that the non-believer should consider the following theses about the nature of bodies in general:

> I. That a body can be moved as well as stay in rest i.e., be not moved; and in both these cases remain a perfect body, and conserve its essence.
>
> II. Whereof it follows that motion does not belong to the essence of body.
>
> III. And one may remark about this that the famous Mr *Newton*, and the commentator of his demonstrations and argumentations, Mr *Whiston* . . . have described or defined *body* rightly; [namely,] that it is an extended and solid substance, not only indifferent to motion and rest but also devoid of any power in itself and only passive (*substantia iners & passiva*)[63]

In the *Essay de cosmologie*, Maupertuis suggests that "We see some parts of matter in motion, we see other parts of it in rest: motion is hence not an essential property of matter; it is a condition in which it may happen to be or not to be, and which we do not see that it could give to itself by itself . . ."[64]

It is questionable whether this offers a conclusive counter-argument, as Toland's thesis is precisely not that in the sublunar world, there is a spontaneous locomotion of extended bulks of matter, but that unconscious natural agents constantly experience physical, chemical, and biological change. They are never completely inactive but keep

62. Cf. Toland, *Letters to Serena*, Epistle V, 202 (in Gawlick edition).

63. Nieuwentyt, *Het regt gebruik der werelt beschouwingen* (1730), 771 (Nieuwentyt's emphasis).

64. Vol. I, 32 (Tonelli edition).

reacting to each other, apparently by their inherent active powers.[65] As this dynamic interaction or mutual commerce of substances is the natural condition of the material universe (as an atheistic hylozoist might argue), it is likely that universal nature engenders all mechanical and biological phenomena with her own power. Nature is hence her own explanation; her endless complexity and eternal perfect order do not require an external determining ground. This (very modern) formulation of the atheistic hylozoistic theory is the true philosophical challenge to physico-theology.

It seems that at this point, we may have recourse to Wolff's physico-theological conception, which is, as we have indicated in section 2 above, grafted onto the Leibnizian cosmological argument from the ontological and qualitative contingency of the world. In Wolff's view, something can be its own explanation (can be its own existential ground i.e., may include existence in its essence) only if it is a necessary thing (*ens necessarium*). Otherwise it will be necessary to posit a cause (*causa sufficiens* or *determinans*) which exhaustively determines a contingent thing to be so and so rather than differently. Since the world is never *necessarily* so as it actually is, it is philosophically justified to conclude that it has an *external* determining ground:

> We would say that the world is necessary if space and time let themselves be filled up also in a different way than the world is filled . . . in that case, we would not need any further ground why there is *this* world and not any other . . . But . . . if other arrangements of the world are also possible . . . then we find no sufficient ground why it is in this way rather than another way, and consequently, we have to look for the determining ground outside it . . . In this manner, the contingency of the world renders it suitable

65. As a matter of historical fact, Toland himself, toward the end of Epistle V, proposed in a somewhat desultory manner that matter had first been invested with motion by God, who keeps directing all natural motion: "*Besides, that God was able to create this Matter active as well as extended, that he cou'd give it the one Property as well as the other, and that no reason can be assign'd why he shou'd not endue it with the former as well as with the latter; is there likewise no necessity that he shou'd ever or rather always direct its Motions?*" (ed. Gawlick, 234). Thus, in a final instance, Toland reserves the direction of the perfectly coordinated motions of matter—among other things, of "the Formation of Animals or Plants"—for God. That nature might produce all these phenomena entirely alone is Diderot's development of the idea. None of the earlier (deistic or theistic) hylozoists (Robinet, Buffon, Maupertuis) went as far as that.

> for revealing to us that there is a God, i.e., that it has an originator and is not by itself.[66]

This argument (which depends not on the ontological but the qualitative or constitutional contingency of the natural world) may be completed with two further, systematic considerations concerning the more specifically biological facet of the atheistic hylozoistic theory: 1) the transition from the *spontaneity* involved in the generation of life to the *necessity* manifest in the conservation of the internal nature of the species; and 2) the time paradox implied in the process of the spontaneous development of an animal towards higher degrees of structural complexity and perfection. Regarding (1), it is logically inexplicable how a process that began *spontaneously* could ever transform into a *necessary* process guaranteeing the uniformity of reproduction, considered (by, e.g., Buffon in the significant article "Animal" of the *Encyclopaedia*) as the *differentia* of the biological concept of "species." In other words, the emergence of thoroughly (by birth) determined internal natures, from an original process which, being fully spontaneous, has no determinative power, remains a mystery.

On the other hand, (2) there seems to be a time paradox involved in the idea that a spontaneous process of development may advance towards ever higher degrees of organization and, therefore, perfection, because it is impossible to progress towards higher complexity or perfection unless an organism knows in advance which condition (structure, constitution) is more perfect than another one, among the indefinitely many possible conditions (structures, constitutions). Such knowledge would imply that the primitive machine the primary molecule is must know the perfection of a stage of development it has not yet reached. In other words, it would have knowledge about a future state in the present, which is a time paradox.

In a final analysis, it seems that the atheistic hylozoistic alternative to physico-theology is ultimately reticent on the principle of life; that it refuses to put the question *why* the primary molecule was formed at all. The physico-theological answer to that query is a philosophically more articulate reaction than reticence. The postulation of final causes and thereby an ordering intellect is a more satisfying explanation to a

66. Wolff, *Vernünfftige Gedancken von den Absichten der natürlichen Dinge*, 9–10 (emphasis added). The Leibnizian Lesser, in the *Theology of the Insects*, vol. I, book I., chapter I, repeats essentially the same argument.

rational being, man, who, conditioned by its intellectual constitution, looks for reasons everywhere. But the physico-theological argument, in order to gain a degree of cogency, still has to overcome Kant's classical criticism.

KANT'S CRITICISM OF PHYSICO-THEOLOGY

Kant, who in *The Only Possible Premise* still accepted that the argument from design carries some though not categorically demonstrative conviction, entirely abandoned the argument in the transcendental theological part of the *Critique of Pure Reason* (*CPR*, 1787[67]), his most famous work. His philosophical criticism directs our attention to the second syllogism of the physico-theological inference, which attributes infinity and supernaturality to the intelligence designated as the cause of natural teleology in the first syllogism. In the *CPR*, Kant essentially accepts the first syllogism, but adds that the argument next performs a philosophically unjustifiable transition as it proceeds from the not fully determined concept of a master-builder or demiurge (*Weltbaumeister*—a concept of relation i.e., a *Verhältnisvorstellung*) to the fully determined idea of a creator God conceived as the most perfect being.[68] For, as Kant points out, the physico-theological argument does not tackle the ontological problem of the *origin* (or creation) of matter. It takes for granted that the notion of the arranger of matter is that of a Creator, who is perfect in every respect. For the mature Kant, however, this is a philosophically illicit identification within the bounds of the physico-theological argument, and a crossing over into the domain of the cosmological argument, which he, in turn, rejects.

The last logical move of the *cosmological* argument, he continues, is an identification of the absolutely necessary being (*ens necessarium*)[69] with the absolutely perfect being (*ens realissimum*). Thus, the cosmological argument essentially presupposes (can be reduced to) the ontologi-

67. The first edition (1781) of *CPR* does not contain a well-developed transcendental theology—this part was upgraded in the second edition 1787. All references are thus to the second edition.

68. "*das allerrealste Wesen*" (B 624, AK III 399); "*die höchste Realität*" (B 628, AK III 402). Cf. also his *Lectures on Metaphysics*, *Vorlesungen über die Metaphysik*, [Pölitz] ed., 49–50: *Of the Real and the Negative* (*Vom Realen und Negativen*).

69. "*Ein Schlechthinnothwendiges*" (*CPR*, B 657, AK III 418).

cal argument.[70] Hence, in the Kantian analysis, the physico-theological argument for the existence and attributes of God performs a primarily a priori, purely notional transition from the concept of an arranger of pre-existent matter, ultimately to the existence of the absolutely necessary being through the mediating concept of the most perfect being. At the same time, we repeat, the existence of at least a demiurge seems acceptable to Kant. He says that "Therefore, the proof could show, at most, a master-builder of the world who would be very constrained by the suitability of the material he handles, and not a creator of the world, to whose idea everything else is subordinated; which is by no means sufficient for the great end one has set here, namely, to prove a self-sufficient primordial being."[71]

A CRITICISM OF KANT'S CRITICISM OF PHYSICO-THEOLOGY

To our mind, the sight of the conspicuous harmonies of nature fairly disproves Kant's conclusion—the arranger of the natural universe has been perspicuously all but constricted by the limited "suitability of the material." On this point, physico-theology seems more realistic (or less prejudiced) as it philosophically recognizes that the degree of perfection the arrangement and coordination of nature reaches gives logical ground more to the conclusion that the Arranger had an indefinitely great (quasi-infinite) power over matter, than to the conclusion that His power was limited by anything. In the Kantian criticism of the physico-theological argument, there is a certain reluctance dialectically to exploit as much as the argument *does* prove. Even Kant himself is apparently ready to accept the argumentation as far as it posits the existence of an arranger of (pre-existent) matter. But this point is very far from being argumentatively unusable material. For it seems reasonable to think, at the sight of the "marvels of nature," that the arranger of universal matter must have been at least pre-constituted so as easily to be able to compel matter into order, and that matter must have been at least pre-constituted so as perfectly to yield to that compulsion. In other words, arranger and arranged must themselves have been perfectly coordinated

70. *CPR*, B 636-637, AK III 406–7. Cf. also *The only possible Premise* . . . , III, 3; AK II 157–59.

71. *CPR*, B 655, AK III 417.

from the very beginning. Now by virtue of Ockham's razor, it seems more reasonable to account for that harmony by the supposition that the arranger has produced or created matter, than by any other supposition (including even the pre-existence of matter). To put it differently, if matter itself (cf. hylozoism) or chance cannot be made responsible for the perfect teleology of nature, then by virtue of the argument from design, we must conclude that an arranger has arranged matter—and an a priori precondition of this conclusion is the assumption that the perfect coordination of arranger and arranged has had a condition or *principle.* At this point, the logically simplest, that is, most reasonable solution of the query is—unless we want to go to infinity—to say that probably, the Arranger Himself has created matter, and that this best explains the perfect coordination of arranger and arranged.

On the other hand, it is flying in the face of reason and experience to maintain that the pervasive order and teleological functioning of the natural universe does *not* teach us anything we may articulate philosophically, about the existence of a (transcendent) efficient cause of order. It is in fact completely *un*reasonable to give up the physico-theological argument because "it can only prove the existence of an arranger." If it can be proved that matter is not self-moving and self-organizing (which implies disproving the entire hypothesis of evolution), that the arranger has had a perfect control of matter, and that He has been guided by the greatest bounty and wisdom possible in the arrangement of matter, then this is enough for the arranger to deserve our religious awe, and it is on this basis philosophically legitimate to set down the concept of a transcendent ordering cause, namely, a God.

CONCLUSION: JACOB'S LADDER AS A PHYSICO-THEOLOGICAL METAPHOR

Having considered all of this, it seems reasonable to maintain that the argument from design conserves at least the degree of cogency attributed to it by the early Kant in *The Only Possible Premise*. This implies that it is reasonable to accept the ("pre-critical" and "critical") Kantian thesis that this argument proves first and foremost the existence and chief attributes of an *arranger* of universal matter. But if we creatively reflect on the above-mentioned principle of the apparently perfect coordination of arranger and arranged, the argument from design will further be able to render the likely thesis that the Arranger *is* the perfect being we call God.

Eighteenth-century classical physico-theology, for its part, never hesitated to draw this conclusion. All authors acknowledged the authority of Scripture, and several of them used other arguments as well to prove the existence and attributes of God, such as, for example, Wolff (in his *Theologia naturalis*, 1736) or the early Kant (who in *The Only Possible Premise* also advocated an upgraded version of the ontological argument as the only *categorical* demonstration of the existence of God). None of these authors seriously disbelieved the harmony of reason and revelation. These historical circumstances, together with the fact that many of them were not professional philosophers, explain that they in a sense jumped to the creationist conclusion in the physico-theological inference.

At the beginning of this paper we claimed that eighteenth-century physico-theology had a tendency to combine terse natural scientific enquiry with spiritual-devotional elaborations (as Trescho says,"effusions"). At the end of this review, we shall give an idea of how physico-theology ascended from wondering at the marvels of nature, through the natural philosophical induction and metaphysical syllogisms, to admiring more immediately the supernatural glory of God, of which nature bears but a faint reflection. The image of Jacob's ladder, suggested by some physico-theological sources, may reasonably be seen as the ultimate metaphysical metaphor of this intellectual movement.

In classical physico-theological spirituality, the contemplation of the open book of the heavens entails a Platonic rapture of the soul, which begins with an admiration of Creation. This admiring contemplation of the "marvels of nature" (*les merveilles de la nature*) leads to an elevation of the soul, which transports it to a vision of celestial glory offering as it were an immediate intellectual view of divine magnificence. Nature is conceived here like Jacob's ladder, on which the soul gradually ascends towards the divinity, every grade of the ladder being a class of things, which—scientifically examined in the physico-theological full induction—guides the pious scholar, disciple of God, step by step towards the moment in which the ladder itself may be dropped. Thus, for Buffon, "Nature is the throne of divine magnificence: the person who contemplates her, who studies her, gradually ascends to the interior throne of omnipotence."[72] Or, as Saint-Pierre put it, "by the sight of the actual harmonies of nature, I ascend towards her maker, and I hope for more blissful destinies in another world."[73] But Wolff is the most explicit of all on this point: "This same description of the world . . . is the ladder on which we may ascend to God and see him as he is, namely, a being of unrestricted freedom, endless knowledge, the highest wisdom, the greatest power, unspeakable bounty and the strictest equity."[74]

This ascension of the soul was further represented by Nieuwentyt as an illumination of the finite understanding through an irradiation of divine light, which removes the obscurity deriving from the finitude of the human intellect, and reveals the existence and attributes of God. In this rhetorical image, the word *verligten* ("to enlighten") was used with the pregnant meaning, "the divinity emits supernatural intellectual light and gives insight into divine existence." Classical physico-theology, then, apparently had the capacity to combine the spiritual enlightenment (*verlichting*) of the soul with the scientific Enlightenment (*Verlichting*) of the mind, as it is expressed by the following passage from Nieuwentyt:

> And may it please the same [the admirable benefactor of all] to enlighten the eyes and the mind of this unlucky person [the atheist], so that he may not only be convinced, with full conviction, of the great Creator's wonderful and inscrutable wisdom, of His immense and unlimited, discretionary power, and of His generously given and undeserved lovingkindness, by virtue of the magnificent framing of this so beautiful world and all the marvellous things that are in it: but may he also experience, together with believing Christendom, the wonders of divine grace, by virtue of

72. Buffon, *De la nature*, vol. I of *L'histoire naturelle*, 179: "La nature est le trône de la magnificence divine: l'homme qui la contemple, qui l'étudie, s'élève par degrés au trône intérieur de la toute-puissance." That the result of this elevation was rhetorically thought to offer a quasi-intellectual vision of the Godhead is shown by the short prayer Buffon inserted in the text of his *De la nature* (vol. I of *L'histoire naturelle*, 185): "Great God, whose presence alone maintains nature and keeps up the harmony of the laws of the universe; you, who from the immobile throne of the empyrean can see roll under your feet all the celestial spheres without collision or confusion; who from your resting place reenact in every moment their immense motions and govern in profound peace this infinite number of heavens and worlds; give back, give at last back the rest to the agitated Earth!"

73. *Paul et Virginie*, ed. Aimé-Martin, vol. IV, 93.

74. Wolff, *Vernünfftige Gedancken*, 3 *verso* (Foreword).

> the steady and immobile grounds of God's Holy Word, so that he may thereby be joyful in the so certainly approaching eternity.[75]

This hopeful sight of the "certainly approaching eternity" opens up the true ultimate metaphysical perspective of physico-theology.

75. Nieuwentyt, *Het regt gebruik der werelt beschouwingen, 30th Contemplation. Of the Unknown.* §17. *Conviction gained from that which is unknown, and conclusion,* 915–16.

PART TWO

Wonder in Seventeenth-Century Europe

4

Descartes on the Excellent Use of *Admiration*[1]

Dorottya Kaposi

THE THEME OF *WONDER* appears in Descartes's French and Latin writings as *admiration* or *admiratio*. In this essay I try to show how these occurrences determine in Descartes' philosophy the essentially twofold character of the concept of *admiration*: on the one hand, it means a "sudden surprise," i.e., an unexpected feeling caused by the perception of some new or unknown object; on the other hand, it can mean "veneration" or "regard" with respect to an unusual, rare, and extraordinary object.[2] In this latter sense, *admiration* can be felt repeatedly and become a long-lasting disposition or habit.

Descartes uses both meanings of *admiration* mentioned above; its use seems to subordinate the first one to the second in favor of which the concept presents a certain progress. This semantic change fits what can be observed in the literature and philosophical writings of the seventeenth and eighteenth centuries,[3] to which Descartes contributes in

1. For the quotation of Descartes's writings, we use the following abbreviations: AT for *Œuvres [de Descartes]*, published by Ch. Adam and P. Tannery, 11 volumes (Paris: Cerf 1897–19091, Vrin, 1964–19742, Vrin 19963) followed by volume number and pagination; CSM and CSMK for the English translation by J. Cottingham, R. Stoothoff, D. Murdoch and A. Kenny: *The Philosophical Writings [of Descartes]*, 3 volumes (Cambridge: Cambridge University Press, 1985–1991).

2. The definition of "wonder"/*admiration* in *The Passions of the Soul* (art. 70) is composed of both aspects in such a way that they cannot be distinguished (AT XI, 380–81, CSM I, 353). It is only in the development of the concept that we can perceive the difference between the two aspects of admiration, especially when the passion of *generosity* is explored in Part III.

3. As explained in the analysis of Sophie Hache: "the rhetorical treatises allow us to support a double hypothesis, which one may resume by noting a progressive detachment of *admiration* from the notion of sublime, correlated to an evolution of the

an active and important way as a prominent author; the modification consists in a certain shifting of stress, in favor of the second sense, *admiration* considered as a feeling of respect which is neither sudden nor unexpected but can become constant and durable.[4]

The distinction between the two meanings of *admiration* turns out to be fruitful in the study of how Descartes's thought is related to two different approaches of wonder in the early modern period, based on Aristotelian sources. It also helps us to understand the philosopher's critical attitude toward the Aristotelian conception of wonder in his epistemological thought, and the refitting of a conception of *admiration* within his moral philosophy.

• • •

According to a widely accepted conception in medieval and early modern thought, based on Aristotle's account in Book A of *Metaphysics*, wonder is the beginning of the philosophical inquiry since it promotes the knowledge of the first causes and principles: "it is owing to their wonder that men both now begin and at first began to philosophize; they wondered originally at the obvious difficulties, then advanced little by little and stated difficulties about the greater matters, e.g., about the phenomena of the moon and those of the sun and of the stars, and about the genesis of the universe."[5] Thus, wonder is, according to Aristotle, the principal motor of philosophy thanks to which we can advance in knowledge little by little; it directs our attention to what we are ignorant of "a man who is puzzled and wonders thinks himself ignorant . . . since they philosophized in order to escape from ignorance, evidently they were pursuing science in order to know, and not for any utilitarian end."[6]

In the Aristotelian account in his *Metaphysics*, wonder is essential in promoting the desire of search for truth. Therefore, wonder has a dynamic aspect which is due to our ignorance rather than to any intrinsic character of the object itself which would be wonderful. Furthermore,

---

meaning attached to it over the course of the century – the *grandeur* progressively takes on the surprise." "Le saisissement de l'âme : sublime et admiration à l'âge classique," in *L'Admiration*, études recueillies par D. Denis et F. Marcoin (Arras: Artois Presse Université, 2003), 121–33, 121. See also Baldine Saint Girons' contribution below.

4. This is also what *admiration* means in current French and English.

5. Aristotle, *Metaphysics*, A, 2, 982 b 13–16.

6. Ibid., 982 b 19–21.

since we seek wonder in order to know what we ignore, it contains something which incites to deliver us from it.

> Yet the acquisition of [Wisdom] must in a sense end in something which is the opposite of our original inquiries. For all men begin, as we said, by wondering that things are as they are, as they do about self-moving marionettes, or about the solstices or the incommensurability of the diagonal of a square with the side; for it seems wonderful to all who have not yet seen the reason, that there is a thing which cannot be measured even by the smallest unit. But we must end in the contrary and, according to the proverb, the better state, as is the case in these instances too when men learn the cause; for there is nothing which would surprise a geometer so much as if the diagonal turned out to be commensurable.[7]

While Aristotle underlines the role of wonder in philosophy in virtue of the recognizing of our ignorance and in virtue of its dynamism in the progress of knowledge, Descartes prevents excessive wonder in the search for truth. In his thought, wonder cannot be considered as the principle of progress in the knowledge which would accompany man in his attention moving from the known to the unknown. Indeed, even though wonder may be at the origin of our first inclination to know, it must disappear as early as possible; otherwise, it would be incoherent with a scientific method to consider objects of our knowledge as wonderful. Descartes seems to maintain that the phenomena of nature are all susceptible to be known by the method and therefore we should not admire them.[8] In the *Regulæ*, Descartes shows that they who hide the causes of phenomena before the others excite wonder in the public, but this is not a legitimate wonder and it is not conducive to knowledge. Descartes's critics direct attention to the fact that the phenomena seem wonderful only for those who cannot apply the simple and easy rules of the method. Once having acquired them, there is nothing among the objects of knowledge that should be considered wonderful.

Now, if we consider at first Descartes's critical attitude toward wonder, we can say that his criticism concerns above all the epistemological context.[9] Inside this, the criticism is mainly addressed toward the admi-

7. Aristotle, *Metaphysics*, A, 2, 983 a 19–25.

8. Cf. L. Renault, *Descartes ou la félicité volontaire* (Paris: PUF, 2000).

9. See especially the critique of *admiration* by Descartes in C. Coutel, "Le paradoxe de l'admiration," in *L'admiration*, 23-30; Hache, "La saisissement de l'âme : sublime et

rable character of the objects of knowledge. This is especially true, not just for the epistemological works on method such as the *Regulæ*, but also for the epistemological context of *The Passions of the Soul.* What Descartes criticizes is the desire to wonder only in order to wonder, which does not allow a more precise knowledge of the object: "This is what prolongs troubles of those afflicted with blind curiosity, i.e. those who seek out rarities simply in order to wonder at them and not in order to know them. For gradually they become so full of wonder that things of no importance are no less apt to arrest their attention than those whose investigation is more useful."[10]

Descartes seems to attack in this way a certain conception of *admiration*, which consists of the view that, since wonder is good (as the beginning of philosophy), we should seek wonderful things and wonder at them all of the time, rather then try to know them. According to Descartes, if we consider the objects of knowledge wonderful we cannot discover their *natural order.*

Thus, in the epistemological context, the understanding of the phenomena of nature requires the knowledge that nature must obey the rules of scientific method and contain nothing that would be worthy of admiration. This is why the Aristotelian *thaumadzein* as astonishment has no place in Cartesian scientific investigation.

The view on wonder in Book A of *Metaphysics*, however, is not the only impact of Aristotle's philosophy on the Cartesian conception of *admiration*. Indeed, there is another doctrine, which, though indirectly originated from Aristotle, influences Descartes and profoundly determines his conception of *admiration*: it concerns the aesthetic and rhetorical developments of the notion of wonder in Aristotle.

• • •

This thought is expressed in Aristotle's *Poetics* as follows: "the wonderful is pleasing" (*Poetics*, chapter 24, 60a); as the philosopher maintains in his *Rhetoric*, "people like what strikes them, and are struck by what is out of the way" (book 3, chapter 2, 1404b). The function of *wonder* in these

---

admiration à l'âge classique"; C. Noille-Clauzade, "L'admiration classique, une passion critique," in *L'admiration*, 145–64. These papers examine the epistemological context of *admiration*, without analysing the occurrences of the related concept of *generosity* as it is exposed by Descartes in Part III of the *Passions of the Soul.*

10. *Passions of the soul*, art. 78 (CSM, I, 355–56).

writings is, first, that it provokes pleasure; and, second, that it makes a strong impression caused by the striking character of the object, which is the reason why people like wonder; and, third, that the striking character comes from the consideration of an object which is "out of the way," i.e., which is different from ordinary things. As a result of these effects, wonder helps retaining in the memory a strong impression of wonderful objects. This effect can be explained by the Aristotelian conception of *phantasia* which constitutes a material support in the soul where the representation or imagination can engrave images.[11] Mnemo-technical writings all underline the importance of the imagination of unusual things and scenes, the usage of which allows to strengthen memory thanks to the *impression*, in its material sense as etching images by engraving marks on the *phantasia*.

The function of wonder in Aristotelian rhetoric seems to inspire the Cartesian conception of *admiration* indirectly, through the Ciceronian art of rhetoric and the classical and renaissance forms of *ars memoriae*.[12] Descartes knows well the mnemo-technical tradition and, though criticizing the way authors use them,[13] he is inspired by all the above characteristics of wonder in this tradition. Admiration has a striking character and it makes a strong impression in the soul by unusual objects. According to its definition in *The Passions of the Soul*, *admiration* brings the soul "to consider with attention the objects that seem to it unusual and extraordinary. It has two causes: first, an impression in the brain, which represents the object as something unusual and consequently worthy of special consideration; and secondly, a movement of the spirits, which the impression disposes both to flow with great force to the place in the brain where it is located so as to strengthen and preserve it there, and also to pass into the muscles which will continue to maintain the impression in the way in which they formed it."[14]

11. Cf. Aristotle, *De anima*, III, 7, 427 b 14–20, 431 a 16–17; *De memoria*, 449 b 31–32, 450 a 20–25.

12. See especially the studies of F. A. Yates, *The Art of Memory* (Chicago: University of Chicago Press, 1984) and P. Rossi, *Clavis Universalis: Arti Della Memoria E Logica Combinatoria Da Lullo a Leibniz* (Sociétà Editrice Il Mulino, 1983).

13. See Descartes's fragment on the *ars memoriae* of Lambert Schenkel: *Cogitationes privatæ*, AT X, 230:3–25 ; cf. the fragment about the "true art of memory" *Cartesius*, AT XI, 649:5–8.

14. *The Passions of the Soul*, art. 70, AT XI, 380–81, CSM, I, 353.

In the same way, article 72 explains that wonder has "considerable strength because of the element of surprise, i.e. the sudden and unexpected arrival of the impression which changes the movement of the spirits . . . Its strength depends on two things: the novelty and the fact that the movement it causes is at full strength right from the start."[15]

This effect of *admiration* is also what allows to keep in our memory the impression caused by such objects. This is precisely what makes admiration most useful and allows us to consider its good, if not *excellent use*: indeed, feeling this passion helps to fortify the impression caused by its object and, while the object is worthy of great consideration, fortifying its impression turns out to be the principal utility of *admiration*.

> Of wonder, in particular, we may say that it is useful in that it makes us learn and retain in our memory things of which we were previously ignorant. For we wonder only what appears to us unusual and extraordinary; and something can appear so only because we have been ignorant of it, or perhaps because it differs from things we have known (this difference being what makes us call it "extraordinary").[16]

Now, let's observe the reasons Descartes gives in order to explain this unusual character of the object of *admiration*: it "differs from what we have known." This could be understood in reference to the first conception mentioned above, which is related to Aristotle's *Metaphysics*, by saying that the difference between what we have known and what is unknown gives the essential dynamic character of wonder, promoting the progress in knowledge by orientating our attention toward ignorance. However, this can also be understood in a second way, by saying that objects which differ from what we have known have something intrinsically different from the *order* of our knowledge; that is why we can consider them *extraordinary*. In this sense, a critical *excellent use* of admiration allows us to point to this difference between objects obeying rules of natural knowledge (which are, as a matter of fact, unworthy of great attention) and the ones which surpass the natural order of knowledge (which are worthy of our special consideration). In other words,

15. Ibid., AT XI, 382, CSM, I, 353–54. We can also find the origin of these reflections in Descartes' correspondence with Elisabeth, in his *Letter* of May, 1646, AT IV, 409–10, CSMK, III, 286–87.

16. Ibid., art. 75, AT XI, 384, CSM, I, 354–55.

the good use of admiration allows us to distinguish the *ordinary* from the *extraordinary*, the *natural* from the *supernatural*.

The utility of admiration is first observed in the description of *The Passions of the Soul*. Whether it is a sudden feeling (of surprise) or a long-lasting regard (of veneration), it appears useful in virtue of its strength by which it fortifies the impression caused by the object. Nevertheless, this is also what turns out to make admiration harmful, in the case in which it fortifies so excessively the impression of an object that it makes it impossible to turn our attention to other details in order to know the object, as explained in the definition of "astonishment."

> This element of surprise causes the spirits in the cavities of the brain to make their way to the place where the impression of the object of wonder is located. It has so much power to do this that sometimes it drives all the spirits there, and makes them so wholly occupied with the preservation of this impression that none of them pass thence into the muscles or even depart from the tracks they originally followed in the brain. As a result the whole body remains as immobile as a statue, making it possible for only the side of the object originally presented to be perceived, and hence impossible for a more detailed knowledge of the object to be acquired. This is what we commonly call 'being astonished.' Astonishment is an excess of wonder, and it can never be other than bad.[17]

Now note that this use can only be harmful if it remains a long-lasting admiration; an instantly felt sudden surprise is caused only by ignorance, i.e., by the fact that the object appears to the mind as new, and not because of any intrinsic characteristic of the object; thus, if *admiration* is a temporary passion it disappears immediately with the disappearance of novelty. Therefore, the sudden and unexpected surprise, susceptible of disappearing, is not only harmless in knowledge but also useless in the distinction between what is worthy and what is unworthy of special consideration. This allows us to narrow down the effect of the use of *admiration*: nothing but a durable and long-lasting admiration can have a good or bad use, depending on whether it serves to fortify the impression of an object which is worthy or unworthy of our special attention.

17. Ibid., art. 73, AT XI, 382–83, CSM, I, 354.

This is, in general, the utility of *all passions*, and this is why such a consideration is made by Descartes in the center of his analysis of *admiration*.

> From what has been said it is easy to recognize that the utility of *all the passions* consists simply in the fact that they strengthen and prolong thoughts in the soul which it is good for the soul to preserve and which otherwise might easily be erased from it. Likewise the harm they may cause consists entirely in their strengthening and preserving these thoughts beyond what is required, or in their strengthening and preserving others on which it is not good to dwell.[18]

• • •

While in the epistemological context Descartes rejects the function of wonder, by referring to the use of rules in the direction of the mind, wonder retains a positive function in the distinction between the objects of our attention which have an ordinary and an extraordinary character. Hence, while the objects of scientific knowledge obey the natural order of knowledge that can be discovered methodically, what is extraordinary turns out to be beyond this natural order. Following Descartes's reasoning, legitimate *admiration* can only concern what is really *extraordinary* ("out of the way," according to the Aristotelian expression).

Furthermore, the distinction between the two meanings of the term *admiration* allows us to see that Descartes's criticism is moving from the consideration of the *feeling* of the admiration to that of the admirable character of the *object* of admiration. This is also what makes *admiration* the most useful passion in moral considerations, based precisely on the rhetorical conception.

This reasoning can already be observed in the young Descartes's reflections according to which science and invention can be considered wonderful (see the fragments on *scientia admirabilis*[19]). They do not constitute objects of knowledge but the foundation of knowledge itself, obtained by the use of an autonomous mind. Since we do not require

18. Ibid., art. 74, AT XI, 383, CSM, I, 354 (emphasis added).

19. "In the year 1620, I began to understand the fundamental principles of a wonderful discovery" (AT X, 216, CSM, I, 3). Cf. Baillet's biography according to which on 10 November 1619 Descartes was "full of enthusiasm, convinced he had discovered the foundation of a marvellous science" (quoted in CSM, I, 4).

that they obey the order of knowledge, actually, they are *beyond the order*. For this same reason, divine creation *ex nihilo*, our free will as well as Man-God count among the *tria mirabilia*, according to an early fragment of Descartes'.[20] Thus, the object of a legitimate admiration must contain: something divine, as God himself; the recognition in the human of having been made in the image and resemblance of God; our infinite free will.[21] We will see that human free will and its good use turns out to be in the center of the development of the concept of admiration in Descartes's moral philosophy.

This positive function is one of the aspects of the semantic change in the use of the concept of admiration. It is no more the novelty but the rare and extraordinary character of the object which provokes admiration. The other aspect of the change is the possibility that admiration becomes a durable and long-lasting disposition of the soul. Thus, the semantic change makes it possible to re-evaluate admiration in moral philosophy. On the one hand, admiration is not a criticized feeling but a passion that has an essentially positive function: that of the distinction between the ordinary and the extraordinary, the natural and the supernatural. On the other hand, the positive use of admiration can become long-lasting just because its object is characterized not as new and unknown but as rare and extraordinary.

Let us develop these two aspects. First, in moral philosophy, Descartes uses admiration in connection with the discovery of the rare and extraordinary character of our free will and its good use. In this way our attention is directed toward free will as the legitimate object of *admiration* since it is worthy of special consideration. This thought is connected to the development of the concept of admiration in *The Passions of the Soul*. Through the concept of esteem, which is the admiration of greatness,[22] Descartes arrives at the most elaborated final form of admiration, that of our own greatness, i.e., our self-esteem, which is

20. "The Lord has made three marvels: something out nothing; free will; and God in Man" (*Cogitationes privatæ*, AT X, 21819–20, CSM, I, 5).

21. "The desire that everyone has to possess every perfection he can conceive of, and consequently all the perfections which we believe to be in God, is due to the fact that God has given us a will which has no limits. It is principally because of this infinite will within us that we can say we are created in his image" (*Letter to Mersenne*, 25th December 1639, CSMK, III, 142).

22. *Passions of the soul*, art. 149 and 150, AT XI, 443–44, CSM I, 383.

expressed in the doctrine of *generosity* as the experience of our free will and its good use according to our best judgment.

> Thus I believe that true generosity, which causes a person's self-esteem to be as great as it may legitimately be, has only two components. The first consists in his knowing that nothing truly belongs to him but this freedom to dispose his volitions, and that he ought to be praised or blamed for no other reason than his using this freedom well or badly. The second consists in his feeling within himself a firm and constant resolution to use it well—that is, never lack the will to undertake and carry out whatever he judges to be best. To do that is to pursue virtue in a perfect manner.[23]

The extraordinary character of the good use of our free will can also be found in the doctrine of the *supreme good* (*summum bonum*), which has, according to Descartes's analysis in his *Letter to Christina*, two forms: in itself, absolutely, or in relation to us, human beings. Taken in the first sense, the supreme good is identified with God; in the second sense, the supreme good cannot be but what is greatest among all things that depend on us so that they are entirely in our power. Thus the notion of the supreme good contains not only our free will but also its good use, considered as "the firm and constant will to bring about everything we judge to be the best, and to use all the power of our intellect in judging well."[24]

Secondly, this concerns the rare and extraordinary character of the object that allows for a long-lasting admiration, which can become a disposition of the soul. Article 161 of *The Passions of the Soul* teaches us that *generosity* is not only a passion, but it can also become a habit that everyone can acquire with a constantly repeated good use of the free will: "if we occupy ourselves frequently in considering the nature of free will and the many advantages which proceed from a firm resolution to make good use of it—while also considering, on the other hand, the many vain

23. *The Passions of the Soul,* art. 153, AT XI, 445–46, CSM I, 384.

24. *Letter to Elisabeth*, 18 August 1645, AT IV 277, CSMK III, 262; cf. *Letters to Elisabeth*, 15 September 1645; 6 October 1645; to Christina, 20 November 1647: "The supreme good of each individual . . . consists only in a firm will to do well and the contentment which this produces. My reason for saying this is that I can discover no other good which seems so great or so entirely within each man's power (AT V, 82–83, CSMK III, 3:24). Cf. P. Frierson, "Learning to Love: from Egoism to Generosity in Descartes," *Journal of the History of Philosophy* 40 (2002) 313–38.

and useless cares which trouble ambitious people—we may arouse the passion of generosity in ourselves and then acquire the virtue."[25]

This means that constantly repeated exercise helps us to develop a disposition which is virtue itself. The essential result of this thought in Descartes's moral philosophy is that the firm resolution to admire our greatness is nothing other then the constitution of a habit, a disposition of the soul; that is why the constant and long-lasting search for the *admiration* of our greatness coincides with the exercise of virtue: "In this sense the scholastics are right when they say that virtues are habits."[26] This is precisely what *The Passions of the Soul* expresses in the development of the doctrine of *generosity*. "It should be noted that what we commonly call 'virtues' are habits in the souls which dispose it to have certain thoughts: though different from the thoughts, these habits can produce them and in turn can be produced by them."[27]

Now, the admiration of our own greatness, which proceeds from the practice of virtue and which can become a habit or disposition of the soul, produces according to Descartes the pleasure which is the natural beatitude, inner satisfaction or, in other words, contentment of the soul. In this way we can see the interrelation of the concepts of virtue, habit, admiration, and pleasure in the ethical context of Descartes's philosophy.

We can say that the Aristotelian rhetorical conception of wonder, as the former quotations make evident, involves the thesis, "what is wonderful is pleasing." This conception would be the worst thing to take into consideration in the epistemological context of Descartes's writings; indeed, this kind of pleasure caused by wonder is not allowed in the search for truth. That is why the search for wonder in order to wonder is criticized in the theory of knowledge. On the contrary, in Descartes's moral thought, this doctrine of wonder explains the relationship between *wonder* and *pleasure*. In Descartes's vocabulary it has its moral equivalent in the relationship between *admiration* and *beatitude*: *admiration* of our own greatness as *generosity* provokes in the generous soul the natural beatitude, i.e., the inner satisfaction which results from the good use of our free will and from the firm and constant resolution to use it well. However, satisfaction is also a kind of pleasure, as Descartes makes clear

25. *The Passions of the Soul*, art. 161, AT XI, 453–54, CSM I, 388.
26. *Letter to Elisabeth*, 15 September 1645, AT IV, 296.
27. *The Passions of the Soul*, art. 161, AT XI, 453, CSM I, 387.

in his *Letter to Elisabeth*, 18 August 1645, in connection with the practice of good actions.[28]

This doctrine surpasses the explanation that would be limited to the consideration of the tradition of the *ars memoriae* and the rhetorical benefit of wonder. The relationship between *generosity* and *beatitude* allows us to see that, if we ask why *admiration* of our greatness as passion is capable of becoming a virtue, we see that it is able to mobilize a firm resolution in order to exercise generosity, i.e., to use well our free will according to our best judgment, and so become more and more familiar with this exercise which can become a habit. As a result, long-lasting and repeated *admiration* has a direct benefit in the recompense that is a pleasure.

In conclusion, we can understand only in the framework of this semantic shift why in Descartes some aspects of the initial definition of *admiration* (*Passions*, art. 53) such as novelty and ignorance go more and more onto a secondary plan and why its development in the theory of passions involves more and more an ethical dimension. The *admiration* which can become habit as a disposition of the soul does not concern an unknown object which can become known, but an object which intrinsically has an extraordinary character. It is in this way that the doctrine of *generosity* represents in Descartes's moral thought the excellent use of *admiration*.

28. "Last, when Epicurus considered what happiness consists in and to what purpose or end our action tends, he was not wrong to say that it is pleasure in general—that is to say, contentment of the mind. For although the mere knowledge of our duty might oblige us to do good actions, yet this would not cause us to enjoy any happiness if we got no pleasure from it" (*Letter to Elisabeth*, 18 August 1645, AT IV, 27620–27, CSMK, III, 261).

5

Admiration, Fear, and Infinity in Pascal's Thinking

Tamás Pavlovits

THE CONCEPT OF *THAUMADZEIN* ("wonder," "admiration") has played an important role in philosophy since Greek Antiquity. Plato and Aristotle claim that people have begun doing philosophy because of admiration or wonder. Normally, one wonders at or admires something unusual. Yet there is a difference between feeling wonder at someone and admiring a natural phenomenon. A person provokes wonder by his or her unexpected behavior. When someone says "I wonder why he did this," one expresses dislike, even disillusionment. There's a need to distinguish from this kind of wonder another kind of "admiration" linked to natural phenomena: in such a case nature produces something unexpected. One knows that natural phenomena follow the order of physical laws. When a phenomenon seems to contradict these laws, one feels admiration. This admiration is inseparable from incomprehension: one does not understand how something could happen, as the admirable phenomenon seems to contradict the logical order of nature. The mind is unable to place these occurrences into its own system, so it is unable to provide reasons and explanation for it. Facing the unusual and admirable event, the mind faces something not-yet-understood, and necessarily feels ignorant. The ignorance that works in tandem with admiration nudges the mind to find out an explanation and provide reasons, and, indirectly, drives it to philosophy: "It is through wonder that men now begin and originally began to philosophize; wondering in the first place at obvious perplexities, and then by gradual progression raising questions about the greater matters too, e.g. about the changes of the moon and of the sun, about the stars and about the origin of the universe. Now he who wonders and is perplexed feels that he is ignorant."[1] According to Plato, it is primarily the philosopher who is able to admire something "for wonder is the feeling of a philosopher, and philosophy begins in wonder."[2] The experience of admiration is triggered by an event that testifies to one's ignorance. As a result, one begins to think. This is why the ancient Greeks consider admiration the source of philosophy.

Looking at this another way, we can say that philosophy aims at diminishing admiration. Benedict Spinoza (1632–1677), somewhat vehemently, writes: "someone who seeks the true causes of 'miracles' and is eager (like an educated man) to understand natural things, not (like a fool) to wonder at them, is denounced as an impious heretic by those whom the people honor as interpreters of Nature and of the Gods. For the denouncers know that if ignorance is taken away and replaced by real knowledge of mechanical processes, then foolish wonder is also taken away."[3] Thus, Spinoza seems to agree with Plato and Aristotle that admiration leads to philosophy, but also adds that the primary aim of doing philosophy is to diminish admiration. In contrast, Blaise Pascal (1623–1662) has an altogether different opinion. In his *On the Geometrical Spirit* (*De l'esprit géométrique*, 1657/8), he writes about the infinity of nature: "those who clearly perceive these truths will be able to admire the grandeur and power of nature in this double infinity that surrounds us on all sides," and adds that "nature has confronted man with double infinity not to make him understand it but rather to make him admire it."[4] According to Pascal, understanding a certain truth leads to admiration, a point where reason has to renounce the intent to understand. Thus, for Pascal, philosophy does not start out from or aim at canceling admiration. On the contrary, philosophy arrives at admiration. Let us see what his reasons for this opinion are.

For Pascal the object of admiration is infinite nature. It is widely known that during the Copernican turn the cosmic enclosure of the world has opened and the world has become an infinite universe. The heliocentric world view had induced Giordano Bruno[5] to claim the

1. Aristotle, *Metaphysics*, 982b.

2. Plato, *Theaetetus*, 155d.

3. Spinoza, *Ethics*, Part I, Appendix.

4. Pascal, "On the geometrical Spirit," in Pascal, *Thoughts, Letters and Minor Works* (New York: Collier, 1910) 443–44.

5. See chapter 1 by Elisabeth Blum and Paul Richard Blum, above.

infinity of the universe, and in this respect Pascal shared his opinion. At the same time, the argument about the infinity of nature provides the perspective of infinity also to scientific and rational inquiry at the dawn of modern thought. In his fragment titled "Foreword to the Treatise on Vacuum," Pascal contrasts the infinite possibility of progress with the closed authoritarian structure of medieval science, and claims that man, because of his reason, has been created for infinity. Yet infinity not only means an infinite possibility but also a serious challenge for thinking man.

Pascal faced the problem of infinity very early in life: at the age of sixteen in his essay on conic sections (*Essai pour les coniques*, 1640). This essay, together with Gerard Desargues's (1591–1661) work, played a pioneering role in the invention of projective geometry. One of the theses explained here runs as follows: let us cut a given cone with a plane in such a way that the plane is parallel to one but only one generator line of the cone. If we draw or project straight lines from the apex of the cone to the plane through the points of the base circle of the cone, the image appearing on the plane will be a parabola. This parabola is infinite as a result of the fact that one point of the base circle, the one where the generator line of the cone parallel to the plane cuts the circle, will never be projectable to the plane, because the straight line between the apex of the cone and the base circle is parallel to the plane. One could also say that this point is projected onto the plane in infinity, where the parallel lines cross each other. This outcome seems intelligible, yet it is highly ambiguous. During the act of projection a finite plane surface (the base circle) gives an infinite image. This is only possible if the finite circle is composed of an infinite number of points. The result forces us to accept this condition. So as a result of the projection, the infinite appears in the finite. But as the finite is absolutely incommensurable with the infinite, the meeting of the two in an apparently finite plane surface remains unintelligible to human reason. In a rigorously based and methodically performed geometrical argumentation, reason only achieves a result incomprehensible to itself. In his paper titled the "*Brouillon projet*," Desargues, Pascal's master, comments on this phenomenon: reason itself gets lost in the thought, because methodical argumentation makes it draw conclusions the existence of which reason is unable to comprehend. The phenomenon of infinity in mathematics (at least in this age) faces irresolvable problems when confronted by reason. Through the problem of infinity reason wanders into directions where it meets its own limits.

Another instance of the incomprehensibility of mathematical infinity is linked to the problem of double infinity in Pascal. This problem is elaborated upon in his above mentioned *Geometrical Spirit*. The principle states that everything can be divided infinitely and everything can be extended to infinity. The principle of infinite division is an axiom for mathematicians: a given straight line AB can be divided infinitely without getting a part that cannot be divided any more. In this sense, there are no atomistic segments in geometry. The points making up a straight line AB are infinitely small and have no extension. But if one sees a straight line AB as a set of such points, then the straight line AB needs to be interpreted as made up of an infinite number of points. In all geometrical forms—similar to the example taken from projective geometry—the finite is linked to the infinite, and this connection again presents an incomprehensible ambiguity to reason. Chevalier Méré, Descartes's disciple, rejects the principle of infinite divisibility because of its incomprehensibility: "It seems to me," he writes, "that the arguments provided by the science of geometry, as they are often unclear and against our senses, make the conclusions drawn from them quite suspicious, mainly when the infinite gets involved in the issue . . . I wish to draw your attention to the fact that as soon as the infinite gets involved in the question to the slightest degree, the question becomes unintelligible right away, since in such instances the human spirit gets disconcerted and loses its way . . . And everything that shocks common sense is false."[6] There was not much need for Méré's warning that the infinite blocks any attempt to make it intelligible, as Pascal already saw. Yet Pascal's attitude to the infinite is basically different from that of Méré. Instead of rejecting the infinite, Pascal turns the incomprehensibility of the infinite into the central problem of his thinking.

For Pascal the phenomenon of the infinite is not limited to the sphere of geometry but also belongs to nature. This position situates him as an opponent to his contemporary, Descartes, who differentiates between the indefinite and the infinite and only accepts the indefinity of

6. "Il me semble aussi que les raisons qu'on trouve en cette science [la géométrie], pour peu qu'elles soient obscures, ou contre le sentiment, doivent render les conséquences qu'on en tire fort suspectes, surtout quand il s'y mêle de l'infini . . . Je vous apprends que, dès qu'il entre tant soit peu d'infini dans une question, elle deviant inexplicable, parce que l'esprit se trouble et se confond." Letter of Méré to Pascal, in Pascal, *Oeuvres Complètès*, III, éd. J. Mesnard (Paris : Desclée de Brouwer, 1991) 354–55.

nature, and not its infinity. In Pascal's thinking, the projection of the infinite from the realm of geometry to the real world is based on the identification of geometrical and physical space. This identification is justified by Pascal's belief in the existence of a vacuum—opposed to by almost all his contemporaries, most eminently Descartes—a phenomenon he tried to prove by scientific experiments. Being devoid of all material, the vacuum has the same characteristics a purely intelligible geometrical space (*solide*) has. Thus the principle of double infinity is valid for physical space the same way as it is valid for geometrical space and objects. As he explains in the famous fragment 230 of *Pensées*, everything can be infinitely divided in the world and everything can be infinitely extended without limits. In this sense everything "floats" between two infinities, the infinitely small and the infinitely big, being as remote from one as from the other. This double infinity is the deepest and most incomprehensible mystery of nature.

At the same time it is important to mention that the incomprehensibility of nature remains detectable only for a strict and methodical use of thinking. It is not easy to think about the infinite. Whilst doing so, the danger of substituting something small or big as you will for infinity and involuntarily imagining it as finite is lurking in the background. Among others, Descartes also draws attention to this danger in connection to God's infinity. Because of this danger, it is only justifiable for reason to approach the infinite in a strict mathematical way, as this is the only method by which the infinity of the infinite can always be remembered and the incomprehensibility of the infinite can be experienced. Thinking can allow one to experience how the infinite, as Méré says, "shocks" reason through thinking about the infinite geometrically.

As a characteristic of the infinite, the shock brings us back to the problem of admiration. As pointed out above, admiration is created in one as a result of experiencing an unusual, not natural phenomenon, something reason cannot explain. For Pascal this unusual experience is the existence of Nature, as in all its particles the infinite is mixed with the finite. Through a mathematical and physical study of nature reason recognizes the double infinity of nature. Recognizing the infinity of nature actually means understanding the most important feature and deepest secret of nature. However, this understanding is ambiguous, as this is an understanding of the fact that nature is eventually incomprehensible, as the infinite rejects all attempts at rationalizing it.

Facing the infinite results of specific border experiences in which reason recognizes its own finite nature, reason has to renounce wanting to understand and doing further research. The world exposes itself for the methodical thinker as something that obviously exists but flaunts understanding. This is the point where one, unable to comprehend, faces one's ultimate ignorance and begins to admire. For those who get to this stage, says Pascal, "his curiosity changes into admiration, he will be more disposed to contemplate them in silence than to examine them with presumption" (S. 230, L. 199).[7]

So the relationship between philosophy and admiration in Pascal is the opposite of Aristotle's or Spinoza's standpoint. While for Aristotle admiration urges one to do philosophy, for Pascal it is philosophy that leads one to admiration. For Aristotle philosophy aims at eliminating admiration, for Pascal philosophy aims at triggering admiration. However, there is a comment to make here. In Pascal's work, one needs to differentiate between astonishment and admiration. This differentiation does not appear in Aristotle's term, *thaumadzein.* In a nutshell, the difference lies in the fact that the concept of admiration signifies an activity in which one participates with one's whole being. Pascal's astonishment and admiration signal the end of philosophy because in them thinking faces its utmost limits, the incomprehensible it cannot rationalise any more by itself, so the only remaining possibility is to renounce understanding and to silently admire what remains incomprehensible. Philosophy leading to admiration is "real" philosophy for Pascal.

At the same time, admiration has an important companion piece: one has to admit one's ignorance. It is scientific research that leads us to ignorance. Pascal claims that "[w]isdom leads us back to childhood" (S. 116, L. 82), and he continues "[k]nowledge has two extremes that meet. The first is pure, natural ignorance of everyone at birth. The other extreme is reached by great souls who run through everything that can be known, only to find that they know nothing and to find themselves in the same ignorance from which they set out; but this is an erudite ignorance that knows itself" (S. 117, L. 83). With this, Pascal joins the legacy of Socrates and Nicolaus Cusanus, according to which the knowl-

7. Citations of the *Pensées* are from the following edition: Blaise Pascal, *Pensées*, edited and translated by R. Ariew (Indianapolis: Hackett, 2005). In parenthesis I give the number of the fragment according to the edition of Sellier (S) and the edition of Lafuma (L).

edge of the highest order is an awareness of ignorance. Yet there is an important difference, namely that for Pascal admiration is linked to the existential act of ignorance. The methodical and geometrical use of reason results in ignorance. As a result, one has to renounce any further research. However, if one is able to contemplate the double infinity of the world in its own incomprehensibility, then this border experience triggers admiration.

Pascal links admiration to contemplation in specific texts. Contemplation (*theoria*, *contemplatio*) played an important role in Ancient and Medieval philosophies. In most cases it signified an activity in which the senses (seeing) and the human spirit both participated, but did not involve a logical, discursive thinking. Above all, contemplation was directed to the world order (*kosmos*). Its sensible and intelligible construction made it possible for a thinker to rise to the one, to its intuitive understanding. For Pascal contemplation has a different objective. The primary difference lies in the fact that for him the object of contemplation is the infinite that cannot be experienced through the senses. As a corollary, the intellect cannot comprehend the infinite, only recognize its existence. As a result, there is only one way for the intellect to be connected to the infinite: not by wanting to understand it but by understanding its incomprehensibility. While doing so, it experiences its borders in admiration. This border experience opens reason up to the infinite, and this opening up is to be understood as the contemplation of the infinite.

Another corollary to border experience and admiration is anxiety or fear (*effroi*). This experience has its roots in the fact that the world, the double infinity of which is detected by reason, is not only the object of scientific research but also the dwelling place of man. The infinite defying the rationalization of the world also questions the purpose of human life. Man's life, the same way as his intellect, is finite: and this quality makes it incommensurable with its world in which it is to lead a meaningful life. This incommensurability leads to the experience of anxiety. "The eternal silence of these infinite spaces frightens me." (S. 233, L. 201)—Pascal writes, and in another fragment he comments: [w]hen I consider the brief duration of my life absorbed in the eternity that lies before and after . . . the small space I occupy and can even see, engulfed in the infinite immensity of spaces I do not know and that do not know me, I am frightened and astonished to see myself here rather than there; for there is no reason why I am here rather than there, why now rather than then. Who put me here? By whose order and direction have this place and time been allotted to me?" (S. 102, L. 68). The infinity and incomprehensibility of the world defies man's attempt to situate his life as part of a bigger rational whole founded by the world—as was normal in the premodern ages. Anxiety and fear result from the comprehension of this basic meaninglessness and baselessness of the world. When one contemplates the infinite, one understands this: such admiration of the infinite is completed by the feeling of anxiety.

• • •

At the final point of philosophy one faces the incomprehensibility of the world and the incomprehensibility of one's own existence. Reason fails when it confronts the final questions and has to renounce any further research. All this seems to prove the final weakness of reason and thinking, although it is exactly the other way around with Pascal. Reaching the point of not comprehending final questions proves the force of reason for him. The force of reason lies in the fact that confronting its limits, it is able to experience its own limits and stabilize this border experience. Reason makes one experience admiration and anxiety at the same time, while one's approach to one's relation to the final; i.e., the infinite turns from rational into existential. In the border experience one becomes open to the final and to the infinite, and awakens to the wonder and anxiety of mundane existence.[8]

—Translated by Ágnes Zsófia Kovács

8. The author would like to thank the Bolyai scholarship of the Hungarian Academy of Sciences and the OTKA research grant for their support of this research.

6

On Thomas Hobbes's Concept of Wonder

Jianhong Chen

Thomas Hobbes (1588–1679) was the most prominent British philosopher in the period between Francis Bacon (1561–1626) and John Locke (1632–1704). As the author of the formidable *Leviathan*, he is among the most famous political philosophers throughout the ages. Hobbes did not write extensively on wonder, but he dealt with it in such a way that his discussion of wonder shed light on a number of issues such as human nature, philosophical reasoning, and religious matters. Therefore an analysis of Hobbes's concept of wonder would help us better understand various vital issues involved in Hobbes's philosophical, religious, and political doctrines. Beginning with some biographical facts, this essay will explore the philosophical and the biblical aspect of Hobbes's concept of wonder, contrast Hobbes's modern idea with the negative attitude toward curiosity typical of Christian thinkers, and conclude with some critical remarks drawing on Heidegger's reflection on the distinction between wonder and curiosity.

LIFE AND WORK

Hobbes was born at Westport, adjoining Malmesbury, in Wiltshire, in 1588. His birth was a premature one. For his mother was frightened by the rumors of the impending invasion of the Spanish Amanda, fearing that Malmesbury would be the natural target of the Spaniards. For this reason, Hobbes notes in his verse *Vita* that his mother gave birth to him and a twin brother, Fear.[1] As this description is often associated with

1. Thomas Hobbes, "The Verse Life," in *The Elements of Law*, edited by J. C. A. Gaskin (Oxford: Oxford University Press, 1994) 254–64, 254.

Hobbes's timid disposition and his sense of insecurity, it may not be entirely irrelevant to the connotation of wonder as Hobbes conceives of it.

John Aubrey, Hobbes's first biographer, reports that Hobbes was "of plebian descent," and his father was "one of the ignorant 'Sir Johns' of Queen Elizabeth's time." Supported by his uncle Francis, Hobbes went to school at the age of four. In the early years of his youth, Hobbes learnt Greek and Latin from a Mr Robert Latimer, who was "a good Grecian." At fourteen years old he went to Magdalen Hall at Oxford, where he stayed for five years undertaking the study of logic and Aristotelian physics.[2] Later in his writings, for instance, in *Leviathan* (1651), Hobbes showed a strong distaste for what he was taught in the university. Yet, the education he received from the university may perhaps have exerted a more positive influence on the formation of Hobbes's thought than he himself tended to admit.

After having obtained his Bachelor of Arts degree, Hobbes was fortunate to be recommended by the principal of Magdalen Hall to the family of Cavendish. He was received into the family as the tutor of William Cavendish, the son of the first Earl of Devonshire. Hobbes served the Cavendish family throughout his life, except a short interval after the death of his patron and friend in 1629. His service for the Cavendish family enabled Hobbes to make acquaintance with distinguished political and philosophical figures in the British Isles as well as with some of the most prominent philosophical minds of the day on the European continent.

Hobbes's first scholarly work was an entire translation of the ancient Greek historian, Thucydides (c. 460 BCE–c. 395 BCE), which was published in 1628. This translation testifies to Hobbes's enthusiasm for "the historians and the poets" of ancient times in the early years of his life. Yet, speaking of Hobbes's "early years," one has to bear in mind that Hobbes was already forty years old in 1628.[3] He was not known as a philosopher until he completed his first original philosophical work, *The Elements of Law, Natural and Politic*, in 1640, when Hobbes was past fifty.

2. John Aubrey, "The Brief Life," in *The Elements of Law*, 231–45, 231–33; see also Hobbes, "The Prose Life," in *The Elements of Law*, 245–53, 245–46.

3. For another interpretation of Hobbes's early years, based on the biography of A. P. Martinich (cf. note 6 below), see Michael Funk Deckard, "A Sudden Surprise of the Soul: The Passion of Wonder in Hobbes and Descartes," *Heythrop Journal* 49 (2008) 948–63.

This work contains the primary source of Hobbes's main ideas presented later in *De Cive* and *Leviathan*.

In Hobbes's understanding, philosophy comprises two principal parts: natural and civil philosophy. Natural philosophy or physics searches after the generation and properties of natural bodies, whereas civil philosophy investigates those of the commonwealth. Hobbes further divides civil philosophy into two subsections: ethics and politics, or moral philosophy and civil philosophy simply. Moral philosophy treats of the dispositions and manners of men, whereas civil philosophy proper deals with civil duties of men and rights of the sovereign.[4] Therefore, Hobbes's philosophical system can also be said to have a tripartite structure: Physics or natural philosophy, ethics or moral philosophy, and politics or civil philosophy. The logical structure of this can be recognized in the philosophical outlook as presented in *De Corpore* (1655), *De Homine* (1658), and *De Cive* (1642). Obviously, Hobbes did not materialize the system according to its logical order but rather presented to the reader, first of all, the head of his system, viz. *De Cive*. *De Cive* was first published in Latin in Paris and received considerable attention among the learned, home and abroad. In 1651, *Leviathan* as well as the English version of *De Cive* appeared in London. *Leviathan* treats of the matter, generation, and nature of an artificially constituted political body, i.e., the state. This work provided a much more detailed analysis of human nature, i.e., of the dispositions and manners of man, than *De Cive*, though Hobbes did not complete his moral philosophy until the publication of *De Homine* in 1658.

In constructing his philosophical system, Hobbes was also actively involved in a number of debates and controversies with his contemporaries. In the 1630's, he offered his objections to Descartes's *Meditations*. In the middle of the 1640's, Hobbes conversed with Bishop Bramhall concerning liberty and necessity and the controversy was published in 1654 and continued with further development.[5] Since 1656 Hobbes had entered a debate with Oxford mathematicians Seth Ward (1617–1689) and John Wallis (1616–1703) over his untenable claim to have successfully squared the circle.

4. Hobbes, *The Elements of Law*, 192–93.

5. See Nicholas D. Jackson, *Hobbes, Bramhall and the Politics of Liberty and Necessity* (Cambridge: Cambridge University Press, 2007).

It is through *Leviathan* more than any other of his books that Hobbes became famous and infamous. In 1666, the House of Commons sought to gather information about *Leviathan* in relation to its bill against Atheism. It was said that some bishops once intended to have him burnt for a heretic. Around that time, Hobbes composed *A Dialogue between a Philosopher and a Student of the Common Laws*. In 1668, Hobbes wrote a posthumously published history of the English civil war between King Charles and his parliament, entitled *Behemoth, or the Epitome of the Civil Wars in England*. In the last years of his life, Hobbes seemed to have resumed his love for the poets of the Greeks and translated Homer. Half in jest and half in earnest, Hobbes said that he translated Homer because he "had nothing else to do."[6] In 1679, Hobbes died at Hardwick Hall at the age of ninety-one, leaving behind a literary and philosophical legacy full of wit, eloquence, and controversy.[7]

PERSPECTIVES ON WONDER

Hobbes does not write much of the passion of wonder, but he does touch on it and think it through. Though he does not present the concept systematically, Hobbes considers wonder as a mental quality peculiar to human nature. How does Hobbes define wonder? What role does it play in Hobbes's philosophical system? What can we learn from Hobbes's discussion? First of all, how does one approach Hobbes's concept of wonder? We may begin with a discussion of different perspectives on wonder: philosophical, biblical, and poetic.

(a) *The philosophical understanding of wonder*. Hobbes claims that his way of doing philosophy marked a new beginning. This implies that he consciously breaks with the traditional ways of thinking, especially with the Aristotelian and scholastic school. His concept of wonder, however, does not deviate essentially from Aristotle's. Taking up Socrates's brief description of wonder as an experience characteristic of a philosopher,[8]

6. Hobbes, "Concerning the Virtues of an Heroic Poem," preface to *The Iliads and Odysses of Homer*, in *The English Works*, edited by Sir William Molesworth, vol. x, iii–x (London: Bohn, 1844) x.

7. For more detail about Hobbes's life and work, see Sir Leslie Stephen, *Hobbes* (Ann Arbor: University of Michigan Press, 1961); for a recent account of Hobbes's life and thought, see A. P. Martinich, *Hobbes: A Biography* (London: Routledge, 2005).

8. Plato, *Theaetetus*, translated by M. J. Levett, in *Complete Works*, edited by John M. Cooper (Indianapolis: Hackett, 1997) 155D.

Aristotle discusses wonder as the beginning of philosophy with more detail:

> For it is owing to their wonder that men both now begin and at first began to philosophize; they wondered originally at the obvious difficulties, then advanced little by little and stated difficulties about the greater matters, e.g. about the phenomena of the moon and those of the sun and the stars, and about the genesis of the universe. And a man who is puzzled and wonders thinks himself ignorant (whence even the lover of myth is in a sense a lover of wisdom, for myth is composed of wonders); therefore since they philosophized in order to escape from ignorance, evidently they were pursuing science in order to know, and not for any utilitarian end.[9]

In this passage, Aristotle states that men begin to philosophize through wonder, that men at first wonder at the obvious difficulties and then the difficulties about the greater matters, i.e., cosmological issues, and that, motivated by an experience of ignorance, men pursue to attain knowledge. From the context of the passage quoted above, we also learn that men overcome ignorance and pursue knowledge through investigating into the causes of things. At the end of this investigation lies the philosophical discovery that God is the cause of all things and the first principle. In brief, the passion of wonder drives men to escape the cave of ignorance and enter into the clarity of knowledge by inquiring into the causes of things.

(b) *The biblical understanding of wonder.* Though being bold and relentless in attacking Aristotle and his school, Hobbes confesses that he "took great care not to write in any way against the sense of sacred scripture."[10] A discussion of the biblical understanding of wonder would help us better understand Hobbes's concept of wonder. What is wonder in the biblical sense? The most wonderful of all things are perhaps the divine word and the divine act. From the biblical point of view, there is nothing so wonderful in this world and "there is nothing new under the sun." The pursuit of much wisdom and much knowledge is "a chasing after

9. Aristotle, *Metaphysics*, translated by W. D. Ross, in *The Complete Works of Aristotle*, edited by Jonathan Barnes, 2 vols. (Princeton: Princeton University Press, 1984) 2:982b12–22.

10. Hobbes, *The Elements of Law*, 248.

the wind."[11] This reminds us of the fact that eating the tree of the knowledge of good and evil is forbidden by God in the biblical account of the creation of the world. Through eating the tree of the knowledge man comes to know good and evil in his own judgment. In another word, this is a kind of knowledge that is attained by the unassisted human mind. According to the biblical understanding, without divine guidance, the philosophical sense of wonder and its end are futile and "chasing after the wind."

Wisdom in the biblical sense is obedience to the divine command, whereas wisdom in the Greek sense is the rational inquiry into the causes of all things. This results in two different images of God. The Aristotelian God is the final cause of all things, the impersonal thought thinking of itself, and the highest end of human reason. In contrast, the biblical God is the source of all things, a personal legislator, and a mysterious and supra-rational being. There is nothing wondrous to the almighty God. God can make possible what man cannot even imagine. For instance, God says to Abraham that his wife Sarah will bear a son for him at the age of ninety. Sarah laughed spontaneously on hearing that. It turned out to be that she did conceive and had a son.

Miracle is the word in the Bible equivalent with the Greek sense of wonder. Wrought by the hands of God, miracles cannot be understood by natural means. God is the real cause of miracles, and miracles in turn manifest the power of God. Nothing can be called wonders except divinely wrought miracles. The biblical wonders invite beliefs in and promote obedience to the mysterious God, who cannot be understood in the light of natural reason. In contrast, wonder in the Greek sense is to inquire by natural reason into the rational causes of what remains mysterious to human reason.

(c) *The poetic or aesthetic sense of wonder*. The poetic sense of wonder can be described as an exclamation of the grandeur or minuteness of something either natural or artificial. We may have a sense of this experience, for instance, in beholding a most delightful scene or appreciating a fine piece of work of art visual or musical. We may also find this kind of experience in reading a poem, for instance, William Blake's verses from *Auguries of Innocence* (1803):

> To see a World in a Grain of Sand

11. Eccl 1:9, 1:16–17.

And a Heaven in a Wild Flower,
Hold Infinity in the palm of your hand
And Eternity in an hour.

The poetic experience of wonder is not to inquire about natural causes of things. Nor does it focus on obedience to the mysterious God. It may have certain similarities with the biblical understanding of wonder to the effect that it can also express men's feeling humble before something great and grandeur. The poetic sense of wonder is nevertheless not exclusively limited to the promotion of obedience to God. What causes it can be natural or artificial or supernatural. Generally speaking, this sense of wonder is not important to understanding Hobbes's discussion of wonder. Therefore, we will limit ourselves to the first two understandings of wonder in discussing Hobbes's concept of wonder in the following pages.

CURIOSITY AS THE PHILOSOPHICAL SENSE OF WONDER

Hobbes mentions three aspects of human life that make men essentially different from animals: the faculty of reason, the phenomenon of religion, and first of all, the passion of curiosity. All these three aspects are tied to wonder in a sense.

The passion of curiosity can be considered the original form of wonder. According to Hobbes, all intellectual phenomena derive their existence from sense, the origin of the thoughts of man, and sense is caused or stirred by the pressure of the external body upon the organs of sense. In consequence of that pressure, there arises within the human body a resistance or counter-pressure. Hobbes uses the word "endeavor" to describe this kind of "small beginnings of motion" within the body of men.[12] This inner endeavor expresses itself outward as the original fancy in different organs of sense: light to the eyes, sound to the ear, odor to the nostrils, savor to the tongue, and other qualities to the other parts of the body. After the object has been removed, the retained image or the decaying sense is what Hobbes calls imagination. Hobbes identifies imagination with memory, and further defines experience as "much memory, or memory of many things." Accumulation of experience leads

12. Hobbes, *Leviathan*, edited by C. B. Macpherson (London: Penguin, 1968) 85, 119 (chapter 1, section 4; chapter 6, sections 1 and 2).

to knowledge, and the expectation of future knowledge is curiosity, a passion characteristic of human nature:

> Forasmuch as all knowledge beginneth from experience, therefore also new experience is the beginning of new knowledge, and the increase of experience the beginning of the increase of knowledge; whatsoever therefore happeneth new to a man, giveth him hope and matter of knowing somewhat that he knew not before. And this hope and expectation of future knowledge from anything that happeneth new and strange, is that passion which we commonly call ADMIRATION; and the same considered as appetite, is called CURIOSITY, which is appetite of knowledge.[13]

Curiosity is thus "the hope and expectation of future knowledge" from what happens new and strange. This is Hobbes's characterization of a fundamental passion peculiar to man. This fundamental passion could further develop itself either into a rational inquiry into the causes or into an obedient awe of God. In other words, the passion of curiosity could grow either into wonder in the philosophical sense or in the biblical sense.

In *Leviathan*, *admiration* and *curiosity* are given separate definitions. Nevertheless, Hobbes deploys them as interchangeable in meaning. Curiosity is defined as "*Desire*, to know why, and how." By the "why" and "how" Hobbes means "the care of knowing causes," "a perseverance of delight in the continuall and indefatigable generation of Knowledge."[14] As Hobbes puts it more clearly elsewhere, curiosity is precisely "love of the knowledge of causes."[15] Therefore, human curiosity can be characterized by two things: (a) men experience novelties; (b) men naturally desire to know what causes novelties. Yet, animals also experience novelties, and they even have a sense of the consequences of one event upon another. But they know these consequences from experience only, and do not naturally desire to determine their natural or rational causes. In this sense, man is both experiential and rational. Thus Hobbes divides human knowledge into two kinds: experiential and conceptual, or historical and rational. The former is "experience of fact," and the latter is "evidence of truth,"[16] according to which there is a distinction between

13. Hobbes, *The Elements of Law*, 57.

14. Hobbes, *Leviathan*, 124 (chapter 6, section 35).

15. Ibid., 167 (chapter 11, section 25).

16. Hobbes, *The Elements of Law*, 41.

prudence and sapience among the ancients. In one word, men are distinguished from animals by their natural desire to know and to determine the natural causes of what happens new to them.

Furthermore, man does not possess rational knowledge by nature but has to acquire it by industry. This implies that man can err in reasoning. For Hobbes, the fact that man is capable of reasoning as well as erring in reasoning further marks man out from animals. Curiosity about causes of things and reasoning of the natural causes of things inevitably leads to a rational conception of God: "Curiosity, or love of the knowledge of causes, draws a man from consideration of the effect, to seek the cause; and again, the cause of that cause; till of necessity he must come to this thought at last, that there is some cause, whereof there is no former cause, but is eternall; which is what men call God."[17]

Hobbes's rational conception of God is in a sense not far from the Aristotelian concept of the unmoved mover, the thought thinking itself. Hobbes differs from Aristotle, however, in recognizing geometry as the paradigm of right reasoning, which enables men to begin their ratiocination from definitions. This purely abstract concept of God is deduced from human inquiry into the causes of things. The capability of ratiocination from definitions enables the passion of curiosity to discover a rational concept of God.

Hobbes's rational theology is in line with Aristotle's natural theology. Nevertheless, Hobbes refuses to endorse the Aristotelian idea that rest is the telos of human life. He declares that life is but motion, ceaseless motion from one goal to another. If there is an end of human life, it is nothing else but death. In this respect, Hobbes is a rationalist and an absolute materialist. Hobbes does not deny the existence of God as the cause of all causes, and yet he implies at the same time that the rational conception of God results from, and is guaranteed by, a methodological deduction. In conclusion, God as conceived of in Hobbes's thought as the name that is either invented by human reason to designate, or supposed by human fancy to be, the final cause of things. In either case, God does not wonder what happens in the world. Man is the only creature that wonders at novelties.

17. Hobbes, *Leviathan*, 167 (chapter 11, section 25).

MIRACLE AS THE BIBLICAL SENSE OF WONDER

Apart from understanding curiosity as the rational desire for knowledge of causes, Hobbes relates the passion of curiosity to the biblical sense of wonder, i.e., miracle. He equates miracles with "Wonders," and defines them as "the Admirable works of God." This understanding of wonder is also characterized by two things: (a) the strangeness or rarity, and (b) what cannot be imagined to have been done by natural means but only by the immediate hand of God. Yet, Hobbes quickly adds that miracle is in a sense subjective or relative. For instance, what seems strange or rare to one man may be just ordinary to another, and what one cannot comprehend by natural means may be due to one's inexperience in reasoning. Therefore Hobbes says that the ignorant and superstitious men make great wonders of the works that more knowledgeable men do not admire at all.[18]

The subjectivity of a miracle goes even further in Hobbes's statement that "a private man has always the liberty" to believe or not believe in his heart what are called miracles, and to judge by their own reason "whether they be Miracles, or Lies." Hobbes asserts that this personal liberty has to submit itself to the public reason whenever the public confession of faith is concerned.[19] Even so, the subjectivity of miracles does not disappear at all. It is only that the personal subjectivity is transformed into a political relativity. Seeing that public reason is in Hobbes nothing but the will of the sovereign, miracles are thus what the sovereign commands his subjects to believe as miracles, and they cease to be miracles when the sovereign declares them no longer to be miracles. Hobbes mentions that the end of miracles is to make manifest to the elect "the mission of an extraordinary Minister for their salvation" and to beget belief, not universally in all men, but in the elect only.[20] Only the political sovereign has the authority to declare what is and what is not a miracle, and to decide on what is to be publicly believed as a miracle. Therefore, Hobbes's concept of wonder in this sense can be said to be a politicized form of the theological concept of miracle.

Considered as a passion peculiar to human nature, curiosity could lead man either to discover by means of right method the true causes

18. Ibid., 469–71 (chapter 37, sections 1–5).

19. Ibid., 478 (chapter 37, section 13).

20. Ibid., 471–73 (chapter 37, sections 6 and 7).

of things or to suppose causes of them "either such as his own fancy suggesteth; or trusteth to the Authority of other men."[21] This ignorance of true causes of things constitutes one of the natural seeds of religion, which is defined by Hobbes as "fear of power invisible, feigned by the mind, or imagined from tales publiquely allowed."[22] "Tales publicly allowed" means precisely tales sanctioned by the public authority. This conception of religion is obviously a civil or political theology. One is entitled to say that the passion of curiosity could lead either to a rational theology or a civil theology. In the former theology, the one God eternal is derived from man's knowledge of the natural causes of things; in the latter, the Gods were created by human fear, which has its origin in man's ignorance of the true causes of things. Civil theology is thus closely connected to the passion of fear, whereas rational theology is an overcoming of fear. We could say that there is a tension between rational theology and civil theology in Hobbes's thought.[23] His rational theology inclines us to conceive of Hobbes as an Aristotelian in this respect, whereas his civil theology bears the character of an Erastian who places religious issues entirely under the control of the state authority.[24] For Hobbes, what is beyond reason is not supra-rational in the traditional sense but political. Hobbes's use of the concept of wonder as miracle is political. This causes suspicion concerning the sincerity of Hobbes's claim to be a Christian.

As I have been analyzing, Hobbes's concept of wonder can be understood from the philosophical as well as from the biblical point of view, as is expressed in curiosity and in miracle. The distinction between these two aspects of wonder is later explicitly described by David Hume as the distinction between "speculative curiosity" and "trembling curiosity."[25] Hobbes suggests this distinction by seeing in

21. Ibid., 169 (chapter 12, section 4).

22. Ibid., 124 (chapter 6, section 36).

23. For a more detailed discussion, see my "On the Definition of Religion in Hobbes' *Leviathan*," *Bijdragen: International Journal in Philosophy and Theology* 67 (2006) 180–94.

24. See Jeffrey R. Collins, *The Allegiance of Thomas Hobbes* (Oxford: Oxford University Press, 2007).

25. David Hume, *The Natural History of Religion* [Introduction by John M. Robertson, London: A. and H. Bradlaugh Bonner, 1889, 9–10 (section 2)]. Elsewhere, speculative curiosity, i.e., "the love of knowledge, which displays itself in the science," is also contrasted to "a certain curiosity implanted in human nature, which is a pas-

curiosity the seed of religion on the one hand and the inner inception of rational knowledge of natural causes on the other. Both rational knowledge and religious beliefs are thought to be derived from curiosity about causes of things. Curiosity can motivate men either to investigate the natural causes of things or to suppose the unnatural or supernatural causes of things. Science and religion agree with each other in their search for causes of things. They differ from each other, however, in their different account of what the real causes of things are. While science has recourse to the natural or rational causes of things, religion appeals to the supernatural or "irrational" causes of things. Seen in another perspective, science is the methodical inquiry into causes of things and religion is the unmethodical search for causes of things.[26] Yet, when one distinguishes science from religion in terms of method, one is already in favor of science. Hobbes describes science as the knowledge of consequences, while he sees the seed of religion in the ignorance of causes of things and associates religion with politics. Hobbes's choice is therefore not ambiguous if one considers Hobbes's positive attitude toward science on the one hand and his emphasis on the dependence of religion on politics on the other.

CURIOSITY CONSTRAINED AND UNCONSTRAINED

For Hobbes, the passion of curiosity leads either to knowledge or to supposition of causes of things. Of the two aspects of wonder, curiosity leading to knowledge is explicitly associated with the philosophical truth, while no philosophical significance is conferred on trembling curiosity that drives men to bow to the divine will that is incomprehensible by reason alone. As Jeffrey Barnouw points out, prior to Hobbes no thinker "had seen curiosity in such a positive light or made it so fundamental to human nature."[27] In order to better understand Hobbes's favorite consideration of the passion of curiosity in general and speculative or cognitive curiosity in particular, I shall briefly introduce the characteristically

---

sion deriv'd from a quite different principle." See Hume, *A Treatise of Human Nature* (London: Penguin, 1969) 499.

26. Leo Strauss, *Spinoza's Critique of Religion*, trans. E. M. Sinclair (New York: Schocken, 1965) 87, 90. For Hobbes's positive evaluation of method, see Hobbes, *Leviathan*, 114–15 (chapter 5, sections 8 and 17).

27. J. Barnouw, "Hobbes's Psychology of Thought: Endeavours, Purposes and Curiosity," *History of European Ideas* 10 (1989) 519–35; 534.

Christian attitude toward curiosity and consider what Hans Blumenberg calls the "trial" of theoretical curiosity and the self-assertion of reason.

Curiosity has generally been negated as a temptation or vice in the Christian tradition. Augustine and Aquinas are good examples. For St. Augustine, curiosity is a form of temptation planted in the soul and "cloaked over with the title of knowledge and science," desiring to acquire new experiences. Augustine considers curiosity as a "vain" and "morbid" desire that motivates men to "search out the secrets of nature, things beyond our end, to know which profits us nothing, and of which men desire nothing but the knowing."[28] In Augustine's view, curiosity is a negative desire rooted in the soul, and is characterized as the desire to take pleasure in experiencing novelties and especially in the knowing itself. Augustine objects to this desire for two reasons. First, it drives men to investigate "the secrets of nature," the secrets that are beyond human understanding. Augustine obviously suggests that it is unfitting and of no use for men to seek the knowledge that is not suited to them. Second, Augustine admits that the desire of curiosity brings forth pleasure. Yet, it is an intellective pleasure in the purely knowing, which profits human beings nothing but knowing only. In other words, Augustine regards speculative curiosity, i.e. pure knowledge for the sake of knowledge itself, as a temptation, as a forbidden knowledge.

Aquinas continues Augustine's negative evaluation of curiosity. In his *Summa Theologica*, in the "Second Part of the Second Part," Aquinas explicitly speaks of curiosity as a vice. In Question 167, He discusses curiosity by raising and answering two questions: (1) Whether the vice of curiosity can regard intellective knowledge? (2) Whether it is about sensitive knowledge? Aquinas analyzes and answers these two questions in the affirmative. That is to say, he thinks that curiosity plays a positive role in men's pursuit of both intellective and sensitive knowledge. Moreover, Aquinas argues that curiosity about intellective and sensitive knowledge may be sinful. At this point, Aquinas is slightly different from Augustine. As Hans Blumenberg contends in his splendid analysis of curiosity in *The Legitimacy of the Modern Age*, the Augustinian idea of the nonreligious inquiry finds in Thomas Aquinas "a very characteristic modification and Aristotelianizing correction."[29] Though he agrees with

28. St. Augustine, *The Confessions of St. Augustine*, translated with an introduction by John K. Ryan (New York: Doubleday, 1960) 264–65 (book 10, chapter 35).

29. Hans Blumenberg, *The Legitimacy of the Modern Age*, trans. Robert M. Wallace (Cambridge: MIT Press, 1983) 332 [orig. pub. 1966].

Augustine in regarding curiosity as a vice, Aquinas does not denounce the goodness of the knowledge of truth itself, nor does he deny lawfulness and worth to the study of philosophy. What Aquinas emphasizes is that the pursuit of knowledge and the study of philosophy should be under the guidance of divine revelation. Otherwise, they can be misused to attack and destroy Christian faith.[30]

In the third part of *The Legitimacy of the Modern Age*, Blumenberg demonstrates that the unleashing of curiosity is characteristic of the modern mind and examines how the passion of curiosity, which is traditionally condemned as a vice, became a cardinal virtue in the modern age, how the "vice" of curiosity became an unconstrained and triumphant desire of the modern man. The peculiar feature of the modern worldview is that unconstrained theoretical reason can freely reconstruct and manipulate the world of nature. As Bernard Yack points out, this understanding of modernity is not novel but in agreement with Heidegger, Adorno, and many German thinkers.[31] What distinguishes Blumenberg from other thinkers does not lie in his understanding of the novelty of modernity but in his account of it.

Since it is not possible to go, in this paper, into Blumenberg's detailed account of the novelty of modernity, I can only indicate in a brief manner one of Blumenberg's major contentions. Blumenberg contends that modern thinkers do not merely secularize theological concepts and transform them into nonreligious concepts for their own use. This is a refutation of the widely accepted secularization thesis: Modernity is conceptually a secularized form of Christianity. Carl Schmitt, for instance, declares "all significant concepts of the modern theory of the state" to be "secularized theological concepts."[32] This statement obviously underlines the conceptual dependence of modernity on Christian theology and hence undermines the novelty of modernity. Blumenberg does acknowledge the importance of Christianity to the modern age.[33] Nevertheless, he entirely disagrees with the secularization theorists. He argues that

30. St. Thomas Aquinas, *Summa Theologica*, vol. 4, trans. the fathers of English Dominican province (New York: Benziger Brothers, 1868–1890) Pt. II–II Q. 167.

31. Bernard Yack, "Myth and Modernity: Hans Blumenberg's Reconstruction of Modern Theory," *Political Theory* 15 (1987) 244–61; 253.

32. Carl Schmitt, *Political Theology*, trans. George Schwab (Cambridge: MIT Press, 1985) 36.

33. Blumenberg, *The Legitimacy of the Modern Age*, 30: "The modern age is unthinkable without Christianity."

theoretical self-assertion against tradition is particularly demanded by the concrete historical situation of the late middle Ages, the situation created by the decay of Scholastic theology into what Blumenberg calls "theological absolutism." Thus, Blumenberg intends to give a historical account of "the singular situation of provocation and self-assertion from which springs the incomparable energy of the rise of the modern age."[34]

To return to Hobbes, it should be emphasized that Hobbes's concept of curiosity exhibits its continuity and discontinuity with the Christian tradition. It continues the traditional idea that curiosity is rooted in human nature and leads to intellective and sensitive knowledge. Yet Hobbes turns the negative concept into a positive one. On the other hand, Hobbes breaks with the traditional idea first by releasing curiosity from the duty of conforming to revelation, and second by subjecting the truth of miracles to the will of the political sovereign. Both steps indicate the modern fashion of Hobbes's idea of curiosity.

CURIOSITY OR WONDER?

Thus far, I have analyzed the philosophical sense and the religious sense of wonder in the thought of Hobbes and contrasted Hobbes's concept of curiosity with the negative attitude toward it typical of Christian authors. As a modern thinker, Hobbes releases the passion of curiosity from any religious or moral constraint. According to Blumenberg, the unleashing of curiosity is precisely characteristic of the modern age. I shall now bring in another perspective on wonder and curiosity, i.e., Heidegger's clear distinction between curiosity and wonder.

Heidegger's attitude toward curiosity is flatly negative, though completely different from the Christian point of view. Curiosity, Heidegger says in section 36 of *Being and Time*, "concerns itself with seeing, not in order to understand what is seen (that is, to come into a Being towards it) but *just* in order to see. It seeks novelty only in order to leap from it anew to another novelty." So curiosity is first of all characterized by the search for novelty. Consequent on its search for continual novelty, curiosity does not seek the leisure of "*tarrying* in the environment with which one concerns oneself, and hence is concerned with "the constant possibility of *distraction*." Then Heidegger describes the third essential feature of curiosity as "the character of *never dwelling anywhere*,"

34. Blumenberg, *Legitimacy of the Modern Age*, 148.

to the effect that "curiosity is everywhere and nowhere."[35] In addition, Heidegger makes a brief distinction between curiosity and θαυμάζειν, the Greek understanding of wonder. Here in *Being and Time*, Heidegger does not explain what distinguishes wonder from curiosity.

In *Basic Questions in Philosophy*, Heidegger writes that the understanding of the origin of philosophy out of *thaumazein* cannot be confused with the idea that philosophy arises from curiosity.[36] Heidegger distinguishes wonder from curiosity by emphasizing that wonder is concerned with the unusualness of the most usual while curiosity focuses on the extraordinary and the unusual beyond the ordinary and the usual. That is to say, curiosity presupposes a distinction between the usual and the unusual, while wonder finds the unusualness in the most usual. Heidegger then further declares "beings *as* beings" are "what alone is wondrous."[37] According to Heidegger, the early Greeks regard *Dasein*, or man as man, as the most unusual or the most uncanny.[38]

Hobbes defines *curiosity* as the desire to know causes of things consequent upon the continual experience of the new. In his discussion of miracle, Hobbes emphasizes strangeness or rarity, which is of course extraordinary and uncanny. This corresponds by coincidence to Heidegger's analysis of curiosity. What is implied in Heidegger's critique of curiosity is that in seeking the most unusual and the most extraordinary man forgets himself as being the most unusual and the most extraordinary. In this perspective, what is first and foremost is to understand man as man himself, to understand beings *as* beings.

Heidegger's critique of curiosity calls our attention to return to the most usual, to beings as beings. To vulgarize Heidegger's thesis, man has to return to the reality of the human world as it is. Politically speaking, Hobbes is realistic enough. Yet, he does not hesitate in supposing the mere state of nature as the starting point of his theory of the state, a theory that optimistically looks forward to a time when the republic of universal light can be eventually established.

35. Martin Heidegger, *Being and Time*, trans. John Macquarrie and Edward Robinson (London: SCM, 1962) 216–17.

36. Heidegger, *Basic Questions in Philosophy*, trans. Richard Rojcewicz and André Schuwer (Bloomington: Indiana University Press, 1994) 135–36.

37. Ibid., 144–45.

38. Heidegger, *Introduction to Metaphysics*, trans. Gregory Fried and Richard Polt (New Haven: Yale University Press, 2000) 156.

7

"Straight toward Heaven"

Natural Theology and Politics in Milton's Paradise Lost

Veronika Szántó

Is there any natural theology in John Milton's (1608—1679) *Paradise Lost* (1674)? The question may sound odd: there is no-one in the epic, Satan included, who needs to be convinced of the *existence* of God, since the Father is virtually a personal acquaintance of each character. However, the way one can obtain knowledge of God as the *creator* of the world is a major concern for Milton, and he addresses this question time to time as the epic unfolds. The problem of natural theology is all the more significant, because it not only is an epistemological and religious question, but also has far-reaching political consequences, as we will see. Thus, the question whether there is a natural theology in *Paradise Lost*, and if there is, what is it like, can help us shed some light on one of the most disputed aspects of the poem, that is, its political philosophy, an aspect that can hardly be separated from the old riddle of Milton's presumed sympathies or antipathies toward his characters, most importantly toward Satan and the Father.

In what follows I use the term *natural theology* distinctly from *natural religion*,[1] in the broad sense of yielding theological beliefs from the interpretation of nature based on experience. Thus, I do not use the term interchangeably with *physico-theology*, which I regard as a subter-

1. For a distinction between natural theology and natural religion, see Robert H. Hurlbutt, *Hume, Newton and the Design Argument* (Lincoln: University of Nebraska Press, 1965) 65, 76–77. Sometimes the term "natural theology" is also used to distinguish the branch of theology based strictly on reason and experience from theology based on revelation. Scholars of early modern thought usually prefer the first kind of usage mentioned above, the one I endorse for now.

rain of natural theology, focusing particularly on purposive and adaptive design manifested in nature. Another subterrain of natural theology, to some degree overlapping with the former, deals with the lawful order of the universe as it is often expressed and argued for in what is usually called the cosmological argument.[2]

In *Paradise Lost* we find two different accounts of the generation of rational beings—that is, of angels and humans. The one is told by Satan in his remarkable dispute with the faithful Abdiel; the other is Adam's account on his creation, related to Raphael, his heavenly visitor in Eden. In fact, both narratives concern the narrators' own peculiar origin, the way they came into being respectively. Otherwise, the two narratives could not be more divergent.[3] Satan's account deals with the process itself whereby he and his fellow angels emerged from animated matter, a process allegedly governed by blind fate:

> That we were formed then say'st thou? and the work
> Of secondary hands, by task transferred
> From Father to his son? strange point and new!
> Doctrine which we would know whence learned: who saw
> When this creation was? remember'st thou
> Thy making, while the maker gave thee being?
> We know no time when we were not as now;
> Know none before us, self-begot, self-raised
> By our own quickening power, when fatal course
> Had circled his full orb, the birth mature
> Of this our native heaven, ethereal sons.
> (v. 853–863)

Adam, in contrast, is concerned not with the creative activity of his Maker, of which he confessedly cannot be aware, but his first experience of the perception of his existence. The relevant passage is worth quoting here in full length:

2. Neal C. Gillespie, "Natural History, Natural Theology and Social Order: John Ray and the Newtonian Ideology," *Journal of the History of Biology* 20 (1987) 1–49; 4. Cf. also chapter 3 above.

3. In this essay I am not concerned with the question of which of the two accounts contains the "truth." Both can be true in a sense, if we differentiate, as evolutionary biologists routinely do, between the *ultimate* and *proximate* causes of creation. See John Rogers, *The Matter of Revolution: Science, Poetry and Politics in the Age of Milton* (Ithaca: Cornell University Press, 1996) 124–25. Then it can be said that both versions are true but both of them deficient; Abdiel's silence on the proximate causes is well meaning, whilst Satan's silence on the ultimate cause is malicious.

For man to tell how human life began
Is hard; for who himself beginning knew?
Desire with thee still longer to conserve
Induced me. As new waked from soundest sleep
Soft on the flowery herb I found me laid
In balmy sweat, which with his beams the sun
Soon dried, and on the reeking moisture fed.
Straight toward heaven my wondering eyes I turned,
And gazed awhile the ample sky, till raised
By quick instinctive motion up I sprung,
And thitherward endeavouring, and upright
Stood on my feet; about me round I saw
Hill, dale, and shady woods, and sunny plains,
And liquid lapse of murmuring streams; by these
Creatures that lived, and moved, and walked, or flew,
Birds on the branches warbling; all things smiled,
With fragrance and with joy my heart o'erflowed.
My self I then perused, and limb by limb
Surveyed, and sometimes went, and sometimes ran
With supple joints, and lively vigour led:
But who I was, or where, or from what cause,
Knew not; to speak I tried, and forthwith spake,
My tongue obeyed and readily could name
Whate'er I saw. Thou sun, said I, fair light,
And thou enlightened earth, so fresh and gay,
Ye hills and dales, ye rivers, woods, and plains
And ye that live and move, fair creatures, tell,
Tell, if ye saw, how came I thus, how here?
Not of myself; by some great maker then,
In goodness and in power pre-eminent;
Tell me, how I may know him, how adore
From whom I have that thus I move and live,
And feel that I am happier than I know.
(viii. 250–82)

A notable element of this account is that Adam, from the instant of his awakening, is engaged by the rich surroundings, and the smooth execution of his capacities grasp his attention immediately after first opening his eyes. This intense absorption in external and internal experiences is, however, preconditioned by the disposition of Adam already present before his exposition to any sensation whatsoever: a disposition called by Milton "wondering." Adam was born wondering, a primary state of his mind that precedes or at least coincides with the beginning of experience. This wondering is ready to grasp any appropriate object. It is notable that the first particular orientation deliberately given by Adam to his wondering attention is "straight towards Heaven." His first experience after finding himself laid on the flowery grass is thus the scenery of the very realm of his creator. Wondering, however, is not just a passive contemplation of the world. Instead, it is a motivating force that gives the impetus to act, to stand up, look around, walk, run, and speak, in a word to seek for novel experiences that further preserve and enhance the wondering itself.

Wonder is a disposition that prevents the wonderer from taking things for granted. If something is not taken for granted then there are open questions to be asked and answered about it. That for the freshly created being everything is amazing and wondrous is not necessarily so. Without wondering, the newborn human adult would be able to understand the world around him and take his place in the great cosmic order. However, what wondering adds to this hypothetical initial indifference towards the creation is a sense of incompleteness. Of course, the contrivance of the human body and the beauty and exuberance of nature are themselves perfect and complete, but wonder necessitates inquiring about the cause of their existence. For as Satan's account shows, the notion of life's emergence by blind forces acting upon matter makes existence a natural necessity in which there is nothing wonderful. Satan thus takes his existence for granted; thereby it does not make sense to him to reflect upon the causes of his creation beyond the fateful operations of chaotic matter. He does not need witnesses of his creation; indeed it does not make sense to him to muse on the world that preceded his own being, his existence being self-sufficient. He even ridicules any attempt that tries to penetrate the mystery of the time when one was not yet born. His version of his life's genesis is more similar to the creation of animals than to that of Adam: Satan emerged from matter just like "the ounce, the leopard and the tiger" or as the bleating sheep that rose from the soil "as moles" or "as plants" (vii. 466–73) on the sixth day of creation.[4] The animals occupy there respective niches without wonder-

4. C. S. Lewis, *Preface to Paradise Lost* (Oxford: Oxford University Press, 1941) thinks that Satan's theory heightens nonsense to an almost improbable degree: "Uneasily shift-

ing at their mysterious generation. As we will see, this is the point where Satan deviates from the animal world: his lack of wonder has far-reaching consequences since it is accompanied by conscious reflection. Contrary to lions and deer, he speculates on his supposed self-emergence, and he comes to the conclusion that there is no superior power to admire and obey so he, unlike the beasts, refuses to accept the rank he is granted in the cosmic hierarchy.

Adam, on the contrary, is led to a rapid recognition of his dependency and insufficiency by the wonder he unavoidably feels whatever he senses or does. He agrees with Satan that it is obviously true that no-one can know one's own beginning in the sense that no-one can have direct memory of it. But he does not find it absurdly futile to inquire about it and he is far from deducing from this futility a self-raising power that had him pulled out of non-existence by his hair. He wants the sun and the earth, the hills and the dales, the woods and the living creatures to bear witness to his creation. Of course, they cannot tell the truth in the literal sense, however they are able to point to a "great maker" who is to be known and adored at the same time. Nature, both human and nonhuman, both organic and inorganic, induces an easy and evident interpretation as the masterpiece of an infinite divinity, when it is contemplated through Adam's wondering curiosity. It is not a proper design or cosmic argument, nor is it a well-formed argument at all, but nonetheless it is a natural theology that provides knowledge of God including his worship. For Adam, the desire to know his creator and the desire to worship him is exactly the same desire. Analogously, Satan's rejection of God's creatorship and his rejection of submission to the Father and the Son is the exactly the same gesture. The presence of wonder in Adam engenders admiration and obedience, the lack of it in Satan results in vanity and rebellion.

It can be argued that the unfallen Adam was in a very different epistemological position from ours. Adam, for example, named the animals according to their nature, because his "sudden apprehension" had been endued with such knowledge directly by God (viii. 351–53). And, after

---

ing on the bed of Nonsense which he has made for himself, he then throws out the happy idea that 'fatal course' really produced him, and finally, with a triumphant air, the theory that he sprouted from the soil like a vegetable. Thus, in twenty lines, the being too proud to admit derivation from God, has come to rejoice in believing that he 'just grew' like Topsy or a turnip" (98). Rogers's invitation to interpret this idea in the context of a radical vitalist theory of organization is more promising.

all, no sooner had he addressed the hills and woods with his question concerning the "great maker" than he met the divine presence face to face, a happily convenient verification of natural theology, hardly accessible for postlapsarian mankind. But we know that, according to Milton, our first parents had perfect knowledge limited to the natural world, a kind of knowledge not altogether unattainable for their fallen posterity.[5] "The end then of our learning is to repair the ruins of our first parents," Milton famously wrote in his 1644 treatise on education.[6] The fallen human species is by no means barred from a certain degree of "knowledge of God and things invisible" that can be reached "by orderly conning over the visible and inferior creature."[7] Exactly the same method is described by Adam in his polite conversation with Raphael:

> Well hast thou taught the way that might direct
> Our knowledge and the scale of nature set
> From centre to circumference, whereon
> In contemplation of created things
> By steps we may ascend to God.
> (v. 508–12)

This description of the proper method by which human beings can ascend to heaven from earth through the multiple interposed grades of the cosmic hierarchy contains nothing specific to the privileged pair of Adam and Eve. It also is sufficiently consonant to Milton's own recommendations regarding the efficient method to be applied in England's reformed educational system. The gradual ascension to the outermost sphere of the cosmos is thus a suitable path from man to God irrespective of at which point one stands in the sacred history.

5. Joseph E. Duncan, *Milton's Earthly Paradise: A Historical Study of Eden* (Minneapolis: University of Minnesota Press, 1972) 153–55. Here Duncan convincingly argues that Milton attributed limited but perfect knowledge to Adam. Stanley Fish argues, however, that Milton might have thought that the linguistic perfection of Eden cannot be restored by our fallen science despite the efforts of Bacon, Wilkins and other champions of the advancement of science. If science's progress depends on a linguistic reform, as these men held, then this impossibility may have a considerable effect on mankind's state of knowledge as well. See Fish, *Surprised by Sin: The Reader in Paradise Lost,* 2nd ed. (Cambridge: Harvard University Press, 1997) 127–30.

6. John Milton, *The Major Works*, eds. Stephen Orgel and Jonathan Goldberg (Oxford: Oxford University Press, 1991) 227.

7. Ibid.

It is remarkable that in the passage cited above from Adam's speech invokes the orthodox view of the universe with the earth at the center and heaven on the periphery. The Ptolemaic system had been for a long time under serious attack with a considerable proportion of the intellectual elite having already been converted to heliocentrism. Though Milton himself was at best ambiguous on this question, he attributed the more obsolete version to Adam intentionally, thereby emphasizing the relative unimportance of the much-debated question from a Christian activist point of view.[8] Adam's inquiries "concerning celestial motions" are only "doubtfully answered" by Raphael, as we read in the *Argument* of Book VIII. Raphael's reluctance to reveal the real structure of the universe and his anticipation of the futility of future astrological debates[9] carries the moral that there are "things more worthy of knowledge."[10] Raphael does not blame humans for their curiosity which is understandable given the cornucopia of God's wondrous works throughout the universe. Nevertheless, it is not their business to "divulge" God's secrets but rather to admire and obey him. The primary aim of the divinely orientated contemplation of nature is not to stockpile experimental or observational data and construct scientific hypotheses or theories based on them but to know God and "him serve and fear" (viii. 168). To gain this insight it is sufficient for humans to recognize their dignified place in the universe: sunbeams are "barren" and become fruitful only by shining the earth, more specifically shining on earth's chief inhabitant, man (viii. 91–99). In this moderately anthropocentric cosmos man is highly ranked among living creatures even if it "is obvious to dispute" that the entire "vast room in nature" is completely "unpossessed by living soul" apart from the earth (viii. 512–18).

Milton's view of natural theology and its relation to relevant knowledge can therefore be justly called idiosyncratic. It is markedly different from two peculiar currents of thought though by no means diametrically opposed to them: the educational reformers and the virtuosi of the early Royal Society. Samuel Hartlib (1600–1662), John Dury (1596–1680), and the notoriously anti-Copernican John Amos Comenius (1592–1670)

8. Milton "wanted to stress that this is not the crucial issue either for the Christian or for the revolutionary politician" (Christopher Hill, *Milton and the English Revolution* [Harmondsworth: Penguin, 1979] 399).

9. See viii. 75ff.

10. Milton, *Major Works*, 507.

were interested in the advancement of learning not for its own sake but for it useful applications. Milton's attitude to learning was pragmatic and inherently hostile to all "disinterested intellectual curiosity."[11] As Charles Webster noted on Dury, the purpose of education according to him was "to produce 'good commonwealth men,' replacing the ornaments of scholastic learning by ideals of public service. The criterion for judging the value of knowledge is its degree of 'usefulness' or 'practical' application."[12] So far Dury's opinion was shared by Milton as well. However, the reformers, unlike Milton and the virtuosi, were insensible to nature's aesthetic beauty. "For them God's power resided in nature for the purpose of practical exploitation through science and technology."[13]

The early members of the Royal Society, such as Robert Boyle, Christopher Wren, Thomas Sprat, and John Ray agreed with Milton that the investigation of the works of the creation is a kind of religious duty and worship that deepen our knowledge of their creator. As Boyle remarked, the further we contemplate creation, "the more footsteps and impressions we discover of the perfections of their Creator, and our utmost can but give us a just veneration of His omniscience."[14] Ray told his readers in his *Wisdom of God* that man is "commanded to consider" vegetables, beast, birds, insects and all animals particularly, "to observe, to take notice of their curious structure, ends and uses, and give God the praise of his wisdom and other attributes therein manifested." Ray even noted that those who, though having "the leisure, opportunity, and abilities," fail to pay nature the due attention "rob God of some part of the glory."[15] Ray thus sanctioned the godliness of the young discipline of natural history, and he linked it tightly to natural theology, hitherto a

11. H. F. Kearney, "Puritanism, Capitalism and the Scientific Revolution," in *The Intellectual Revolution of the Seventeenth Century*, edited by Charles Webster (London: Routledge and Kegan Paul, 1974) 218–42, 236. The quotation is from Arthur Lovejoy, cited by Kearney.

12. Charles Webster, *Samuel Hartlib and the Advancement of Learning* (Cambridge: Cambridge University Press, 1970) 11.

13. Ibid., 14.

14 Quoted in Richard Westfall, *Science and Religion in Seventeenth-Century England* (Ann Arbor: University of Michigan Press, 1973) 42.

15. John Ray, *Wisdom of God Manifested in the Works of Creation* (London: Ray Society, 2005) 155.

distinct strand of thought, so that from now on they seemed "to future generations virtually a joint endeavor."[16]

However, as implied by Ray's remark just quoted, the natural scientists tended to despise superficial, uneducated wonder and they claimed that only the systematic research of the natural philosopher can discover the *real* wonder that lies in nature's secret depths and they are the only ones who can report authentically the wondrousness of the universe and the greatness of its creator. Wren, for example, wrote that "who can better magnify the arm that expanded heavens, than he who tells you that seven thousand miles will fall short of the diameter of the earth, and yet that this diameter repeated a thousand times will not reach the sun, or this distance between the sun and us repeated a thousand times reach the nearest fixed star."[17] An ordinary man may have a vogue impression about the immensity of the universe, however it is the professional astronomer who is entitled to give a full meaning of this immensity by carefully quantifying it, and consequently he is privileged to formulate and communicate the true nature of the wonders to the laity. Thomas Sprat put it even more sharply:

> [God's] works receive but little glory from the rude wonder of the ignorant, and there is no wise man that values the applauses of a blind admiration . . . The tribute of praise that we owe our maker is not a formal slight confession that His works are wonderful and glorious, but such an acknowledgement as proceeds from deep observations and acquaintance with them.[18]

The wondering of Adam would not deserve much respect from Sprat since according to his principles it is deemed incurably facile: in the blink of an eye Adam leaps to the creator without lengthy immersion in contemplation and experimental observation. His admiration is blind, and his wonder is rude from the virtuoso's point of view.

We saw that Milton in *Paradise Lost* did not attribute too much significance to such "nuances" as the relative position of the sun and the earth. For the virtuosi nothing could have been more important. In their opinion the devil, or more precisely God, hid in the details. This justified

16. Gillespie, "Natural History," 46. Gillespie mentions other earlier efforts to combine the two disciplines in certain writings of, among others, Thomas Browne and John Wilkins (see ibid., 9-11).

17. Quoted in Westfall, *Science and Religion*, 28.

18. Ibid., 47.

more meticulous observations than Milton would have endorsed. Milton might have regarded Ray's demand to investigate vegetables and insects in minuscule detail as exactly the kind of disinterested intellectual curiosity he rejected as mere hairsplitting.[19] Closer to the actual text of *Paradise Lost*, we can say that for Raphael it was relatively unimportant to acquire certainty regarding the central problems of astronomy. Many leading scientists were far from this indifference, and they placed much more weight upon heliocentrism; in their efforts to redirect anthropocentrism in the post-Copernican world, they forged a positive argument from the sun-centered scheme that proved the dignity of mankind bestowed on it by God, who in his benevolence chose earth's orbit to be the central one.[20] For Milton, this dignity and God's greatness was already guaranteed wherever truth actually lay concerning the structure of the solar system, and that was what really mattered.

The knowledge that is the end of learning is the knowledge of God. Human beings may not conceive of God's true nature until he will be all in all, but he revealed himself enough to provide humankind with the possibility of salvation: God has "adjusted his words to our understanding," so we have "in ample quantity those things which we need to understand about him for our salvation."[21] Out of this knowledge, whether obtained through Scripture or through the contemplation of the universe, we come "to love [God], imitate him and be like him."[22] The purpose of this kind of learning is not so much a certain cognitive state or erudition as a disposition well attuned to virtuous conduct.

The reader of the poem has the occasion to observe how wondering admiration and virtuous action are linked together, how the former is canalized into the latter. The spontaneous worship performed by Adam and Eve at the beginning of Book V shows exactly that. God's "glorious works" are so "wondrous" that they implicate the unspeakable wondrousness of its author (v. 153ff.). The lower creatures can show only dimly but nonetheless sufficiently the creator's goodness. As nature was able to

19. See, for example, Alexander Marr, "Introduction," in *Curiosity and Wonder from the Renaissance to the Enlightenment* (Aldershot: Ashgate, 2006) 1–20, esp. 4–6.

20. John Hedley Brooke, "'Wise Men Nowadays Think Otherwise': John Ray, Natural Theology and the Meanings of Anthropocentrism," *Notes and Records of the Royal Society of London* 54 (2000) 199–213, esp. 202–3.

21. "On Christian Doctrine," in Milton, *Major Works*, 727.

22. Ibid., 227.

bespeak God, the great master, on Adam's first day, without words, similarly Adam and Eve regard them now suitable to join and amplify the Edenic pair's praise and glorification of him. Having finished the prayer, they hasten "on to their morning's rural work" (v. 211), which, along with nighttime rest, had been "set" by God to dignify the human species and emphasize its superiority to the rest of living creatures, which "all day long / rove idle unemployed" (iv. 612–13, 616–17.) Pious worship and fulfillment of the duties spring from the awe and admiration at the works of nature that continuously advertise their creator in a harmonious chorus with Adam and Eve.

As wonder begets worship and obedience, the doctrine of "self-begetting" leads to rebellion. Although our general ancestors live in the prepolitical milieu of Paradise, *subjection* and *obedience* are sufficiently political terms to contrast them to Satan's revolt. His doctrine is the ideology of self-sufficiency and autonomy, and at its foundation there is the outright rejection of dependency on any external, intervenient operation that could be a necessary and inevitable cause contributing in the process of coming into being. Though this theory does not logically exclude wonder or awe at the spectacular process of self-generation[23] (Satan however does not seem to express any feeling of this kind), any hypothetical wonder induced by material evolution of the angels would at best be directed at the self-organizing matter, that is, at the angels themselves, since the very essence of the process is that the result of it needs nothing apart from itself. So although the spontaneous generation can in theory induce a kind of wonder, it is certainly incapable of inducing a kind of wonder that in turn would engender adoration, worship, and voluntary obedience to an external, invisible authority. Indeed, Satan's undeniable despotic allures and self-congratulatory manners, which mingle with his more praiseworthy characteristics such as self-sacrifice all the time, might partly derive from that the logic of his theory of self-sustaining

23. Consider the rather fruitless debate between creationists and evolutionists: the latter often argue that the view of life as an evolutionary process is more wonderful (or it has *grandeur*, to use Darwin's term) than the biblical Genesis. In our present context, see William Empson's flip on C. S. Lewis, who in his *Preface* "treats creation by a personal God as the only theory which is not mere jabberwocky," whereas Empson and in his view the majority of Lewis's audience hold the Darwinian account "more numinous, more imaginatively stirring." See Empson, *Milton's God* (Cambridge: Cambridge University Press, 1981) 89. For Lewis's scornful judgment on Satan's theory, see his remark on the turnip, cited above, n. 4.

existence forces him to attribute the overwhelming power alone deserving respect to himself. This may also explain the much debated inconsistency of his rhetoric that defends the angels' equality trampled by the Father, while at the same time he tries to preserve the angelic hierarchy of "Thrones, Dominations, Princedoms, Virtues, Powers" (v. 772ff).[24] It is the tension between the metaphysically grounded equality of all the angels as "ethereal sons," and Satan's own political ambitions and convictions that surface in Satan's "verbal paradox."[25]

As John Rogers remarks, there is a consensus between the pro- and the anti-Satanist camps regarding the centrality of Satan's claim of self-begetting.[26] As wonder at the marvelous works of nature can establish a particular way of quasi-political conduct in the case of the unfallen Adam and Eve, so also the significant lack of this wonder gives rise to a totally different politics of the fallen angels. It follows that the collapse of Satan's theory has far-reaching consequences too. It is agreed to by critics with diverse sympathies that on the top of the Niphates after his conversation at the end of Book III with Uriel, witness of the creation of the earth, Satan is forced to acknowledge God's creatorship.[27] He has no reason to doubt Uriel's account, whom he deluded by transforming himself into the shape of a young cherub. More important is that the beautiful panorama from the mountaintop to the young earth and sun prompts Satan to reconsider his whole enterprise and thrusts him into despair and a nervous breakdown. Although Milton remains silent about the exact cause of Satan's conversion, we learn that "His face / Thrice changed with pale, ire, envy and despair" (iv. 114–15). It is a turning point in the development or rather degeneration of the archfiend's character that from now on keeps "rapidly rotting away."[28]

Milton's omission of the emotional changes in Satan that must have occurred between the end of the third book and the beginning of

24. Rogers, *The Matter of Revolution,* 129; Empson, *Milton's God,* 73–74.

25. Cf. *Paradise Lost,* i. 245ff., x. 460ff.

26. Rogers, *The Matter of Revolution*, 122–23.

27. Michael Bryson, *The Tyranny of Heaven: Milton's Rejection of God as King* (Newark: University of Delaware Press, 2004) 83–84; Empson, *Milton's God,* 61–63.

28. Empson, *Milton's God,* 71; Lewis, *Preface to Paradise Lost,* 99. The chief point of divergence between Empson and Lewis is not the fact of degeneration rather than the initial moral quality of Satan. Empson insists that initially Satan is praiseworthy while Lewis thinks he is contemptible from the beginning (Empson, *Milton's God,* 88; Lewis, *Preface,* chapter 13, passim.).

the fourth is usually explained by the fact that Satan's soliloquy on the mountaintop was composed earlier for other purposes. However, the gap left by the lack of an explicit narrative can be easily filled in by folk psychology. Envy results from the recognition how mistaken he had been when he had contested God's creatorship and attributed a similar faculty to himself. Ire accompanies this discovery as natural when one discovers that one's greatest enemy was right. He naturally falls into despair since he is forced to find himself in the middle of an enterprise that he had built on false premises. The beauty of the creation constitutes a painfully sneering background to Satan's struggling.[29] What Satan had been trying to avoid with all his efforts now seems to happen all of a sudden: he is no longer able to withhold wondering.

Although at this point of the narrative the role of nature's beauty in Satan's submission to God's creatorship can only be inferred from the context, the perfection of human beings triggers explicit awe:

> My thoughts pursue
> With wonder, and could love, so lively shines
> in them divine resemblance, and such grace
> The hand that formed them on their shape hath poured.
> (iv. 362–65)

Later, just before the temptation, Satan gives voice to his admiration of the newly created world and the scale of nature: "O earth, how like to heaven, if not preferred / more justly . . . With what delight could I have walked thee around" (ix. 99–100; 114). In order to be able to continue his compromised project it is necessary to him to persuade himself that God is incapable of creation or that the world is in reality unworthy of praise or at least cast some aspersion on God's creatorship time and again (ix. 135, passim). Otherwise he has no choice but to confess that it is not his high-flying republican ideas rather than a mere desire for destruction that motivates his actions: "For only in destroying I find ease / To my relentless thoughts" (ix. 129–30).

We can thus see how unchained wonder at the works of creation induces obedience and true worship and how willfully suppressed admiration is crucial to contribute to rebellion and disobedience. It is important that wondering itself is not a guarantee for lawful obedience. Although the immediate object of wondering is the creature, the won-

29. Empson, *Milton's God*, 62.

dering nonetheless must not stop there: its task is to point to an invisible reality that is the fountain of all wonders, the most wonderful thing of all. The things created can only be wondrous inasmuch as they borrow their wondrousness from God. Wondering that does not transcend its proximate object results in idolatry. Milton finds it necessary to note that Adam "bow[s] low" before Raphael without "awe" but "with submiss approach and reverence meek" (v. 358–60). However, when at the day of his creation, the "presence divine" appeared before Adam, he "fell submiss" at his feet "with awe and adoration" (viii. 313–16). Milton's cosmos is inherently hierarchical, and citizens of each sphere owe respect to the inhabitants of respectively superior spheres.[30] Wondrousness is distributed quite evenly throughout the different levels from the most inferior to the uppermost level. But the proper effect of wonder (that is, awe and admiration) must be directed exclusively toward its invisible source. When people bow to the fallen angels who in human history appear as biblical idols,[31] or, for that matter to the Stuart king, and thereby fail to worship their true God, they act in diametrical opposition to the unfallen Adam, who withheld awe when he met Raphael, but let it overflow in the presence of the Divine. The Royalist sin[32] of Milton's age had its origin partly in a misdirected awe that replaced the invisible superior with the visible equal. At the core of Milton's lifelong political project and his epic works lies the relentless exhortation and warning that in order to regain

30. This is not to deny the presence of an alternative, potentially democratic cosmology in *Paradise Lost*, as John Rogers has shown in his *The Matter of Revolution*. However, the two rival schemes can be, if not in their political implications, synthesized. I diverge from Rogers's view in that I regard the official hierarchical cosmos as a default cosmology rather than a reflection of Milton's late, increasingly aristocratic political philosophy. At this point I can agree with Tillyard without accepting his much-contested opinion that Milton was altogether "orthodox." See E. M. W. Tillyard, *The Elizabethan World Picture* (Harmondsworth: Penguin, 1972) 31, 40–41. I also doubt that Milton hid his more egalitarian views in an esoteric cosmology. What Rogers overlooks is that egalitarianism is overtly asserted (e.g., xii. 69–71), but it is valid only *within* spheres, such as the human sphere, which is at the macrolevel only a component in the cosmic hierarchy.

31. i. 369 passim.

32. Milton and the Puritan revolutionaries may have also been guilty in conflating the secular and the sacred. *Paradise Lost* and *Paradise Regained* can be interpreted as recognition of this mistake hitherto imputed by them only to the Royalist camp. See Blair Worden, "Milton's Republicanism and the Tyranny of Heaven," in *Machiavelli and Republicanism*, edited by Gisela Bock et al. (Cambridge: Cambridge University Press, 1990) 225–45, esp. 243–45.

Paradise on earth our awe and wonder must be redirected where it was originally aimed: straight toward heaven.

8

Malebranche on Restlessness and Curiosity

Roland Breeur

INTRODUCTION

WAS IT ON THE "quai des Augustins" or in the "rue Saint Jacques"? In any case, Malebranche was twenty-six and purchased a copy of Descartes's work, *Traité de L'homme* (*Treatise on Man*), which provides a mechanical account of human physiology. It is told that "the joy of learning such a great number of discoveries caused him such violent palpitations of the heart that he was hourly forced to put aside his book and stop reading in order to catch his breath."[1] Reading this book changed the entire life of this young Oratorian, born in Paris on August 6, 1638:[2] it converted him not only to Cartesianism but probably to philosophy as such.

What made him so breathless about that work? We may guess that he became particularly interested in Descartes's radical critique of Aristotelian thought, which he had found so sterile. Descartes's book at least stimulated him to devote ten years of his life to studying mathematics and (Cartesian) philosophy. This thorough study resulted in his major work, *The Search after Truth* (*De la recherche de la vérité*) (published in 1674–1675), a work that was read and commented on by very important

1. Father Yves André, in *La vie du R. P. Malebranche, prêtre de l'Oratoire, avec l'histoire de ses ouvrages*, quoted by Th. Lennon in his *Philosophical Commentary* at the end of his translation of *The Search after Truth* (abbreviated LO), translated by Thomas M. Lennon and Paul J. Olscamp (Cambridge: Cambridge University Press, 1997) 835.

2. For history's sake, it is interesting to know that he was born and died in the same years as Louis XIV (1638–1715).

thinkers and intellectuals of his time and beyond. This book also forms the background for what will follow.

The aim of this work is very explicit: it is, as a good Cartesian, to investigate the origin of our errors in order to help us to devote more interest to what is really important and true about ourselves, about our world, and about God. But at the same time, it is more, and in its objective it goes far beyond a strict Cartesian background: Malebranche wants finally to demonstrate God's active presence and role in all the aspects of the world he has created. It is not a coincidence that this treatise represents a kind of synthesis of the thoughts of Descartes and Augustine:[3] it reformulates and rethinks purely Cartesian topics (about the body, the world, and the mind) against the background of religious and theological concerns, from Augustinian inspiration. Hence Malebranche's typical and famous doctrines of, for example, "occasionalism" and the "vision in God."

• • •

It would lead us too far to enter in detail into these two central notions of Malebranche's metaphysics.[4] It is enough for the present purposes to point to the intertwinement of the religious and philosophical aspects that compose these notions. Concerning the latter (vision in God), heavily criticized by Antoine Arnauld (1612–1694), Pierre Sylvain Régis (1632–1707), and others, it offers a good illustration for what we just called the "revision" of the Cartesian sources: Malebranche shares with Descartes the idea that every perception and knowledge is accomplished by way of immaterial entities, ideas, or mental representations. But for Malebranche, it would be inconceivable that what stands for the essences of reality would depend upon our own mental and finite activity, as in Descartes's view. In the "Elucidation 10" of his *Search after Truth*, he defends his opinion by insisting precisely on the eternal character of what we conceive and apprehend as necessary and objective truths. The necessity, immutability, and infinity of the ideas we conceive make

3. For that synthesis, cf. H. Gouhier, *La philosophie de Malebranche et son experience religieuse* (Paris, Vrin, 1926); in his appendix, he provides a list of passages of Augustine quoted by Malebranche; see 411–20.

4. For a clear presentation of both issues, cf. the chapters of Steven Nadler ("Malebranche on Causation") and Tad M. Schmaltz ("Malebranche in Ideas and the Vision in God") in *The Cambridge Companion to Malebranche*, ed. Steven Nadler (Cambridge: Cambridge University Press, 2000).

them irreducible to our own finite mind: hence our own apprehension depends on the access we have to the eternal archetypes existing in the divine understanding.

The doctrine of the "vision in God," laying stress on the dependence of our cognitive faculty upon the divine understanding, exemplifies on an epistemological level the far reaching modification the Cartesian system undergoes. But this religious prominence becomes even more overwhelming in his influential doctrine of "occasionalism." *Occasionalism* is the "view that the finite created beings have no causal efficacy whatsoever."[5] They are in need of what Kant disapprovingly called "supernatural assistance."[6]

"A true cause," as Malebranche understands it, is such that the mind perceives a necessary connection between it and its effect, i.e., a kind of connection that, as Hume will stress later, no one can encounter between created entities, objects, or events in nature. All events between creatures ("natural causes" between bodies, minds, etc.) can only be regarded as providing the "occasion" (or "the condition," as the twentieth-century philosopher, Husserl, says) for a real, efficient cause to occur. Our sense experience can only reveal to us a succession of events; they never disclose necessary relations between these events. "God alone is the true cause of all that occurs in the world." Hence, only his divine assistance is able to bring about these kind of relations and the events that appear at these occasions. In Malebranche's words: "natural causes are not true causes; they are only occasional causes that act only through the force and efficacy of the will of God."[7]

This immediately makes obvious the fact that contrary to what is often stated, *occasionalism* is more than an "*ad hoc* solution" to the Cartesian mind-body problem. It applies to all kinds of causal relations between created entities: hence for relations between two material objects, between a volition and a thought, between a thought and a bodily movement, etc. This theory follows from "theological premises concern-

5. Cf. Nadler, *Cambridge Companion to Malebranche*, 115.

6. Immanuel Kant, *Critique of Pure Reason*, translated and edited by P. Guyer and A. W. Wood (Cambridge: Cambridge University Press, 1997) 436. This text was originally published as *Kritik der reinen Vernunft* in 1781, A 390. See chapter 14 below.

7. LO, 449.

ing the essential ontological relationship between an omnipotent God and the created world that he sustains in existence."[8]

However, the relevance of it is undeniably apparent in the specific context of mind-body dualism: how can pure matter interact with the mind and vice versa? Between the mechanical functioning of our body and the experiences in our mind, there is indeed an "explanatory gap," such that even a Darwinian biologist such as Thomas Henry Huxley (1825–1895), known as "Darwin's bulldog," wrote, so many years after Malebranche: "How is it that anything so remarkable as a state of consciousness comes about as a result of initiating nerve tissue, is just as unaccountable as the appearance of the Djin, where Aladdin rubbed his lamp in the story."[9]

To be honest, *occasionalism* as conceived by Malebranche seems less a real "solution" to that problem, as a beautiful symptom for what is so problematic about it. Although it is clear, indeed, that our arm is moved when we will it, there is no "necessary connection between our will to move our arms, for example, and the movement of our arms."[10] Even worse, since the soul already lacks the knowledge of all the physiological and neurological processes that intervene in the motion of a little finger, "how, then could men move their arms?"[11]

At one level, we neither experience nor live in our body "as a sailor in a ship," but our mind is "united with this body."[12] Consequently, our ignorance concerning the processes responsible for bodily movements does not prevent us to behave and perform all kinds of actions. And in doing so, I never have the impression to navigate a "vessel" or to command an engine. Although my body is nothing more than "a kind of machine equipped with and made up of bones, nerves, muscles, veins," I am closely joined to it. As Descartes already stressed in *Meditation VI*, this machine and my soul form a "unit": I feel pain, hunger, thirst and so on, at the "occasion" of physiological and neurological movements. But these sensations, or "subjective" experiences of my own body, arise from

8. Nadler, *Cambridge Companion to Malebranche*, 116.

9. Quoted by Colin McGinn, "Can We Solve the Mind-Body Problem?" in *The Nature of Consciousness*, eds. N. Block, O. Flanagan, and G. Güzeldere (Cambridge: MIT Press, 1997) 529

10. LO, 449.

11. Ibid., 450.

12. René Descartes: "Nature also teaches me . . . that I am not merely present in my body as a sailor is present in his ship, but that I am very closely joined and, as it were, intermingled with it, so that I and the body form a unit." Meditation VI, in CSM ii. 56.

a *union* of my mind with the body that as such is established by God. For this reason, Malebranche opens his treatise on the search after truth with the statement that the union between the mind and the body is dependent upon one another and that this union is dependent on a more fundamental union: that of our soul with God.

This doctrine of the two unions forms the core of his philosophical explanations in the *Search after Truth*.[13] What we have just said about his theory of the "vision in God" and *occasionalism* already provides us with some outline of what really is at stake in Malebranche's metaphysics. In what follows, we will spend some time enlarging upon the idea of *the two unions* in order to have a better grasp on what he means by *curiosity*.

THE TWO UNIONS

One of the epochal characteristics of modernity is named "ruthless curiosity" ("rücksichtlose Neugierde") by Nietzsche.[14] The tendencies to justify that cognitive appetite coincide with the general question about the legitimacy of the modern cognitive will as such, and more precisely, of the new theoretical pretensions and ambitions. Since Descartes postulated the necessity of a radical new foundation for the criterion of certainty in knowledge, and since he claimed the importance of a method in the progress of our cognitive acts, curiosity seemed to liberate itself from the coercion of dogmatic systems and enclosures. Whereas in the late renaissance, that lust for science was still associated with the image of a Faustian hubris and self elevation, it gradually claimed autonomy. One of the reasons of this self-affirmation of curiosity could have been the fact that the Cartesian concern about the possibility of secured knowledge involved a general leveling off and homogenization of the objects of theoretical inquiry.[15] For the new scientific drive, there is no preferred and no unworthy object of cognition. In the eyes of the new scientific appetites and its method, no object has a dignity on its own. Everything can be assimilated into an inexhaustible inquiry.

13. Denis Moreau, *Malebranche. Une philosophie de l'expérience* (Paris, Vrin, 2004) 36.

14. Friedrich Nietzsche, *Jenseits von Gut und Böse*, VII, § 188.

15. Cf. Hans Blumenberg, *The Legitimacy of the Modern Age*, trans. Robert M. Wallace (Cambridge: MIT Press, 1983) 361–75 ("Interest in Invisible Things within the World"); *Die Legitimität der Neuzeit* (Frankfurt: Suhrkamp, 1999) 422–39 ("Das Interesse am innerweltlich Unsichtbaren").

However, it may appear a little bit curious that precisely this unrestrained appetite for knowledge was condemned by the "founder" of modernity. In a letter to Marin Mersenne (1588–1648), Descartes raises the objection against Galileo's style of inquiry that it makes us oblivious of the more basic questions about the foundations of our cognition, presuppositions, and the like: he writes that his error is that "he continually digresses and he does not take time to explain matters fully. This, in my view, is a mistake. It shows that he has not investigated matters in an orderly way, and has merely sought explanations for some particular effects, without going into the primary causes in nature; hence his building lacks a foundation."[16] Hans Blumenberg refers in this context to the rise of a sort of "intrascientific morality, a rigorism of systematic logic"[17] in the eyes of which uninhibited curiosity was suspect. Now, we find the same suspicion in Malebranche's thought, though with different reasons. Against the medieval idea of curiosity as one of the vices of an individual in need of redemption, we will see that for Malebranche, curiosity is itself the expression of a fundamental unrest or restlessness of the individual inclination toward the Good. What is condemnable in it is not reducible to the Cartesian "intrascientific morality," the metaphysical quest for foundations. As a true Augustinian, Malebranche deplores the reality of the boundless, hectic cognitive drive as being the mark of our "original sin."[18] Original sin has upset and violated the order of nature.

As became clear from the introduction, Malebranche reframes Cartesian notions in a religious context. Hence, he will stress the idea that next to the union our mind experiences with its body, there exists a more essential one: the union with God. Through the unions with the body, the mind is related to sensible objects of the world, and through the latter, it receives its "life, its light, and its entire felicity": this union raises man "above all things."[19]

Hence, our mind is by its very nature situated "between its Creator and corporeal creatures."[20] The aim of his work, in this "search after truth"

16. Descartes to Mersenne, 11 Octobre 1638, AT II, 380, in CSMK iii. 124.

17. Blumenberg, *Legitimacy*, 397; *Legitimität*, 466.

18. For Augustine's condemnation of curiosity, cf. his *Confessions*, Book V, Chapter III; and Blumenberg, *Legitimacy*, Part III, V ('Curiosity Is Enrolled in the Catalog of Vices'); *Legitimität*, III, VI ("Aufnahme der Neugierde in den Lasterkatalog").

19. LO, xix.

20. Ibid.

is to re-establish the original hierarchy between these unions, by investigating all the things that are absolutely necessary "for the perfection of the mind." Because God, says Malebranche, "made minds to know and love Him rather than for *informing* bodies."[21]

Of course, the union with the body as such, contrary to the Lutheran and Calvinist idea of an integral corruption of human nature, has nothing corrupt or sinful. Cartesians agree on its positive function for the preservation of our being. On that level, we may conclude "that senses and passions are very well suited to the end for which they are given to us."[22] Moreover, what deceives us in relation to sensory perceptions, are not our senses, but the way we judge precipitously on the basis of what we experience. However, since there is original sin, this union has become so loud that we almost cannot hear God speaking to us anymore. Since the Fall, human nature has become victim to a disproportion between the two unions. "The first man's sin" has "weakened" our mind's union with God and has "strengthened our soul's union with our body."[23] Or better, the sin has enslaved us to our senses and passions in such a way that, says Malebranche, "we are led to believe that our body is the more important of the two parts of which we are composed."[24]

But although the union with God diminished, proportionally to the way in which the union with the body increased, it is impossible that it would be entirely broken "without the destruction of our being." Even for those who are intoxicated with bodily pleasures and who might be insensible to truth "they are yet joined to it": "The truth does not abandon them, it is they who abandon the truth."[25]

On the other hand, the same holds for the union with the body: it may diminish as the union with God increases, but it can only be broken by our death. "For even if we were as enlightened and as detached from all sensible things as were the Apostles, there is still the necessity stemming from Original Sin that our mind should depend on our body, and that we should feel the law of our flesh resisting and constantly opposing the law of our mind."[26]

21. Ibid., xx.
22. Ibid., 19.
23. Ibid., xx.
24. Ibid.
25. Ibid., xxii.
26. Ibid., xxiii.

Since that Sin, we not only have to endure that union, but *we are condemned to it*. Through it, the mind strives constantly to reinforce its union with sensible objects, to do things according to the relation things have to our body, i.e., to the preservation of our lives. We become enslaved by that union, which "infinitely debases man and is today the main cause of all his errors and miseries."[27]

Malebranche provides a nice example of that misery, that is, of the corruption in the natural equilibrium or hierarchy between the two unions, referring to "the first man" (Adam). After his Sin the pleasures and pains that initially served to advise and inform him about what the body needed for its preservation, no longer had the same meaning. They finally troubled his felicity, his love for God, and his senses and passions "revolted against him" without obeying him any longer. In a strange passage, half theological, half "proto-historical," Malebranche describes what happened as follows:

> Adam, then, had the same senses as we do, by which he was advised of what was necessary for his body, but without being distracted from God. Like us, he sensed pleasure and even pains, or involuntary and prevenient displeasure. But these pleasures and pains could neither enslave him nor make him unhappy, as they do us, because as absolute master of the motions generated in his body, he stopped them, if he so wished, as soon as they had performed their advisory function (and no doubt he always wished to do so with regard to pain). Happy would he, and we, have been had he done the same thing with regard to pleasure, and had he not voluntarily turned himself away from the presence of his God by allowing his mind's capacity to be exhausted by the beauty and anticipated sweetness of the forbidden fruit, or perhaps by the rash joy excited in his soul by the contemplation of his natural perfections, or finally by his natural fondness for his wife and the inordinate fear of displeasing her, all of which apparently contributed to his disobedience.[28]

Had he not been so fond of his wife, and had he been able to resist that "inordinate fear of displeasing her," he certainly would not have plunged us into the misery, i.e., to have strengthened the union with the body at the expense of our intimacy with the Creator. Consequently, we lost the direct control over the motions generated in our nerves: we have to

27. Ibid., xix.

28. Ibid., 22. Cf. also his further comments in Elucidation 8, ibid., 579–606.

suffer and go through our affliction and pains beyond their advisory, "prevenient" function and be continuously distracted from God.

But not only the suffering. The pleasures also exceed their natural function: we enjoy them for their own sake. Concupiscence, which is "nothing other than a natural effort by the brain traces to attach the mind to sensible things" grew victorious with original sin.[29] As a consequence, we are constantly obsessed by things related to our body, and we neglect perpetually our soul's "noble part." We debase ourselves by diversions like "hunting, dancing, gambling and good food" or care only about our health, glory, and grandeur. Even the learned, "faux savants," he says, "and those who pride themselves on their intelligence spend more than half their life in purely animal actions."[30] Nonetheless, we do not endure this humiliating debasement of our nature as a fatality: our sinful disregard of or blindness for the true nature of our soul is something we accomplish in freedom. We are responsible for our concupiscence in the sense that the objects we run after are freely chosen. Malebranche, in some sense, could even say that original sin, since it predetermines us to a strengthened union with our body, condemns us to freedom. This is indeed what he suggests in his exposition of our inclinations and the restlessness of the will.

THE RESTLESSNESS OF THE WILL

Hence our freedom to make choices is like a symptom revealing the latent omnipresence of original sin, i.e., of the corruption of the relation between the two unions. Our union with God manifests itself as inclination in our minds. Indeed, he says, inclinations are motions of the mind that are impressed by God in us. And since the Creator can have no other end for His operations than Himself, our natural inclinations have no other end than His Glory: "He cannot create a creature without directing it toward Himself and commanding it to love Him more than all things."[31]

However, although we cannot suspend that impression and perpetual motion towards the Good in general, we still can deviate it in the direction of what pleases our body, "and thus cause all the disorder

29. Ibid., 120.

30. Ibid., xxi.

31. Ibid., 266.

found in its inclinations, and all the miseries that are the certain and necessary results of sin." That deviation is, in other words, the result of the free choices we make in function of our bodily needs.

It means that for Malebranche, one always has to keep in mind a crucial distinction between the will and freedom. About that distinction he writes: "I propose to designate by the word will, or capacity the soul has of loving goods, *the impression or natural impulse that carries us toward general and indeterminate good*: and by freedom, I mean nothing else but *the power that the mind has of turning this impression towards objects that please us so that our natural inclinations are made to settle upon some particular object*."[32]

Our freedom consists in the power we have, since our sinful intimacy with the body, to determine that divine Love in us towards particular goods, our inclination towards "the good in general" towards particular goods. Our freedom is, in that sense, a sign of our fallen nature. It is the expression of a deficiency. It spoils a pure power of loving, that neither comes from us, nor depends on us, in a tendency to love what we should not love at all. In that sense, it is far from being a positive property of human nature. It is the expression of his weakness.

But how can we choose and dishonor that pure inclination towards particular goods? From the discrepancy or disproportion between the general goal of our will and the particular goods to which we deviate from it. Never satisfied by these finite goods and unable to silence the general will, we are continuously condemned to chase after new things. We are free because we are never satisfied by the things for which our desire is hunting.[33] It is precisely the dissimilarity between the real goal and the accidental ones that sets us free. And that freedom, the necessity to run from object to object, is what Malebranche calls the restlessness of our will. In a famous passage, he writes:

> Because it was made only for a good that contains all other goods within itself, the will's vast capacity for all goods in general cannot be satisfied by all the things the mind represents to it; and yet this continual impulse toward the good God impresses upon it cannot be arrested. This impulse, never ceasing, necessarily imparts a continual agitation to the mind; the will, searching out what it desires, obliges the understanding to represent to itself

32. Ibid., 5.

33. Cf. chapter 11, below.

> all sorts of objects . . . but the soul does not enjoy them; or if it does, it is not content with them . . . because nothing can arrest the soul's impulse except Him who has imprinted this impression upon it.[34]

The finite goods can only momentarily distract the pure love or will of the soul, and it is condemned to look for novel, extraordinary objects that contain the illusionary promise to contain something of the infinite. Hence, always mislead and disillusioned, "disgusted" or frustrated, the soul remains "always restless, because the will is constantly driven to search after what it can never find," moved by the hope of finding the true good among these particular objects. Moreover, the more we possess, the more we desire. However great the goods may be that kings and princes enjoy on earth, kings and princes are not less restless and unhappy than others. Those "on top of the wheel of fortune, they are bound to be infinitely more agitated and shaken by its motion than those lower and nearer the center. For when they fall, it is always from a great height."[35]

No doubt that Malebranche's work is of a remarkable coherence, and one can discover the application and illustration of his model of the unions and the restlessness at any level of his explanation. That restlessness affects man in his love, his self-love (or *amour propre*), in his moral behavior or intellectual endeavors. For example, the restlessness always distracts us as the will continuously drives our attention to those objects that have some relation to our bodily being. It turns our understanding toward things that appear useful or pleasurable, that appear as novel, extraordinary, or exceptional. But for that same reason, given the fact that we tend only to know about things that are related to us, that come to us through our senses and imagination, we remain extremely ignorant of abstract things or of what is really important. Here lies the origin of all the errors Malebranche wants to refute and dismantle, the errors of the "pagan philosophers" who trusted their senses and believed in the reality of mysterious entities like "substantial forms,"[36] all the errors generated by our ignorance of the real distinction between the mind and the body,[37]

34. LO, 269.

35. Ibid., 284.

36. Cf. Book VI, chapter 3, about "the most dangerous error of the philosophy of the ancients" (ibid., 446ff.)

37. Cf. Book I, chapters 10ff. (ibid., 48ff.)

the vanities of the counterfeit scholars[38] obsessed by erudition, anatomy, or by "sublunary objects" that give them the illusion to participate in a sublime and noble science, while, in their "perversissima caecitate," as Augustine would say, they remain ignorant of the "Truth" that is really at stake and is useful to them. "And let no one say that this is to recognize the grandeur of Him who has made all these great objects. The smallest fly better manifests the power and wisdom of God to those who consider it with attention, and without being prejudiced by its size, than everything the astronomers know about heavens. Yet man are not made to spend their lives studying flies and insects . . ."[39]

CURIOSITY

Unfortunately, as Malebranche warns us, "it must be carefully noted that . . . the number of fools, or those who allow themselves to be mechanically led by sensible impressions, is infinitely greater than those who have some breadth of mind." Hence our natural tendency for example to judge about the qualities of certain objects in function of the quality of our sensations (for example, our judgment about the purity of matter[40]), or to disseminate its attention over an infinite amount of things, to believe blindly what "scholars" tell us, etc.

However, one of the more fundamental examples of the intellectual perversion is curiosity. In the third chapter of Book IV, Malebranche immediately relates the idea of curiosity to the restlessness of the will. He writes:

> So long as men have an inclination for a good that surpasses their strength and they do not possess, they will always have a secret inclination for anything that seems novel and extraordinary. They will constantly pursue things to which they have given no consideration, in the hope of finding what they search for in them; and since their mind can never be entirely satisfied except through the perception of Him for whom they were made, they will always be in a state of restlessness and agitation until He appears to them in His glory.[41]

38. Ibid., 295ff.

39. Ibid., 296–97. For the "faux savants," cf. Book IV, chapters 7 and 8.

40. See ibid., 83.

41. Ibid., 278.

As such, curiosity is not bad. It is "very consonant" with "our faith": indeed, it is better to search restlessly after the truth and the good, than it is to linger in our illusions. Instead of remaining in false satisfactions about dubious judgments and opinions, it is good to be aroused by what is new in philosophy and science. And although these new discoveries do not contain the true good, our curiosity about them "should be permitted": it stimulates cognitive repudiation of our vague ancient illusions.

But since the cognitive drive of curiosity rests upon the natural inclination toward the infinite, it is nothing but the intellectual specification of it, and it condemns us to a restless theoretical excitation by novelty. This excitation is as such the infallible sign that we have that restless inclination because we "have not seen the evidence for the truth of the things" we desire to know, or better, because we do not possess "the infinite good" that forms the real object of our cognitive drive.

Consequently, Malebranche does not condemn the existence of curiosity as a human vice but interprets its reality as one of the manifestations of the human sinful state (i.e., the distortion of the paradisiacal equilibrium between the two unions). In his two chapters on curiosity,[42] Malebranche goes on to prescribe three rules for controlling our curiosity. The first rule is, "that men must not love novelty in matters of faith." The second is, "that novelty is not a reason that should lead us to believe that things are good and true." The last one is, we should not permit ourselves to be seduced by novelties the truths of which we know will be impossible to grasp, or novelties that are "so slight and insignificant that they cannot satisfy us."

It would lead us too far afield to comment upon these rules in detail: the importance of what Malebranche says about them lies in their presuppositions about the nature of human knowledge. For example, concerning the second and third rules, they both illustrate what he had previously said about our inclinations: particular new goods urge us to examine new things carefully, but it would be erroneous to believe that these things contain what we really hope and wish for.

One of the suppositions is, in the domain of reason, the Cartesian difference between absolute and partial evidence. To the former belong our "clear and distinct ideas," to the latter our sensory perceptions and ideas of our imagination. Curiosity can be helpful, insofar as it urges us to leave the realm of vague opinions and ordinary goods. But since

42. See ibid., 278ff. (Book IV, chapters 3 and 4). For the three rules, cf. ibid., 279.

its liveliness rests upon the lure of something special and uncommon, the soul is tempted to connect the idea of novelty and truth. Everything new bears the mark of something true. And the other way round: some interesting ideas become boring and are soon to be forgotten.

Hence, it is necessary to resist the intellectual desires exciting a vain interest for earthly goods and to remain critical towards what some philosophical experts present as "difficult questions," "paradoxes," impressive terms, subtle distinctions, etc. All of these things agitate us and make us run from one insight to another: "we dissipate ourselves," and in abandoning ourselves to vain passions, we "lose God and everything else with Him."[43]

In what he says about curiosity, the opposition between the "two unions" once more comes undisguisedly to light. This opposition doubtlessly forms the background of what he finally remarks in the first rule. Our union with God not only permit access to the ideas of reason ("vision in God"), but they finally determine the nature of the kind of truth human desire is chasing after. God is the final goal of our desires and of our different kinds of intellectual zeal. And if the ideas of reason can help us on the way to Him, the light they offer is reliant upon the insights offered by faith. As he suggests in his first rule, "that men must not love novelty in matters of faith," it supposes the idea that the truths in matters of faith do not simply overlap with intellectual truths. The latter are true insofar as they rely on what faith makes clear about God. But it would be displaced and even a sign of "stupidity" to confuse the domain of reason with the domain of faith.

Malebranche's argument refers to a basic difference between truths in philosophy and truths in religion. The problem with curiosity is that it tends to confuse both by applying cognitive and skeptical scrutiny to a domain transcendent and external to it. Indeed, "we must submit to the authority of the Church, because she can never err."[44] For Malebranche, the truth of the Church is incompatible with rational inquiry. Any critical attitude is in this context misplaced or inappropriate and makes us behave like "heretics and philosophers." Worse, he says, it is a "crime," an "intolerable conceit," to search for the truth in matters of faith following our reason.

43. Ibid., 284.

44. Ibid., 280.

Or alternatively: it is better to regard philosophical truths with the kind of devout respect and piety reserved only to truths of religion. In this context, Malebranche unmistakably refers to the "pagan" thinkers of antiquity or the Scholastic tradition, and reproaches their refusal of new sciences and philosophies. These "faux savants" imagine a necessary connection between their own opinions about reality and the soul and the truths of the faith: hence their confusion. They venerate, for example, Aristotle as a saint, and every time some truth is discovered, it must be shown that the master already saw it. This is nothing difficult, since "in certain of his works, one can make him say anything one wishes because, although he makes a great deal of noise, he says nearly nothing, just as children interpret the sound of bells in any way that pleases them because bells make a lot of noise and say nothing."[45] Moreover, if the first kind of "stupidity" was the expression of misplaced arrogance, this kind of blind reverence for matters of reason "is also frivolous, and a contemptible meanness of spirit."[46] It is the attitude of those who believe whatever is told to them blindly and indiscriminately. They project onto natural questions the attitude proper to questions of faith.

CONCLUSION

Within the frame of his religious-philosophical approach to human reality,[47] Malebranche offers an exhaustive methodology conceived to combat all the errors that are generated by our condition as "incarnated" mind. As he said in his introduction, original sin impedes and disappoints our tendencies to break off the union with our body. But philosophy can at least help us on our way towards the "perfection" of our soul, in favor of the absolute truth about our own union with God.

45. Ibid., 281.

46. Ibid., 280. "Nonetheless, most of those we call learned in this world can be said to have acquired this reputation only because they know from memory the opinions of Aristotle, Plato, Epicurus, and some other philosophers, blindly agree with their views, and obstinately defend them. In order to have some degrees and external marks of learning in the universities, it is sufficient to know the views of a few philosophers. Provided one is willing to swear *in verba magistri*, with a bit of memory he soon becomes a doctor" (ibid., 280–81).

47. And as Malebranche claims: "The most beautiful, the most pleasant, and the most necessary of all our knowledge is, undoubtedly, the knowledge of ourselves. Of all the human sciences, the science of man is the most worthy" (ibid., xxv).

Which truth? That this relation is "natural, necessary and absolutely indispensable." Hence, what this method tends to achieve is made very explicit: in order to present the truth about our divine union, "it would be necessary to overthrow the fundamental principles of pagan philosophy, to explicate the disorders of sin, to combat what is falsely called experience, and to argue against the prejudices and illusions of the senses."[48] And this is precisely what he does in the *Search after Truth,* offering "a thousand unquestionable proofs and experiences" to convince us of the obligation to resist the spell conjured up by our bodily existence. He even anticipates Henri Bergson, the twentieth-century French philosopher, when he demonstrates how a great deal of our errors derive from an alteration of a purely functional or practical meaning of our bodily existence into a theoretical one. What was conceived of as useful for the preservation of our everyday lives becomes in itself the source of cognitive representation and convictions about ourselves, the world of material objects, and God. Summarized in his own words: "In this work I combat several errors and especially those most universally received or those that cause a greater disorder of the mind, and I show that these errors are almost all consequences of the mind's union with the body. In several places I try to make the mind realize its servitude and dependence relative to all sensible things so that it might be awakened from its somnolence and make an effort to free itself."[49]

And one must confess that he achieves this with great philosophical insightfulness and sharpness. The "thousand unquestionable proofs and experiences" he deals with in the context of his analysis of the influence of our confusion of the mind and the body in the formation of "false sciences," or his detailed analysis of the origin of optical illusions, of the significance of images in our life, etc. remain of an inestimable, priceless value: a still fresh and inexhaustible source of inspiration. However, as we suggested previously, his ultimate end remains, like a good Augustinian, a religious one. One may distort somewhat the inner finality of his concerns in isolating his purely philosophical insights from his religious purposes.[50] No surprise then if at the end of his introduction, he advises

48. Ibid., xx.

49. Ibid., xxv.

50. One of the great merits of D. Moreau's study on the debate between Malebranche and Arnauld is precisely his concern to reintegrate the objective of their mutual dispute into a religious preoccupation. See *Deux cartésiens. La polémique entre Antoine Arnauld*

those unable to resist the "prejudices of our youth or the continuous impressions made on our imagination by the body" to have "recourse to prayer."[51] Although one can also try to justify the religious finality of his philosophical undertaking in reference to what G. K. Chesterton once said, "But a man is not really convinced of a philosophic theory when he finds that something proves it. He is convinced when he finds that everything proves it."[52]

et Nicolas Malebranche (Paris: Vrin, 1999).

51. LO, xxix.

52. G. K. Chesterton, *Orthodoxy* (New York: Doubleday, 1986) 83.

9

Wonder in the Age of the *Saeculum*

Spinoza

Gábor Boros

WONDER IS TO BE investigated here within the framework of my analyses concerning the transformation of some particular emotions, called *religious* emotions in the early modern period. Habitually, the term "secularization" has come to be used when characterizing the philosophical-cultural processes that took place in the early modern period. In this paper I can only refer to other parts of these analyses in order to clarify what I would like to summarize now by saying I am using this term in the sense the German twentieth-century philosopher Hans Blumenberg established.[1] Earlier, I analyzed the concept pair hope and fear, as well as love, felicity, and generosity. Now I am turning to *wonder*, which is the word that translators of Descartes (1596–1650) and Spinoza (1632–1677) made use of when rendering the Latin-French expression *admiratio-admiration*. The investigation will not extend to the whole of the seventeenth century—I will confine myself to the passage from Descartes to Spinoza—nor will I refer to medieval interpretations of wonder—as I did in earlier papers in connection with other emotions.

It is not a matter of course either that wonder is to be investigated among the passions (emotions), or that it is to be set in the context of religious emotions for that matter. Even Descartes had to overcome considerable difficulties before he was able to assure his reader of some

1. See, for example, my paper "The 'Secularization' of Religious Emotions in Spinoza: Hope and Fear, Love, and Generosity," in *Mededelingen vanwege het Spinozahuis* 89 (Damon, 2007) and "The 'Secularisation' of Religious Emotions: Hope and Fear, Love, and Felicity in Spinoza, Descartes, and Hobbes," in *Religious Emotions*, edited by Wim Lemmens and Walter Van Herck (Cambridge: Cambridge Scholars, 2008) 87–104.

sort of legitimacy for his placing wonder among the passions. His basic problem was his own general definition of passion that includes the reference to bodily changes as efficient causes of passion,[2] whereas he did owe an explanation either of why there are no bodily changes connected to the "passion" of wonder or else why are they so tiny that we are inclined to believe they do not exist at all. In the end he resolves the problem maintaining that there are bodily changes but their proper place is the brain, where bodily processes cannot virtually be observed.[3] Although this solution is given only afterwards, when Descartes has already started enumerating and describing the particular passions, he obviously presupposes this when declaring wonder as "the first of all the passions."[4] And what is even more he does this on the basis of precisely that property from which it follows that no observable bodily change accompanies it. For it is *not the well-being of the body or the body-mind composite* which wonder aims at—as in the case of the passions in general—but *the knowledge of the thing* we wonder at. As the bodily organ of obtaining knowledge, the brain is and remains effectively concealed from us.

As is well known, Descartes enumerates six primary passions—wonder, love, hate, desire, joy, sadness—and it is within this group of passions that wonder is given first place. However this may be, its state as a passion continues to be uncertain, and in my view this fact explains that its primacy is not so much that of the *first passion* within the system of passions but rather that of something *preceding* the passions, taking *priority* over them. The reason for its precedence is that if the novelty of an object does not capture our attention, we'll simply contemplate it without any passions; anachronistically speaking, wonder is the general condition of possibility for the passions. This explains the fact that what Descartes accentuates is the fundamentally positive role wonder plays instead of calling our attention to its negative aspects, although the negative aspects are clearly present in his texts: if we wonder too much at an object, this can compromise the use of reason. There cannot be but one antidote against this negativity: knowing everything that can be known.[5]

2. See Descartes, *The Passions of the Soul* (*Les passions de l'âme*, 1649) art. 27.

3. Ibid., art. 70–71.

4. Ibid., art. 53.

5. Even if Descartes himself does not refer explicitly to this layer of meaning, given the general trends of the age, it is obvious that we have to look here for the foundations

From among the positive aspects of wonder there is one that excels the others by far: within the series of the passions deduced from or connected—at least partly—to wonder we arrive after some small steps at what Rodis-Lewis famously called "the late fruit" of Descartes' philosophy, i.e., the passion—and, for that matter, virtue—of generosity (*générosité*). Already in the second article Descartes makes use of distinctions within or in connection to the concept of wonder. There, we find an element that leads our gaze in that direction: veneration (*vénération*) and scorn (*dédain*) are distinguished from esteem (*estime*) and contempt (*mépris*) by the introduction of the point of view of their respective objects that are or are not—more precisely, that we regard or do not regard—"a free cause capable of doing good and evil."[6] Clearly, this point of view guides us already forward to the concept of generosity. We can recognize this immediately as soon as we relate these same passions to ourselves, i.e., we reflect upon the possibility of veneration or contempt of *our own main merit*. For one of the principal parts of Cartesian wisdom (*sagesse*) is being—or at least becoming—aware of what can render just our self-esteem or—so could we translate this tenet in the terminology of wonder—what we are entitled to wonder at in ourselves the most. We cannot really expect from Descartes other sorts of answers than what he in fact gives us to this question: the object of our just self-esteem can only be what renders us capable of doing good and evil: "the exercise of our free will and the control we have over our volitions,"[7] recognizing and wondering at this capability in us is what renders us generous.

---

of the religious meaning of wonder too. For wonder as the religious phenomenon of the miracle is almost trivially connected to wondering as a mental property: we take for wonder what we wonder at, because the thing in question does not happen in a law-like manner, i.e., according to the laws of nature. In this sense the argument of Descartes that wavers between the positive and negative aspects of wonder can well be projected behind the context of the cognition in the strict sense of the term to the context of the religious cognition and religious emotions as well: if you have less-than enough inclination to wonder, you will not be sensible to the specific religious phenomena; if you are inclined to wonder more-than-enough, you'll possibly be superstitious and be ignorant of the laws of nature governing natural events. When Descartes calls our attention to the importance of cognition as a remedy against this extreme he suggests a general critical attitude concerning the objects evoking wonder: "here is no remedy for excessive wonder except to acquire the knowledge of many things and to practice examining all those which may seem most unusual and strange" (ibid., art. 76).

6. Ibid., art. 55.

7. Ibid., art. 152.

Before coming to Spinoza it is worthwhile to briefly draw attention to the astonishing fact that wonder and freedom of the will had already been connected in the thought of the youngest Descartes at the very beginning of his career. What is especially astonishing in this connection—true, only one singular fragment witnesses this—is that the role of mediator is played by the authentic religious concept of wonder: "The Lord has made three marvels: something out of nothing; free will; and God in Man."[8] I call this an authentic religious concept because this is not colored, so it seems to me, with any critique from the point of view of a skeptical theory of knowledge. In no way can we rightly interpret this thought as if Descartes regarded the mentioned works of God as wonders because they are not observing any laws, and therefore we can only see rarely creation out of nothing, freedom of the will, or even God in man as pure effects of processes regulated by natural laws. I would favor an interpretation according to which this fragment stands on the one side of the threshold of the period ["Epochenschwelle," in Blumenberg's terminology] within the modern history of the concept-pair wonder (i.e., miracle) and wonder (i.e., astonishment), whereas evidently, on the other hand, we have the religious skepticism of Hobbes and Spinoza based on a highly critical theory of knowledge.

• • •

As for the theory of affects in Spinoza it can well be supposed that the basic definitions and descriptions of Descartes's *Passions of the Soul* were carved in Spinoza's mind, and that Spinoza's text presenting the theory was written with at least one eye on Descartes' text—perhaps even more than the other parts of the *Ethics*. Therefore, to understand the lines of thought of the theory of affects in the *Ethics*, we need to discover the way Spinoza accentuated particular theorems differently from Descartes, and what his reasons were for that differing. One of the first important differences of accent can already be seen at the beginning of the first treatise dedicated to the affects, i.e., the third part of the *Ethics*. Spinoza's first affect is not wonder, which provides the condition of possibility for the *Cartesian* theory of passion *in the sense of the theory of knowledge*, but desire, the conscious appetite, which provides the condition of possibility of the theory of affects *anchoring it in the ontology* as far as it takes

8. *Tria mirabilia fecit Dominus; res ex nihilo, liberum arbitrium et hominem Deum.* The fragment has been preserved in the collection called *Cogitationes privatae*; see AT X, 218.

over from it the concept of man's essence, i.e., his striving to persevere (*conatus*) in its being. This consideration yields the conclusion that it is not a sort of cognition that provides the point of departure for the Spinozean theory of knowledge, but a highly particular sort of being. At the same time we are given only three basic affects—desire, joy, sadness—instead of Descartes's six primary passions. But why should wonder be left out from the system of "basic emotions," the passion that Descartes may have regarded as the foundation, the condition of possibility for the other passions, however doubtful it might have been if wonder is to be recognized as a real passion?

When looking for Spinoza's reasons for leaving out wonder from his group of primary passions we have evidently to start with the view shared by many that Spinoza in a way rendered Descartes's respective theories more consistent than they were in the works of his predecessor. As for the theory of affect this claim can be accepted without much ado. The proposed geometric order does not really allow for six primary passions, whose inner relations among themselves are far from being elaborated and presented systematically. The ideal of a geometric order invites its author to sketch a theory like that of chapter 6 in Hobbes's *Leviathan*: what we are given here is *stricto sensu* just one basis for the theory, which is called *appetite*, and which expresses and embodies something like a general action-readiness, that its subject is motivated. This basic appetite, in its turn, is capable to function in two opposite directions, to- and from- something. This furnishes us with the essential formula of a theory built on three basic affects, and it is from within this theory that Spinoza begins thinking, even if it remains true that the essential difference of his fundamental metaphysical idea from that of Hobbes hindered him from taking over all that was to be found there.

Concerning the pair of affects, love and hate, as Cartesian primary passions it must have been obvious for Spinoza that they can be deduced from joy and sorrow. Consequently, only wonder remained there to be eliminated. Now we can suspect that Spinoza was not really convinced of Descartes's arguments in favor of wonder's being a passion, and in this way wonder finally fell out by itself as it were. This thesis gains support from the series of the definitions of affect, which close the third part of the *Ethics*, where wonder surprisingly comes to the fore: it is right after the definitions of the three basic affects that wonder comes up together with the definition of disdain deduced from—or better: connected to—

wonder. The definition itself does not help us much in understanding why wonder is given this distinguished place: "Wonder is an imagination of a thing in which the Mind remains fixed because this singular imagination has no connection with the others." But the explication goes a step further toward the clarification of the issue: Spinoza denies the character of an affect to wonder: "So the imagination of a new thing, considered in itself, is of the same nature as the other [imaginations], and for this reason I do not number Wonder among the affects." This is precisely what we mean when saying Spinoza rendered Descartes's theory more plausible. At the end of the explication Spinoza himself refers to Descartes to the effect that the reader can think the reason for the high ranking of wonder is nothing but the fact that, after Descartes, this is the usual way of enumerating the affects. "I recognize only three primitive, or primary, affects: Joy, Sadness, and Desire. I have spoken of Wonder only because it has become customary for some to indicate the affects derived from these three by other names when they are related to objects we wonder at."[9]

I myself think this explication is basically right. At the same time, I can also see another reason why Spinoza treated wonder as well as other affects connected to it in such an ambivalent manner. This reason leads us to deeper layers of Spinoza's system, his concept of religion, will, and generosity. On the one hand, he denied that wonder and its connected affects had the character of a basic affect—in fact he denied they were affects at all; on the other hand, he appreciated the role they play in human life. For in the remark to proposition 52 we see wonder and the whole subsystem of the affects deducible from it, and—as already mentioned—in the summary of the theory of affect, wonder, even before love, obtains the first place after the primary affects. Therefore we have good reasons to think that Spinoza was uncertain concerning how to interpret wonder: in one layer of the text he attributes an important role to wonder, veneration, and generosity, whereas in another there emerge some considerations that urge him to cut off the thread leading from wonder and veneration to generosity—because generosity is not based on the freedom of the will regarded as illusory—and to reassess wonder. Strangely, the main motivation behind this will be that wonder receives

9. Spinoza, *Ethics*, Book 3, AD4 Expl., in *The Complete Works of Spinoza*, vol. 1, ed. E. M. Curley (Princeton: Princeton University Press, 1985).

a religious dimension for Spinoza—just like Hobbes—and a miracle is what is wondered at.[10]

It was proposition 52 in the third part of *Ethics* that gave occasion for Spinoza to make a remark on wonder and its subsystem. The proposition reads: "If we have previously seen an object together with others, or we imagine it has nothing but what is common to many things, we shall not consider it (*contemplabimur*) so long as one which we imagine to have something singular."

One of the important elements of the devaluation of wonder—which is, of course to be understood only in relation to the importance it had in Descartes—can immediately be seen: the dominant mode of cognition in the text of the proposition to which the *remark* with the definition of wonder is attached is the *imagination.* This is the sort of cognition to be blamed for the inadequate ideas—however it may be true, on the other hand, that as human beings, i.e., finite modes, this must be our first access to the world, and this must even remain the case, imagination must remain our continuous point of reference, even if we possess already some adequate ideas as well. In any case, whenever we wonder, we grasp the world in an inadequate manner. The framework for introducing wonder—be it an affect or not—in Spinoza's referential universe is the question whether we can see something particular in the considered object. But the trouble with wonder is that whether or not we can see something particular in the considered object, the foundation for the whole "consideration" (*contemplatio*) is inadequate cognition: for every object of our imaginative consideration remains true that we always only *think* we see, hear, etc. the object; our idea may be true but it may also be false.

What is common, what is not proper to something, i.e., what is *not* like the aspect of the thing considered in wonder, are the universals forged by the imagination,[11] and not the "common notions" that provide the basis for the reason.[12] With a bit—but really only a bit—of exaggeration we can also say that in the default setting of Spinozist wonder we

10. See my "Schmitt, Hobbes, Spinoza: Die Frage nach dem Wunder," in *Ethik, Recht und Politik bei Spinoza. Vorträge des 6. Internationalen Kongresses der Spinoza Gesellschaft*, ed. M. Senn and M. Walther (Zürich: Schulthess, 2001) 225–38. See also the contribution by Jianhong Chen above.

11. See Spinoza, *Ethics*, Book 2, P40 Expl. 1.

12. See Spinoza, *Ethics*, Book 2, P40 Expl. 2.

see as proper to the thing in question, which is in reality *not*, even if we cannot so easily answer the question concerning what in general could be really proper to a thing, totally different from everything else. We see this aspect as proper to the thing only because this does not harmonize with our expectation concerning this sort of thing, an expectation based on our habitual beliefs on that sort of thing originating in our life experience from our childhood on. Typically, we will wonder when we see a human whose hair, skin, etc. is differently colored than those humans we are usually in contact with. The soldier will wonder when seeing the vestiges of horses in the sand; he must realize there were peasants who went there with their plough horses, instead of soldiers with their battle horses. However, wonder will arise in another way too, and in this sense we will not be more entitled to maintain the soldier was right than to prefer the peasant. Although we can imagine that this circumstance will bother us so much that we launch a scientific project on it, Spinoza thinks this is far from being a typically human reaction. Wonder is the typical affect of the man of resignation—in the *Tractatus Theologico-Politicus* (1670) Spinoza adds superstition to resignation—and not that of someone always on the quest; it is rather to be avoided than recommended.

That is, even if the remark to proposition 52 does resume something of the Cartesian tree of wonder as primary passion, and even if the ramification of the different affects is characteristically Spinozean,[13] this

13. Characteristically Spinozean, since even in this scholium has artfully been smuggled the distinction, which seems to be the most important characteristics of the Spinozean theory of affect. Following Moreau ("Spinoza et les problèmes des passions," in *Les passions à l'âge classique*, edited by P.-F. Moreau [Paris: PUF, 2006] 147–51), we can speak about "objective" and "imitative" affects, where an affect is "objective" if it is evoked in us by the emotion's normal object, while the "imitative" affect is evoked by the fact that we see others in the grasp of an affect: this fact is enough to make an affect be aroused in us: which affect will be aroused depends on whether or not we have emotionally been preheated, and if yes, in which way. In the Scholium to proposition 52, Spinoza projects these two concepts of affect onto each other, namely when treating disdain (*contemptus*), which is introduced polemically, like wonder itself, and he links it to a similarly branchy tree of connected affects as to wonder itself, i.e., this would be a good argument in favor of treating wonder as a basic affect. "To Wonder is opposed to Disdain, the cause of which, however, is generally this: because we see that someone wonders at, loves or fears something, or something appears at first glance like things we admire, love, fear, etc. (by P15, P15C, and P27), we are determined to wonder at, love, fear, etc., the same thing; but if, from the thing's presence, or from considering it more accurately, we are forced to deny it whatever can be the cause of Wonder, Love, Fear, etc., then the Mind remains determined by the thing's presence to think more of the things that are not in the object than of those that are (though the object's presence usually

fact must not let us forget that the whole story deploys within the realm of the passions *in the Spinozean sense*: the affects linked to wonder—which may be regarded as an affect or only an imagination, as we are told in the 4th definition of affect—are based on inadequate ideas. At the same time the affect-tree of wonder is so branchy that we can even ask ourselves if Spinoza ought to pick out wonder and place it among the primary affects, even if his reason for that move would be different from Descartes's reason. For it is no argument against wonder that it cannot be built on inadequate ideas. Considered from a purely systematic point of view the inadequate ideas deserve as much to be treated by the philosopher of the value-neutral way of looking at the world as the adequate, and the passions as much as the actions: sorrow as primary affect can also exclusively and essentially originate from inadequate ideas, i.e., this affect remains always and essentially a passion. Moreover, if we are planning to describe philosophically how our everyday life is dominated by passions the branchy tree of wonder can prove to be at least as useful as that of sorrow: according to Spinoza, wonder permeates everyday life so thoroughly that we do not even possess enough words to name all the derivative sorts of it: "In this way we can also conceive Hate, Hope, Confidence, and other Affects to be joined to Wonder, and so we can deduce more Affects than those which are usually indicated by the accepted words. So it is clear that the names of the affects are found more from the ordinary usage [of words] than from an accurate knowledge [of the affects]."

It is worth mentioning that Spinoza differs fundamentally from Descartes when in the middle of the scholium to proposition 52, at the beginning of his polemical introduction he refers to those who regard disdain (*contemptus*) as the opposite of wonder, since Descartes himself thinks wonder has no opposite at all. The Cartesian passion parallel to Spinozean Disdain is *mépris*, which is the opposite of *estime* but they are both subsorts of wonder. In Spinoza, however, Disdain has an affect tree similar to and equally extended than that of wonder: "Finally, we can conceive Love, Hope, Love of Esteem, and other Affects joined to Disdain, and from that we can deduce in addition other Affects, which we also do not usually distinguish from the others by any single term." This can even be regarded as an argument in favor of the view that

---

determines [the Mind] to think chiefly of what is in the object)." (*Spinoza Opera*, edited by Carl Gebhardt, vol. 2, 180–81.)

Spinoza would better render wonder and disdain primary affects just like joy and sorrow.

In Descartes, as we have seen, there is a high and short road to generosity, i.e., such consideration of ourselves, in which we see *correctly*, which we can rightly appreciate in ourselves: this is our free will and its staying under the control of reason. Now it is fascinating to recognize that there is a line in Spinoza as well that leads from the scholium on wonder up to propositions 58–59 where the affects called actions are introduced, one of which is precisely generosity.

Proposition 53 that follows the definition of wonder in proposition 52 speaks embarrassingly about the *distinct* functioning of *imagination*. Notwithstanding this, its main point remains the all-important human gesture of drawing our attention to ourselves. We consider what even the adequate ways of cognition can acknowledge as something particular, proper to each of us, i.e., our own power of acting. "When the Mind considers itself and its power of acting, it rejoices, and does so the more, the more distinctly it imagines itself and its power of acting." This is a claim problematic enough but does not belong to our topic now, so we can skip it. According to proposition 54, "[t]he Mind strives to imagine only those things that posit its power of acting," whereas proposition 55 formulates the inverse statement: "When the Mind imagines its own lack of power, it is saddened by it." When reading these statements we cannot resist the temptation to mean that the power of acting in Spinoza takes over in some way the function Descartes attributed to the free will, namely, to be the basis of our just self-esteem. This feeling seems to be reinforced in the first corollary of proposition 55 where being blamed by others—the opposite of esteem—is linked to sorrow, the opposite of joy: "This Sadness is more and more encouraged if we imagine ourselves to be blamed by others." The joy stemming from the opposite cause will be called in the scholium to the corollary, "Self-love or Acquiescence in himself [*Philautia vel Acquiescentia in se ipso*; Curley's rendering is "self-esteem," which takes us too swiftly to my goal]." This affect is not very far from what the first aspect of Cartesian generosity is: just self-esteem. The second corollary can be understood as referring to the inverse of the second aspect of this generosity: "No one envies another's virtue unless he is an equal." Envy, Descartes thinks, can in no way be worthy of the generous, who esteems everyone else for the same reason as he esteems himself: for the free will and its being controlled by reason. In a similar manner, Spinoza reckons envy finally to sorrow, which means envy is "an affection by which a man's power of acting, or striving, is restrained." And so it becomes clear that Spinoza's treatment of wonder also prepares for the introduction of the concept of generosity, at least indirectly. There is, however, a big difference: instead of referring to the Cartesian free will, Spinoza employs the concept of the power of action, which becomes a quasi-normative concept in Spinoza's hands.

The treatment of those affects that are not based on the inadequate ideas of imagination but are built on the adequate ideas coming from reason or intellect will be introduced in proposition 58 of part 3, and their main branches are given in the scholium to proposition 59. This classification reflects the one comprising all the affects laid down in the main body of part 3 in the sense that in both cases we have one original affect containing all the other—desire and strength of character, respectively—and these basic affects will be ramified into two—joy and sadness here, tenacity and generosity (Curley's rendering is "nobility") there. Tenacity and generosity are both sorts of joy, for, as we saw, sorrow cannot but be based on inadequate ideas; therefore, this affect cannot but be a passion. This classification reflects also the tree of wonder, insofar as Spinoza stresses in all two—or three—cases that what he can provide the reader with are only starting-points of a theory of affect much broader than his own, since there are lots of derivative affects arrived at by way of combination of the primary ones and some derivatives.

In the following I will highlight some important feature of the Spinozean concept of generosity, because, as I mentioned, it will be instructive to see the difference between the Cartesian, clearly stated dependence of generosity on wonder and the Spinozean hidden connection between the same affects. Perhaps only those who look for such a connection will find the approach of his *Ethics* based on Descartes's *Passions of the Soul*:

> By Generosity I understand the Desire by which each one strives, solely from the dictate of reason, to aid other men and join them to him in friendship . . . Those actions, therefore, which aim . . . at another's advantage, I relate to Generosity. So . . . Courtesy, Mercy, etc., are species of Generosity.[14]

14. Spinoza, *Ethics*, Book 3, P59, Schol.

First of all, we have to stress that the basis for joining other human beings to us is what reason tells us, *dictamina rationis*. This means that the basis has nothing to do with the imaginative way of cognition, and therefore neither to wonder. What we wonder at is what cannot be connected to the imaginative universals, i.e., what affected our bodies in an order different from the order of affection in other men's bodies. However, wonder may or may not be connected to these universals; it would not be capable to build up harmonious communities among people, because the contents of the particular universals are different in the respective particular people. In contrast to this, the function of generosity is precisely to take into account the level of affectivity of the community that is to be found among people knowing things adequately, on the basis of the common concepts of reason.

Let us elucidate the differences with an example. Suppose there lives someone somewhere whose vision is extremely sharp, and she can assess the distance between the Moon and the Earth without any computation, purely on an empirical basis, just like normal people assess distances found in everyday life. Her idea of the distance will be inadequate, although it will be true on supposition, because it originates from experience (imagination in Spinoza's sense); therefore, the inner certainty of an adequate idea is missing in it. Her idea would be adequate only if she deduced the Moon-Earth distance from the laws of the movements of the celestial bodies as from common concepts of reason. On the other hand, her fellow human beings with normal visual abilities will wonder at our superhuman being, since she maintains something contrary to all everyday observation that yield in them in the same sort of inadequate ideas—although false, on supposition—as in the supervisionary being. This is not difficult to imagine, and indeed necessary that the normal and the supervisionary people will have an argument with each other about the right distance, since they will be unable to reconcile the different experiential-imaginative bases that has in the same way stiffened in them long before as unalterable habits. The only way to reconciliation, or the formation of a common conviction, would be if all of them became independent from their experiential-imaginative bases taking as their point of departure the law-like common concepts of reason for their computations leading inevitably to adequate and, to be sure, true ideas. One could argue in a similar manner also in the practical philosophy, as in the case of mercy, whose one form, alms-giving—being built on inadequate ideas—Spinoza refuses—just as in the later Kant—but whose other form, built on adequate ideas and therefore capable to create unity, community, Spinoza enumerates as Generosity among the active affects.

Summing up my argument I maintain that Spinoza devoted much attention to the Cartesian concept of wonder, both in the sense of being the primary passion mentioned in the first place in Descartes, and in the sense of becoming the capstone of the whole building of the system of passion. In a final analysis, however, his manner of thinking philosophically did not allow for him to adopt either of these solutions: wonder as the foundation gives a gnoseological flavor to the whole theory whereas Spinoza took as his foundation the originally and inalienably individual, proper way of *being*. Moreover, wonder is always based on inadequate ideas. Consequently the affects stemming from it will always and necessarily belong to the realm of passions, and what is even worse, they will provide the basis for the political-theological attitudes of the superstitiously religious people, which makes them the supporters of monarchic political order and false religion. From all of this it follows that none of the active affects—one of which being generosity—can be connected to wonder. This is one consideration against wonder's having a positive role in Spinoza's philosophy, but there is another one as well. Whereas Descartes thinks the basis of our own legitimate self-esteem as well as that of others is the free will controlled by reason, in Spinoza this basis will be the—always individual—power of action realized in actions—in the strict sense of the term within the Spinozean theory of affects—that follow from adequate ideas. The freedom of the will that Spinoza thinks is illusory will be replaced by the cognition and recognition of the necessary order of things. The new, the extraordinary loss of all their charm, the disenchanting of the "secularized" period, becomes perfected, and wonder is born anew.

PART THREE

Wonder in Eighteenth-Century Europe

10

Berkeley's Wonderful Divine Language

Apology and Biblical Realism

Péter Losonczi

"... we have first raised the dust,
and then complained we cannot see."

INTRODUCTION

In this essay I intend to demonstrate that the passion of wonder has a special, multifaceted function in the architectonics of the Berkelian philosophy. Although Berkeley does not provide a detailed analysis in which this function would be clarified, the significance of this question is evinced by central arguments of the Irish bishop-philosopher where the mention of this passion, as well as the corresponding thesis concerning the wonderful character of the works of nature turn out to be decisive constituents. This latter question, however, touches upon the very essence of the Berkelian philosophy as an attempt to develop an intellectual position which can meet the conditions of Christian apologetics. If we read Berkeley from this perspective and take his works, in line with David Berman's interpretation, as the last theologically fruitful synthesis, it becomes evident that this positioning of the passion of wonder is a considerable gesture. However, what renders the problem really interesting is that this metaphysico-theological system is evolved in the form of the well-known *divine language-argument*, according to which the created reality is a visual language through which the Creator communicates with His noble creatures, human beings. As I will show, it is the passion of wonder as a "hermeneutical tool" that is the means by which the addressees are capable of an authentic understanding of the divine message.

Due to this central role of the architectonic function of wonder, it is possible to argue that the *Alchipron: or, The Minute Philosopher* (1732), a text whose passages will gain equal importance with the earlier works of Berkeley in my study, carries important additional information about the Berkelian program. This late work is, as its subtitle reveals, one of primarily apologetic importance rather than purely philosophical-theoretical importance. However, in disagreement with the opinion shared by important interpreters of Berkeley, I hold the *Alchipron* to be an elemental part of the Berkelian *oeuvre*. Its apologetic character does not so much as weaken its place as strengthen it. This is not about Berkeley, but rather about his modern—or, strictly speaking, *modernist*—interpreters that this view is not generally accepted. The philosophical preparation of this apologetics in the earlier works, and the application of these philosophical arguments in the course of the development of the train of thought of the *Alchipron* itself is in harmony with the general apologetic strategy of early modern religious thinkers. To paraphrase Jean-Yves Lacoste, this apologetic scheme proceeded from a *general apology for religion* towards a *general Christian apologetics*, which laid down the bases of *the defense of the truth of a specific Christian confession.*[1] This scheme can be detected even in the thought of Schleiermacher whose *Speeches on Religion* is a general apology for religion while this work itself already contains important comments that point towards Christianity, whose priority and relative fulfillment in Evangelicalism is explained in his *The Christian Faith*. If one consults the table of contents of the *Alciphron*, she will easily discover this pattern in the Berkelian construction. As it will be shown, the passion of wonder is operating in the explanation of a properly theistic position, the relevance of which can be better understood if we take into consideration the fact that Berkeley's semantics of wonder goes back to its medieval conception. That is why we need to make a further preparatory note, which concerns questions of atheism and authentic religiosity.

The label of "atheist" or "atheism" was applied not merely to persons or groups who explicitly denied the existence of God. It was rather a name for those whose conception about God or beliefs concerning other

1. Jean-Yves Lacoste, "Athéisme," in *Dictionnaire critique de théologie* (Paris: P.U.F., 2002) 96–102, 99.

religiously significant matters were taken as improper and erroneous in the light of the related convictions of somebody who actually applied this designation. As Buckley writes, "[i]n those stirring times, 'atheism' was a disparaging epithet [that is to say atheists] were what your enemies were—albeit faithful believers by their own profession."[2] This means that it is not enough to accept a religious position in general. The real rationale of an apologetics is to show how the principles of *the* proper and true form of religion can be defended on rational grounds. Against this background, it is necessary to interpret as fundamental all those arguments by the means of which Berkeley intends to lead the reader to the embrace of an authentic religious stance, that is, *theism*. Nevertheless, leading this apologetically saturated use of the passion of wonder back to the earlier writings where its occurrence seems partial at first glance, the general importance of this passion becomes evident. This is why we can rightfully speak of the centrality of the passion of wonder in the Berkelian philosophy if its aforementioned "hermeneutical" function becomes clarified.

THE *ETHOS* OF BERKELIAN THOUGHT: BIBLICAL REALISM

In the closing sentences of the *Principles of Human Knowledge* (1710), Berkeley declares that "what deserves the first place in our studies is the consideration of God, and our *duty*," and continues by saying that "having shown the falseness or vanity of those barren speculations, which make the chief employment of learned men," the reader of the book would "be more disposed to reverence and to embrace the salutary truths of the Gospel, which to know and to practise is the highest perfection of human nature."[3] It may seem striking to a late modern or postmodern eye that these sentences can really be understood as quintessential to Berkelian thought. An essay concerning the principles of human knowledge—that is, a book about epistemology and metaphysics—is concluded by declarations of a religious and moral nature. It is also maintained that the essence of his *philosophical* program is to show the *falseness* or *vanity* of those speculative theories that are held by the *learned men* and to propose a way that leads us to the fulfillment of the highest perfection of

2. Michael J. Buckley, *Denying and Disclosing God: The Ambiguous Progress of Modern Atheism* (New Haven: Yale University Press, 2004) 30.

3. George Berkeley, *Principles of Human Knowledge* (New York: Penguin, 2004) 113.

human nature, the *reverent* embrace of the salutary truths of the Gospel. In fact, the gnoseological problems are intimately linked together with a religious-moral *ethos*. The speculative theories of the learned men are proclaimed not merely as being false—that is, scientifically *improper*—but they are labeled also as being in vain—namely, pointing to a major defection in the religious and moral sense. It is not a value-free matter what sort of principles are leading human beings in the attainment of knowledge, since the ways of knowledge influence not merely the theoretical-scientific horizons of his/her activities but also have essential effects on the general perception of the world as a whole and, consequently, on the *ethos* of the knowing subject. Originally, the Greek term, *ethos*, dwelling, designates a "sense of *place* or habitation" and in this manner "signifies an environment that makes possible and sustains moral living, establishing the direction and parameters of human conduct."[4] The environment or place, which Berkeley wanted to reveal to his readers, is a world that exists *in the presence of God*. Besides the earlier quoted statements, in the concluding passage of the *Principles* the philosopher-bishop also asserts that explaining his philosophical standpoint he wanted to "inspire [the] readers with a pious sense of the presence of God." This divine presence is where people are dwelling and it is a defect that they do not recognize it. Berkeley's purpose is to make this divine presence *seen*. In order that this presence be recognized by his readers he needs to teach the genuine way of seeing and, consequently, thinking. Without the genuine mode of seeing, people are unable to think properly, and this results in improper action. We can take it verbatim that for Berkeley the *world as the presence of God* is that the "place" where human beings live and move, and, consequently, should be a moral life which is in harmony with this "Divine environment" or *ethos*. It is also evident from his conclusive remark that it was the opinion—strictly speaking, the *barren speculations*—of the *learned men* that made him fundamentally dissatisfied as far as the parameters of their moral life was concerned. These learned men were the different representatives of the then-contemporary philosophical and scientific community, first and foremost those who explained the workings of nature in virtue of mechanistic principles and materialistic philosophy. As he asserts in the *Principles*, those who pretend to "explain things in terms of corporeal causes—a claim that seems to have too

4. William P. Brown, *The Ethos of the Cosmos: The Genesis of Moral Imagination in the Bible* (Grand Rapids: Eerdmans, 1999) 11.

much estranged the minds of men from God, that active principle, that supreme and wise spirit, 'in whom we live, move, and have our being.'"[5]

If we understand rightly the Berkelian ambition for revealing the world as divine presence, we can understand why the thinker renders the just mentioned biblical passage to be a quasi-philosophical principle. His ambition was to provide an alternative for these divergent speculative philosophical systems relying on which, as the motto of this paper says, we first raise the dust, and then complain we cannot see.[6] In place of the speculations that can lead only to vagueness, Berkeley's purpose was to show how human beings can acquire *discernment*. This meant for him the admission to a specific scientific and philosophical stance *together with* a religious view of seeing. In fact, Berkeley represents *common sense* thought and biblical religiosity not merely as complementary viewpoints, but identifies them. This is a very peculiar stance, which I call *biblical realism*. Biblical realism is the expression of the Berkelian ethos, and, consequently, the leading principle of the bishop's philosophical endeavors. The complete development of this biblical realism is the precondition of Berkelian apologetics—namely, the disclosure of the essential identity between a philosophy of discernment and the biblical religion. This position should be interpreted against the background of that general scientific and philosophical situation which Berkeley intends to challenge. The materialistic and mechanistic world view of his immediate predecessors and his contemporaries rendered possible the conception of the world that was more and more independent from God. The deists were the representatives of a "religious"—that is, not expressly atheistic—standpoint, which nevertheless could divorce the world and God in a radical manner without any hesitation. This perspective was extremely hazardous in Berkeley's eyes precisely because it meant evident jeopardy for the *ethos* of a world as the presence of God. Any version of the improper conception about God, or, as the other side of the same coin, about the world could lead to the devastation of this *ethos*.

It is interesting to quote a passage of the *Principles* in order to picture this element as the driving force of Berkelian biblical realism. At a point he uses the passage from Isaiah, "thou art a God that hides thyself," and emphasizes that this biblical declaration of divine hiddenness should not be understood in an improper way. He emphasizes that

5. Berkeley, *Principles*, 77–78 (§66).

6. Ibid., Introduction (§3).

"though God conceal himself from the eyes of the sensual and lazy, who will not be at the least expense of thought; yet to an unbiased and attentive mind, nothing can be more plainly legible, than *the intimate presence* of an all-wise spirit, who fashions, regulates, and sustains the whole system of beings."[7] In an era when the thesis concerning divine concealment is distorted and implies an increasing distance between God and the world, Berkeley finds it necessary to make it evident that divine presence is an even more fundamental principle. I think that it is not only his religious commitment in general that should be born in mind if we want to understand his writings correctly, as Stephen R. L. Clark correctly assumes,[8] but this is the case also with biblical realism, the expression of the Berkelian *ethos*.

BIBLICAL REALISM AND THE *LIBER MUNDI* TRADITION: THE DIVINE-LANGUAGE ARGUMENT

As Costica Bradatan persuasively demonstrates in his *The Other Bishop Berkeley*,[9] the philosophical program of the Irish thinker can evidently be related to the medieval *Liber Mundi* tradition. As Bradatan explains, "*Liber Mundi* is one of the most complex and fascinating cultural-philosophical topics of the medieval universe."[10] This scheme of understanding, or *episteme* in Foucaldian terms, implied "a consideration of the whole visible world in symbolic terms: namely, as a coherent system of signs, as a sophisticatedly encrypted message that God is continuously sending to his creatures."[11] The human being was destined to orientate itself taking two books into account: the Bible and the Book of Nature. Bradatan detects certain antique prefigurements of the metaphor but emphasizes that the metaphor of the book becomes relevant in the Middle Ages. Naturally, the significance of the metaphor is fundamentally attributable to the status of the Book. In the Scripture itself Bradatan reveals two passages where the elements of the *liber mundi* metaphor can be

7. Ibid., 111 (§151, emphasis added).

8. Stephen R. L. Clark, "Berkeley on Religion," in *The Cambridge Companion to Berkeley*, ed. Kenneth P. Winkler (Cambridge: Cambridge University Press, 2005) 369–404.

9. Costica Bradatan, *The Other Bishop Berkeley: An Exercise in Reenchantment* (New York: Fordham University Press, 2006).

10. Ibid., 57.

11. Ibid.

identified. The one is the well-known *logos* doctrine from John 1:1–4. Here the creating divine Word is the source of intelligibility, so declaring the act of creation, as it were, to be a "linguistic/textual gesture" having a similar status to that of revelation. The other biblical passage mentioned by Bradatan is from the First Letter to the Corinthians (13:12–13): "what we see now is like a dim image in a mirror; then we shall see face to face. What I know now is only partial; then it will be complete – as complete as God's knowledge of me. Meanwhile these three remains: faith, hope, and love; and the greatest of these is love." According to Bradatan, both passages can be interpreted according to a "text perspective."[12] This interpretational horizon is, in my view, even more radical if considered according to the following two factors.

First, we should consider what William P. Brown makes evident in the passage quoted earlier. It is, in fact, inappropriate to understand the biblical image of the world in terms of a *cosmos* since the Scripture speaks not merely about the calling into being of a harmony in the natural order but shows the divine act of creation as being a law-giving gesture which makes matters of harmony and obedience equal elements of the question. In fact, the two biblical texts referred to by Bradatan make this reading unavoidable. In the case John, the divine Word is the manifestation not merely the intelligible order of the created world but also refers to the person of Christ as the fulfillment—and the *re*-creation—of the reign of the law of God. Similarly, the Pauline text speaks not merely about the mysterious character of the created world as such but also about the enigma of human life as a moral and religious endeavor in which the best "epistemic tools" are faith, hope, and love (charity), all of which appear in medieval moral theology as the theological or infused virtues. Thus, when we try to interpret Berkeley's *ethos* as biblical realism we are not merely in harmony with the moral stance of the biblical vision of creation but this is also an issue that is significant regarding the *liber mundi* tradition.

Another important question is the exact definition of the status of the two Books in the medieval development of the theme of *liber mundi*. As the detailed investigations of Jos Decorte shows,[13] the Bible has prior-

12. Ibid., 58.

13. Jos Decorte, *Raak me niet aan: Over middeleeuws en postmiddeleeuws transcendentiedenken* (Kapellen: Klement, 2001); a translation of this text into English will be forthcoming from Pickwick Publications.

ity over the Book of Nature—and, consequently, over the ways of thinking related to "text"—due to multiple reasons. Before the rise of biblical criticism in the seventeenth century, the Book had the status of an *unambiguous* source of divine message. As the Word of God as *Verbum* and *Veritas*, the Scripture was taken as the "the true divine language in an absolute sense."[14] As God is incapable of erring or lying, the Bible is a language that is impossible to use in an improper way. Moreover, the Bible as a revealed text is the *direct* word in virtue of which God addresses the human being. Hence its evidence follows, which makes it fundamentally unnecessary to "decode" the text of the Bible. At the same time, the Book of Nature carries certain ambiguities. A theologian or an exegete's job is to show that there is no contradiction between the two books in spite of the apparent inconsistencies. This work can be carried out by showing the proper way of seeing and the real order of the text of nature in comparison with the revealed text. Many thinkers of the medieval period hence take the function of guidance, *manuductio*, as the core of their intellectual activity.[15] Bradatan, of course, is aware of the importance of the Bible but he seems to neglect the absolute priority of scripture.

It has been important to make these two corrective comments on Bradatan's reading of the Berkelian philosophy as a case of the *liber mundi* tradition because the stance I call biblical realism is not merely a repetition of the traditional conception but, in my understanding, a *radicalized version* of it. This radicalism is implied by the earlier quoted dilemma regarding divine hiddenness. The different philosophical theories and scientific principles of the seventeenth and eighteenth centuries—like the denial of the applicability of teleology, the mechanistic principles as the sole tools of natural philosophy, and the accompanying materialistic position—equally weakened the applicability of a relational scheme within which the two books could be exhaustively read together. Additionally, the emerging influence of biblical criticism and also the general rejection of historical authority as a tool of scientific work reversed the priorities concerning the two books. The role of the two books underwent a radical change; instead of convergence a fundamental divergence characterized their relation. The *translation* between the two

14. Ibid., 75.

15. See Gergely T. Bakos, *On Faith, Rationality, and the Other in the Late Middle Ages*, forthcoming from Pickwick Publications.

"texts" proved to be more and more complicated. In fact, as for the questions of the natural world the senselessness of such a translation became more and more dubious. Berkeley's program aimed at countering these tendencies that equally destroyed the possibility of understanding the world as a divine *milieu*. Contemporary theories seriously jeopardized the intellectual mediation of a biblical ethos from a traditional Christian point of view. In my interpretation, a quasi-assimilation of the two books is carried out as a consequence of the radicalism of the Berkelian program. The theoretical instruments of the quasi-assimilation are the *esse est percipi* thesis and the divine-language argument, the central elements of Berkelian metaphysics and natural philosophy.

ESSE EST PERCIPI: IDEAS AND SPIRITS

As we quoted earlier, according to Berkeley, the materialistic philosophy estranged the minds of men from God, while his purpose was to reveal the intimate presence of God. But it is not just the materialistic position which endangered theism. The general method of the corpuscularians applied in the understanding of the natural realm was also insufficient according to him. Ever since Galileo expounded that the Book of Nature is written in exact mathematical characters, philosophers of nature were engaged with the explication of these universal and abstract rules of nature. Berkeley himself was unconvinced about the appropriateness of this method. He writes in the *Principles* that this scientific attitude itself misses the point of any serious philosophical endeavor. It is worth quoting this passage in length not only because it explains the essence of the Berkelian account of science, but also because it presents one of the clearest Berkelian references to the *liber mundi* tradition.

> As in reading other books, a wise man will choose to fix his thoughts on the sense and apply it to use, rather than lay them out in grammatical remarks on the language; so in perusing the volume of nature, it seems beneath the dignity of the mind to affect an exactness in reducing each particular phenomenon to general rules, or showing how it follows from them. We should propose to ourselves nobler views, such us to recreate and exalt the mind, with a prospect of the beauty, order, extent, and variety of natural things: hence, by proper inferences, to enlarge our notions of the grandeur, wisdom, and beneficence of the Creator: and lastly, to make the several parts of creation, so far as in us lies,

> subservient to the ends they were designed for, God's glory, and the sustentation and comfort of ourselves and fellow-creatures.[16]

Instead of the overstatement of abstraction and the construction of rigid rule-theories, Berkeley proposes a nonfoundationalist method. He does not deny that by "a diligent observation of the phenomena within our view, we may discover the general laws of nature, and from them deduce the other phenomena."[17] But he adds: "I do not say *demonstrate*; for all deductions of that kind depend on a supposition that the Author of Nature always operates uniformly, and in constant observance of those rule we take for principles: which we cannot evidently know."[18] Later we will come back to these problems. At the moment we need to turn to another important aspect of Berkeley's philosophy, which is, at the same time, in close connection to questions in the scientific methodology. The Berkelian stress upon human ignorance regarding the activity of God has, of course, a general context in Christian theology. However, in the Berkelian philosophy this thesis attains a specific import, in consequence to the above-mentioned idea about the *intimate* divine presence in the world. This intimate presence is disclosed by the famous divine-language argument. This theory can be understood against the general background of Berkeley's immaterialistic metaphysics. According to Berkeley, the entity—namely, *material substance*—on which the main principles of the then-contemporary mechanistic and materialistic theories were based is a fictional being. This theory is stemming from an epistemological-metaphysical argument whose main elements are as follows.

Berkeley proceeds from a general epistemological principle according to which "the objects of human knowledge . . . are either ideas actually imprinted on the senses, or else such as perceived by attending to the passions and operations of the mind, or lastly ideas formed by help of memory and imagination."[19] Besides the ideas there is another sort of entity that knows or perceives the ideas and exercises different operations, and this active being is what Berkeley calls "*mind, spirit, soul* or *my*

16. Berkeley, *Principles*, 93 (§109).
17. Ibid. (§107).
18. Ibid.
19. Ibid., 53 (§1).

self."[20] The ideas exist *in the mind that actually perceives them*. This statement introduces the first fundamental Berkelian philosophical principle. In Berkeley's opinion, the term *exist*, when it is applied to sensible things, means nothing else than the experiential operation is present for the perceiver. "The table I write on, I say, exists, that is I see and feel it; and if I were out of my study I should say it existed, meaning thereby that if I was in my study I might perceive it, or that some other spirit actually does perceive it."[21] The existence of a nonmental entity is equivalent to its being perceived by a spirit or mind—*esse* is *percipi*—and Berkeley finds it *perfectly unintelligible* to speak about the mind-independent, absolute existence of things.[22] It is a misconception based on the doctrine of abstract ideas that, according to him, generated the mistaken distinction between the existence of things being perceived and their existence in an unperceived state. The different perceptual qualities, like the colors, heat, cold, extension and so on, that is the "things we see and feel [are but] so many sensations, notions, ideas or impressions on the sense," which are impossible to separate from perception. This distinction can be made according to Berkeley as easily as a thing *can be divided from itself.* [23]

From all this Berkeley draws the conclusion that leads him forward to the metaphysical basis of his philosophy. He insists that "all the choir of heaven and furniture of the earth, in a word all those bodies which compose the mighty frame of the world, have not subsistence of the mind," and since *esse est percipi* "so long as they are not actually perceived by me, or do not exist in my mind or that of any created spirit, they must either have no existence at all, or else subsist in the mind of some eternal spirit."[24] The only substance in the Berkelian world is "*spirit*, or that which perceives" and since—as the parallel thesis of the *esse est percipi* declares—"to have an idea is all one as to perceive" it is declared by him that there can be no "unthinking substance or *substratum* of these ideas."[25] The possible counterargument which would present the general corpuscularian stance would be that the perceptual qualities like sound are stemming from a unthinking substance in which they are

20. Ibid. (§2)

21. Ibid., 54 (§3).

22. Ibid.

23. Ibid., 54 (§5).

24. Ibid., 55 (§6).

25. Ibid. Cf. ibid., 80 (§73).

inherent—that is, the material substance—and the qualities which we are acquainted with are *copies* or *resemblances* of these and, consequently, the ideas are effected by the material substance via a causal relation. Berkeley's answer for this argument is that "an idea can be like nothing but an idea: a color or figure can be like a color or a figure."[26] A thing of a certain nature, Berkeley says, cannot mediate the property of a thing belonging to another class of entities. For admitting the corpuscularian option one needs to explain how an idea, a being essentially belonging to mental existents in its nature could *resemble* a material substance that is nonmental?[27]

It would surpass the limits of this chapter to analyze the details of this epistemological-metaphysical position and to point to its possible deficiencies. What is important for us is what Urmson points out, namely, that Berkeley's epistemology makes "a very plausible effort to show that his view gives common sense everything it needs."[28] As I mentioned earlier, Berkeley develops a strategy in virtue of which common sense turns out to be a view that can easily be harmonized with Christian apologetics. The epistemological-metaphysical grounding of his standpoint and its direct passage to the proof for the existence of God is in harmony with this strategy. The thesis concerning the nature of ideas as mind- or spirit-dependent entities plays a crucial role in this case for the existence of the divine Spirit. As he explains, he experiences that he can freely produce ideas, as well as change or modify them, from which it follows that the mind can be defined as an *active principle*.[29] However, in connection with the actually perceived sensual ideas he describes these differently, since they have no dependence on his will. It is not optional whether he sees a particular sensual object that presents itself to him, and this is the case with all the senses: "the ideas imprinted on them are not creatures of [his] will." So, he concludes, "[t]here is . . . some other will or spirit that produces them."[30] This proof is a special version of the cosmologi-

26. Ibid., 55 (§8).

27. One may wonder whether Berkeley here is reliant on the Aristotelian-Scholastic principle "*omne agens agit simi simile*." The representative scheme of early modern epistemology goes back to this thesis, and it seems that Berkeley corrects the mistaken application of it.

28. J. O. Urmson, *Berkeley* (Oxford: Oxford University Press, 1982) 45.

29. Berkeley, *Principles*, 63 (§28).

30. Ibid., 63 (§29).

cal argument. The world consists of the ideas perceived by God, but it is also important to emphasize that unlike Malebranche's intellectualist scheme, Berkeley's is a voluntaristic one. In Urmson's terms, "[w]hat makes the natural world and its order is the will of God that men should obtain these ideas, not God's own perception of them from all eternity."[31] But it is important to emphasize that this immediacy of God through the world does not yet exhaust the sense of that *intimate* presence which Berkeley wants to involve. The full meaning of it can be understood if we examine how Berkeley defends the idea of divine providence. From this it becomes evident that intimacy here means not merely a spatial closeness but rather a real *personal relation*. Berkelian apologetics cannot be understood without this element. However, as we shall see, the passion of wonder plays its role in connection with this issue.

THE TRUTH OF THEISM: WONDER, PROVIDENCE, AND ADORATION

Immediately following the passage where the existence of God is demonstrated, Berkeley develops the basic principles of his natural philosophy. As he writes, "[t]he ideas of sense are more strong, lively, and distinct than those of the imagination; they have likewise a steadiness, order, and coherence, and are not excited at random, as those which are the effects of human wills often are, but in a regular train or series, the admirable connexion whereof sufficiently testifies the wisdom and benevolence of its Author."[32] The rules or established methods in virtue of which the Creator excites the ideas of sense in us we can call the *laws of nature*.[33] By observation we can attain sufficient knowledge of these rules and in this way we are to a great extent capable of foreseeing the course of events which "enables us to regulate our actions for the benefit of life."[34] As we have seen, it is not the abstract knowledge of necessary connections that makes us capable of being conversant with the world and all the things significant for our well-being. Nevertheless, this design of the world in which the ideas of the senses are, in fact, *real things*,[35] the consistent uni-

31. Urmson, 63.
32. Berkeley, *Principles*, 63 (§30, emphasis added).
33. Ibid.
34. Ibid. (§31).
35. Ibid., 63–64 (§32–33).

form working of nature "evidently displays the goodness and wisdom of that Governing Spirit whose Will constitutes the laws of nature" do not necessarily lead our thoughts to the Author Himself. Rather, we wonder about secondary causes and try to establish causal relationships between the bodies that constitute the natural realm. However, any speculation about the material agents in nature is the best harbor for atheists.[36] In Berkeley's view, "the being of a spirit infinitely wise, good, and powerful is abundantly sufficient to explain all the appearances of nature."[37] In fact, the connection between ideas is not explainable as the relation of cause and effect but only as that of the marker or sign with the thing signified.[38] This *universal system of signs* implies the fulfillment of the Berkelian theology and apologetics.

The signs—that is, the ideas—in virtue of which God is communicating with human beings are the expressions of divine will. As we have seen, unlike the ideas attributable to capricious human volitions, these ideas appear and are linked together consistently. This well-ordered semiotic system makes up the *visual language* of God. As Berkeley writes in *A New Essay toward a New Theory of Vision*, "the proper objects of vision constitute an universal language of the author of nature."[39] The workings of nature are not phenomena attributable to the bare material substance as materialistic philosophers put it, but expressions of a mutually functioning divine language addressed to men. The Berkelian *manuductio* resides in the disclosure of the *intimate, personal divine presence*: the world as a divine milieu. Berkelian biblical realism is the expression of the *ethos* of such a world, which takes a radical form of the *liber mundi* tradition in so far as the sequence of the ideas imprinted on the senses is an immediate visual divine revelation. This makes the book of nature very similar to that of the Book in so far as their revelatory function is concerned. In consequence of a proper orientation as for the right way of seeing—or "reading"—the divine message attainable through this visual language makes God's providence evident. The course of things are such that it becomes clear that they "were designed for . . . the sustentation and comfort of ourselves and fellow-creatures."[40] For this reason, Berkeley

36. Ibid., 65 (§35).

37. Ibid., 79 (§72).

38. Ibid., 77 (§65).

39. Berkeley, *Essay on Vision* (New York: Bobbs-Merrill, 1963) 147.

40. Berkeley, *Principles*, 93 (§109).

holds it absolutely vain to seek abstract rules in the workings of nature in a foundationalist manner. In a similar vein to the above quote, he explains that this endeavor is as unintelligent as reading books to speculate about grammatical matters of language. It is a possible but ridiculous thing to do in the case of *liber mundi* when there is a much nobler alternative: "to recreate and exalt the mind with a prospect of the beauty, order, extent, and variety of natural things: hence, by proper inferences, to enlarge our notions of the grandeur, wisdom, and beneficence of the Creator."[41]

As we have seen, in different arguments for the existence of a *provident* God, Berkeley speaks of the *wonderful* or *admirable* qualities of the natural world. We mentioned that the regular train of series of the ideas that constitute the world, "the admirable connexion . . . sufficiently testifies the wisdom and benevolence of its Author."[42] In the *Principles* it is also expounded that there is a "curious organization of plants, and the animal mechanism in the parts of the animals" and also that "the clockwork of nature . . . is so wonderfully fine and subtle as scarce to be discerned by the best microscope"—and it is added that in spite of minor difficulties as for the details of the problem all this sufficiently reveals the administration of the Providence.[43] In the concluding arguments of the treatise, Berkeley emphasizes that "[l]ittle and unreflecting souls may indeed burlesque the works of Providence, the beauty and order whereof they have not capacity, or will not be at the pains, to comprehend." Nevertheless, he says, "those who are masters of any justness and extent of thought, and are withal used to reflect, can never sufficiently admire the divine traces of Wisdom and Goodness that shine throughout the Economy of Nature."[44] Wonder and admiration are used as synonyms in these texts. But by means of reading the arguments developed in the Fourth Dialogue of the *Alchipron*, we can determine both the function and the relation of wonder and admiration.

In this text the conception of the visual divine language is the central instrument of a specifically *theistic* argument. In fact, the title of this part is *The Truth of Theism*. Berkeley is fundamentally reliant on the visual language-argument. At a point, Euphranor, a figure who represents the simple sincerity of Berkeley's Christian identity alongside

41. Ibid.

42. Ibid., 63 (§30).

43. Ibid., 74–75 (§60).

44. Ibid., 112 (§154).

Crito's knowledgeable and sarcastic character,[45] explains that "it seems the proper objects of sight are light and colours, with their several shades and degrees; all which, being infinitely diversified and combined, form a language wonderfully adapted to suggest and exhibit to us the distances, figures, situations, dimensions, and various qualities of tangible objects." It is important to add that this is carried out "not by similitude, nor yet by inference of necessary connexion, but by arbitrary imposition of Providence, just as a word suggests the things signified by them."[46] We are practicing this language from the first moment we enter the world.[47] As Crito's sophisticated comments explain, "this optic language hath a necessary connexion with knowledge, wisdom, and goodness [as well as it is] equivalent to constant creation, betokening an immediate act of power and providence." It is also declared that it "cannot be accounted for by mechanical principles, by atoms, attractions, or effluvia."[48] It is also Crito who presents the decisive argument regarding the theological and apologetic import of the visual language-argument. As he explicates,

> Some philosophers, being convinced of the wisdom and power of the Creator, from the make and contrivance of organised bodies and orderly system with all its parts and contents well adjusted and put in motion, as an artist leaves a clock, to go thenceforward of itself for a certain period. But this Visual Language proves, not a Creator merely, but a provident Governor, actually and intimately present, and attentive to all our interests and motions, who watches over our conduct, and takes care of our minutest actions and designs throughout the whole course of our lives, informing, admonishing, and directing incessantly, in a most evident and sensible manner. This is truly wonderful.[49]

It is evident from this passage that the idea of divine intimacy is meant to express the provident attention of God. The *wonderful* workings of the natural world are witnessing that God is constantly addressing human beings with a *message whose sense, not its grammar*—as emphasized in Berkeley's *Principles*—is the signal for the presence of a benevolent

45. See T. E. Jessop, *The Works of George Berkeley* (London: Thomas Nelson, 1951) vol. 3, 15.

46. Ibid., 154 (*Alciphron* IV.10.10).

47. Ibid., 155.

48. Ibid., 159.

49. Ibid., 160.

governor. The demonstration of the existence of a *provident* God is for Berkeley *the* proper form of religiosity as is apt for revealing the *ethos* driving his biblical realism. The argument goes on with Euphranor's comment according to which men basically are encompassed by such a wonder without reflecting on it. Although this language addressed to our eyes mediates something of *divine* or *admirable* that "may well awaken the mind and deserve its outmost attention"[50] for certain reasons it often remains unattended. It is also explained why the admirable and divine nature of this visual language of divine providence is neglected by men. Following Alchipron's note according to which "these strange things affect men but little," Euphranor gives the following answer:

> But they are not strange, they are familiar; and that makes them be overlooked. Things which rarely happen strike; whereas frequency lessens the admiration of things, though in themselves ever so admirable. Hence, a common man, who is not used to think and make reflexions, would probably be more convinced of the being of a God by one single sentence heard once in his life from the sky than by all the experience he has had of this Visual Language, contrived with such exquisite skill, so constantly addressed to his eyes, and so plainly declaring the nearness, wisdom, and providence of Him with whom we have to do.[51]

In his editorial note Jessop emphasizes that it was unusual, in Berkeley's days, to "[find] Nature herself a miracle in the twofold sense of being wonderful and being the direct product and instrument of God."[52] However, this extreme stance of what I have labeled biblical realism seems to suggest that *the world is a permanent miracle* performed by the provident God "in whom we live, move, and have our being." The text of the *Alciphron* in fact expresses dissatisfaction with those common men who are not "used to think and make reflexions, would probably be more convinced of the being of God by one single sentence heard once in his life from the sky than by all the experience he has had of this Visual Language," which also implies aversions to the role of miracles. In the *Principles* Berkeley admits that on some occasions it may prove to be necessary "that the Author of Nature display His overruling power in producing some appearance out of the

50. Ibid.

51. Ibid., 161.

52. Ibid.

ordinary series of things. Such exceptions from the general rules of nature are proper to surprise and awe men into an acknowledgement of the Divine Being: but then they are to be used but seldom, otherwise there is a plain reason why they should fail of that effect."[53] The criteria for the effectiveness of miracles as events diverting from the normal course of things is the same as in the case of the natural proceedings of divine activity: to be surprising they must be *rare*. However, according to Berkeley, "God seems to choose the convincing our reason of His attributes by the works of nature . . . rather than to astonish us into a belief of His being by anomalous and surprising events."[54] Miracles, in harmony with the traditional definition, are described as events that diverge from the natural course of events. It appears that Berkeley's purpose was not to deny the validity of miracles or even to assimilate the natural movements of nature with miraculous happenings. His prior aim was to show that the natural world itself is the *immediate* expression of a divine plan that is also explained in the Bible.

The Berkelian distinction between the grammar and the sense of the divine language and its apologetic significance renders it important to make comments on Michel Foucault's interpretation of Berkeley's theory of signs and its general connection with the problem of the premodern form of knowledge. In chapter three of his *The Order of Things*, Foucault discusses the logic of signature in the Classical age in comparison with what he describes as the premodern *episteme*. In this regard he mentions Berkeley as one of the representatives of a general trend in the course of which the premodern form of knowledge he labels as the prose of the world turned into the body of knowledge as an articulated system of *mathesis*. According to Foucault, the most important consequence of this shift was that the so far dominant epistemic status of *resemblance* as an infinite series of similitude among the diverse forms of existents was replaced by a new attitude that was reliant upon a specific form of knowledge. As he writes,

> [f]rom the seventeenth century onward, the whole domain of sign is divided between the certain and the probable: that is to say, there can no longer be unknown sign, a mute mark. This is not because men are in possession of all possible signs, but because there can be no sign until there exists a *known* possibility

53. Berkeley, *Principles*, 76 (§63).

54. Ibid.

> of substitution between two *known* elements. The sign does not wait in silence for the coming of a man capable of recognizing: it can be constituted only by an act of knowing.[55]

This attitude, Foucault says, demolished knowledge's "kinship with *divination*,"[56] which is evidently a relevant remark for our investigations. By this he means that instead of presupposing signs anterior to it, and growing in parallel with the opening up of these signs, "it is within knowledge itself that the sign is to perform its signifying function; it is from knowledge that it will borrow its certainty and probability."[57]

The relational character carried by the signs enclosed in the premodern *episteme* ceases to function as an enigmatic factor in the unveiling of the link between the Creator and the creature. In Foucault's reading the God of early modern metaphysics is using knowledge instead of signs: he communicates with human beings through "the relations that are set up between our impressions, in order to establish in our minds a relation of signification."[58] He argues that both in the cases of Malebranche and Berkeley, "the sign arranged by God is the cunning and thoughtful superimposition of two kinds of knowledge" and, consequently, we cannot speak about any involved *divinatio*.[59] The function of a sign is not to refer to something other than itself but becomes "a brief and concentrated kind of knowledge."[60] Referring to Berkeley's discussion of the question of causality,[61] he declares that in the place of the random divined signs that characterized the horizon of premodern *episteme*, in the Irish philosopher the treatment of signs developed in a form that involves its assimilation into the knowledge of what is probable.

In my view Foucault's reading of Berkeley can be challenged by means of the Berkelian distinction between the grammar and the sense of divine language. The working of nature as the expression of a *spiritual* causality that in the form of the visual expresses the intentions and *will* of

55. Michel Foucault, *The Order of Things: An Archeology of the Human Sciences* (New York: Vintage Books, 1973) 59. See also Decorte, *Raak me niet aan*, for a discussion of this text in comparison to medieval ways of knowing.

56. Foucault, *Order*, 59.

57. Ibid.

58. Ibid.

59. Ibid., 59–60.

60. Ibid., 60.

61. See Berkeley, *Principles*, 77 (§65).

a spiritual being, a divine person whose inmost personal reality *an Sich* is otherwise absolutely transcendent and inaccessible. And vice versa: the visual refers to another dimension; that is to say, to the sphere of the personal intentions and volitions of the divine person. This reference at the same time expounds the working of divine providence—the content of the divine volitions. The distinction between the *grammar* and the *sense* of the *same* language is meant to clarify the difference between the visual-natural expression of a spiritual causality and the volitional dimension of the same spiritual agent. The Berkelian message about the matter of the sign and the signified is not exhausted by the theory of divine language as a natural expression of the supernatural. This is why we can speak of *biblical* realism in the case of Berkeley. It is very important to emphasize that the conclusion of Berkeley's treatment of the "natural" is concluded in the reader's relegation to the Book which *really and fully* reveals this providential content: to the "salutary truths of the Gospel." The entire philosophical discussion of the workings of the natural realm as a divine language is meant to lead us beyond the grammar and to unclose the sense to which it carries a relation: the existence of divine providence whose exact content is revealed *in the Gospel*. The Berkelian reference to our duties is not the atomic piece of any kind of naturalistic morality but its function is to orient the human being towards the embrace of Christian morality explained in the *narratives* of the Holy Scripture.

The radicalism of this program lies in the introduction of immaterialism and, by the same token, the pattern of pure spiritual causality. This radicalism can be reasoned by the challenge made by the then-contemporary trends of natural philosophy in which the natural processes were more and more radically diverted from direct divine activity and intentionality. As these trends dismantled the *ethos* of the cosmos, Berkeley needed to choose a similarly radical means to lead readers to the reverent embrace of the truths of the Gospel.

Therefore God shows soberness regarding the application of miracles in his communication with the human being. In fact, as the divine-language argument intends to show, the whole functioning of the natural world itself is a communicative gesture, an expression of the divine will. The apologetic purpose is to make people look properly at what they, in fact, always already see. The apologist's work resides in disturbing the gaze that follows from the permanent presence of the otherwise striking phenomena of the skills and purposefulness of nature: to present the familiar as unfamiliar again and again. Thereby engendering *wonder* and *adoration* in his fellow-beings the apologist can help her peers in seeing, reading, and hearing in an appropriate way. We can say that the semantics and the syntax of the visual language altogether produce the *sense of the book of nature*, whose familiarity covers its wonderful, admirable, and divine character. The repeated gesture of wonder, at the same time, is able to disclose the world as a divine milieu implicating a definite ethos. It is evident that the passion of wonder, as well as adoration, belongs to the most central constituents of Berkeley's theology and apologetics. The significance of this fact is strengthened if we consider the significance of *emotive meaning* in the philosophy of Berkeley. Berkeley tightly interconnects the two books, the Bible and that of Nature, in such a way in which it is really unprecedented in the context within which he is working. In so doing Berkeley hopes to show not merely the truth of theism on a *common sense* basis, but also to show, *contra* Toland, that Christianity is through and through *mysterious*.[62]

62. This research was undertaken in the framework of the research project OTKA-67798 of the Hungarian Scientific Fund.

11

"Of Curiosity, or the Love of Truth"

David Hume on Wonder in A Treatise of Human Nature

Michael Funk Deckard

> There is implanted in us,
> doubtless for a good end,
> an insatiable desire for truth;
> we spare no pains in search of it;
> and when we have found it,
> or at least its appearance,
> we are again as uneasy to communicate
> the fruit of our labours to others.
>
> —Edmund Burke (1745)

In David Hume's *Treatise of Human Nature* (1739–1740), there are three books, "Of the Understanding," "Of the Passions," and "Of Morals." How these books link together is at stake in much of the secondary literature on Hume.[1] In other words, can one have an epistemology, anthropology, or a moral theory autonomous of each other? In order to gain an understanding of the text of the *Treatise* as a whole, one must figure out how these three topics—understanding, passions, and

1. For the discussion of this claim in the literature, see, most recently, *New Essays on David Hume*, eds. Emilio Mazza and Emanuele Ronchetti (Milan: FrancoAngeli, 2007); *The New Hume Debate*, Revised Edition, eds. Rupert Read and Kenneth A. Richman (London: Routledge, 2007); Christopher J. Finlay, *Hume's Social Philosophy: Human Nature and Commercial Sociability in* A Treatise of Human Nature (London: Continuum, 2007); Henry E. Allison, *Custom and Reason in Hume: A Kantian Reading of the First Book of the "Treatise"* (Oxford: Clarendon, 2008).

morals—tie together in Hume's mind as well as in one's own life. None of these can be entirely separated from the others: ideas, feelings, actions. They each influence the others at every moment, sometimes without us even realizing it. In this article, I hope to show how all three 'moments' are necessary for Hume's theory of human nature and how these moments teach one about how thought, feeling, and action are intimately related to each other. Due to the fact that they are not cut off from each other in our daily lives, they should not be cut off from each other in a philosophical reflection on or theory of human nature.

Whereas many commentators have discussed the chapter at the end of *Treatise*, book 1 (1.4.7),[2] in which Hume's melancholy is the result of his skepticism, almost no one has discussed the end of book 2 (2.3.10), 'Of curiosity, or the love of truth.' I wish here to read the *Treatise* in light of this short chapter on curiosity and describe Hume's philosophy of human nature as one driven by wonder. This view disagrees with one made by Lorraine Daston and Katharine Park, who write: "Wonder [i]s conspicuously absent in Hume's analysis of the passions in *A Treatise of Human Nature*."[3] In order to do this, I will first analyze the early parts of the *Treatise* in which Hume mentions synonyms for wonder (i.e. surprise and admiration) alongside Adam Smith's "History of Astronomy."[4]

2. In this article, when I quote from David Hume, *A Treatise of Human Nature*, eds. David Fate Norton & Mary J. Norton (Oxford: Oxford University Press, 2000), page numbers from this edition will be followed in brackets by the book, part, section, and paragraph numbers when available. For a discussion of 1.4.7, see, for example, Annette C. Baier, *A Progress of Sentiments: Reflections on Hume's* Treatise (Cambridge: Harvard University Press, 1991) chapter 1.

3. Lorraine Daston and Katharine Park, *Wonders and the Order of Nature, 1150–1750* (New York: Zone, 1998) 326. Although admittedly the word *wonder* is not used as a section heading in the *Treatise*, it is surprisingly present behind what Hume says, as will be seen in this article and in Hume's later *Dissertation on the Passions*, first published in 1757. Cf. David Hume, *A Dissertation on the Passions; The Natural History of Religion: A Critical Edition*, ed. Tom L. Beauchamp (Oxford: Oxford University Press, 2007), henceforth abbreviated as *Dissertation*. Although wonder and admiration are also discussed in Hume's *Enquiry Concerning Human Understanding*, I save the discussion of that text as well as the topic of Hume on religion for another occasion. For a discussion of Hume's first *Enquiry*, see Eric Schliesser, "Miracles and Science: David Hume and Adam Smith on Wonder, Admiration, and Surprise," unpublished presentation to the APA (December 2006). I am also in disagreement with the claim of Annette C. Baier, *Death and Character: Further Reflections on Hume* (Cambridge: Harvard University Press, 2008) 81–82, who did not think wonder a central or important topic in the *Treatise*.

4. For an excellent discussion of Adam Smith's "History of Astronomy" similar to our discussion here, see Eric Schliesser, "Wonder in the Face of Scientific Revolutions:

With these synonyms in mind, I will, second, show the importance of the passions for Hume's theory as a whole. Third, I will concentrate on the chapter on curiosity in particular, and try to show how it provides the transition from passions to morals.

OF THE UNDERSTANDING

In the first book of the *Treatise*, Hume expresses that all ideas are based on impressions. This is the origin of what may be called Hume's empiricism in which there are originals and copies, the former being the impressions and the latter ideas, "so that all the perceptions of the mind are double, and appear both as impressions and ideas."[5] One can never go beyond experience, says Hume. The basis of every perception then, whether it be of colour, sound, touch, etc., "arises in the soul originally, from unknown causes . . . an impression first strikes upon the senses, and makes us perceive heat or cold, thirst or hunger, pleasure or pain of some kind or other."[6] These original impressions form within the mind, whether in memory or in the imagination, but first and foremost there is no more original origin than sensation/impression itself.[7] However, although Hume does not describe precisely why we have sensations, what he does emphasise from the outset is the force, vivacity, and liveliness of impressions.[8]

Thus, at the very origin of sense perception, the mind and body act with immediate force or vivacity. For example, some perceptions are easier to remember than others. This is due to the fact that they have greater force or vivacity from the outset. For Hume, the bodily impression itself will always have a greater force and vivacity than an idea in

---

Adam Smith on Newton's 'Proof' of Copernicanism," *British Journal of the History of Philosophy* 13 (2005) 697–732.

5. Hume, *Treatise*, 8 (1.1.1.3).

6. Ibid., 11 (1.1.2.1).

7. For a discussion of memory and imagination in Hume's *Treatise*, see, e.g., David Owen, *Hume's Reason* (Oxford: Oxford University Press, 1999) 72–73; Claudia M. Schmidt, *David Hume: Reason in History* (University Park: Pennsylvania State University Press, 2003) 20ff.; see also Herman De Dijn, "Hume's Nonreductionist Philosophical Anthropology," *Review of Metaphysics* 56 (2003) 587–603.

8. Cf. Baier, *Progress of Sentiments*, 29.

our minds.[9] In order for the understanding to act at all, impressions must become ideas.[10]

Just as the tongue contains different papilla, or taste buds, so the mind (i.e., soul) has different ways of measuring metaphorical 'taste' and these can be cultivated. All simple ideas are connected by means of some principle into complex ones—the ways in which association between ideas occurs is by means of resemblance, contiguity in time or place, and cause and effect. "[O]ur imagination runs easily from one idea to any other that *resembles* it."[11] Just as ideas relate to each other they also relate to the impressions that they resemble (and their corresponding pleasure/pain). Among these three principles, resemblance, contiguity, and cause and effect, according to Hume, "there is no relation, which produces a stronger connexion in the fancy, and makes one idea more readily recall another, than the relation of cause and effect betwixt their objects."[12] For Hume, there is a kind of *force* that holds ideas together, analogous to blood relations. For example, a child and her parents are closer in terms of this force than cousins. This force is called by Hume "attraction, which in the mental world will be found to have as extraordinary effects as in the natural, and to show itself in as many and as various forms."[13]

Underlying Hume's view of the understanding are two kinds of relation. One is the connection within the imagination just described; this is the "natural" relation. The second is a philosophical (or "artificial") relation in which two disparate ideas are connected arbitrarily by means of the imagination.[14] Hume's delineations of the headings under which natural or philosophical relations may be considered is simply another

9. Cf. *Treatise*, 7 (1.1.2.1); Dabney Townsend, *Hume's Aesthetic Theory: Taste and Sentiment* (London: Routledge, 2001) 88–89.

10. I am inclined even to agree with Finlay when he claims that a passion must intercede here for an impression to become an idea (*Hume's Social Philosophy*, 69).

11. *Treatise*, 13 (1.1.4.2).

12. Ibid.

13. *Treatise*, 14 (1.1.4.6). See the Nortons' edition of the *Treatise*, where Hume writes in his Abstract, "if any thing can entitle the author to so glorious a name as that of an *inventor*, 'tis the use he makes of the principle of the association of ideas, which enters into most of his philosophy. Our imagination has a great authority over our ideas; and there are no ideas that are different from each other, which it cannot separate, and join, and compose into all the varieties of fiction" (416).

14. For a discussion of natural and artificial relations, see Eric Schliesser, "Two Definitions of 'Cause,' Newton, and the Significance of the Humean Distinction between Natural and Philosophical Relations," *Journal of Scottish Philosophy* 5 (2007) 83–101.

way of abstracting what the imagination already does without our always recognizing it. The first kind of relation occurs whether the individual is a philosopher or not. The second requires a more "philosophical" mind to make abstract connections between objects. In the language of the eighteenth century, one might call the first connection a "common sense" one, and the second a specialized one. The imagination is thus free to "transpose and change its ideas" as well as to separate any differences among ideas.[15]

It is this discussion "of the connection or association of ideas" that lies behind both Hume's analysis in the rest of the *Treatise* and, as will be seen shortly, Adam Smith's "History of Astronomy." Although Hume says it is "the love of truth" that is "the first source of all our enquiries," at the end of book 1, there he is at wit's end. After Hume has found himself in an "*intense* view of these manifold contradictions and imperfections in human reason" and his brain is so heated that he is ready to reject all belief and reasoning, he writes the following:

> Where am I, or what? From what causes do I derive my existence, and to what condition shall I return? Whose favour shall I court, and whose anger must I dread? What beings surround me? and on whom have I any influence, or who have any influence on me? I am confounded with all these questions, and begin to fancy myself in the most deplorable condition imaginable, inviron'd with the deepest darkness, and utterly depriv'd of the use of every member and faculty.[16]

This is the state that many philosophers find themselves in after a great deal of thought. It is a frustrating business to try to prove things like knowledge, being, beauty, goodness, etc. It is not solely by means of the understanding or "abstract ideas" that one comes to truth, morals, etc. Thus, this section of Hume's *Treatise* shows why one needs passions. It is by means of passions that most people accept the common sense belief that external objects exist and that causes are real.[17] When Hume is ready to throw all of his books into the fire and he is full of melancholy, he concludes book 1 of the *Treatise*, "Of the Understanding." This is how far *pure understanding* gets one. Nevertheless, Hume is naturally inclined to keep thinking and cannot give up this business of the understanding.

15. Cf. *Treatise*, 12 (1.1.3.4).

16. Ibid., 175 (1.4.7.8).

17. See ibid., with annotation on 489.

He writes, "I cannot forbear having a curiosity to be acquainted with the principles of moral good and evil, the nature and foundation of government, and the cause of those several passions and inclinations, which actuate and govern me."[18] These are the subjects of books 2 and 3 of the *Treatise.* Even though we leave Hume at the end of book 1 in the "deepest darkness," he cannot keep himself from wondering about morality, politics, and the passions.

It is surprising that the end of book 1 *and* the end of book 2 both describe the 'origin' of philosophy to be in curiosity or the love of truth, and that this curiosity is precisely a passion, a desire for further knowledge. But is this merely a vain and prideful pursuit of self-glorification? On the one hand, ambition leads one to pursue objects that most of the human race desires, for example, money, fame, glory, etc. On the other hand, one can attempt to recognize why one ambitiously pursues these objects and how one acts while in pursuit of these. Do we choose to make money or achieve fame *at the expense of* our family, friends, or neighbors, or do we strive after truth for the sake of truth, without losing sight of our nearest and dearest?[19]

The search for truth is not the same for all people, especially the way philosophers make "abstract" connections between objects in order to overcome this deepest darkness. In using the word abstract here, I am referring specifically to philosophical ways of thinking, for example calling something a substance. Some will attain truth in much more "common" or "natural" ways. Hume says explicitly that he does not wish to make everyone in England philosophers nor does he expect them to follow his philosophy. "They do well to keep themselves in their present

18. See ibid., 176 (1.4.7.12). He writes further in this paragraph, "I am uneasy to think I approve of one object, and disapprove of another; call one thing beautiful, and another deform'd; decide concerning truth and falsehood, reason and folly, without knowing upon what principles I proceed. I am concern'd for the condition of the learned world, which lies under such a deplorable ignorance in all these particulars. I feel an ambition to arise in me of contributing to the instruction of mankind, and of acquiring a name by my inventions and discoveries. These sentiments spring up naturally in my present disposition; and shou'd I endeavour to banish them, by attaching myself to other business or diversion, I *feel* I shou'd be a loser in point of pleasure; and this is the origin of my philosophy."

19. As Terry Eagleton writes, "Society for Hume is composed of ever-expanding concentric circles of what Hutcheson calls 'weak love', and the affective air grows thinner the further we move from our blood relations" (*Trouble with Strangers: A Study of Ethics* [Oxford: Wiley-Blackwell, 2009] 53).

situation; and instead of refining them[selves] into philosophers, I wish we cou'd communicate to our founders of systems, a share of this gross earthy mixture, as an ingredient, which they commonly stand much in need of, and which wou'd serve to temper those fiery particles, of which they are compos'd."[20] By examining human nature, both "naturally" and "philosophically," according to a principle of curiosity that strives for truth, Hume hopes to "compose my temper from that spleen, and invigorate it from that indolence, which sometimes prevails upon me." If readers of philosophy or science find themselves despairing or indolent, allowing melancholy to penetrate their humors, then they would do well to read Hume and learn from him in order to find a "good humour."

In following this advice, whether one strives to be a philosopher or not, it is by reading the *Treatise of Human Nature* that one finds wonder reflected in Hume's thinking. As early in the work as 1.2.1 ("Of the infinite divisibility of our ideas of space and time"), following Hume's reflections on the "origin of our ideas," there are two analyses from which philosophy begins. The first is "whatever has the air of paradox" and is most remote from "vulgar conception" (i.e., the common people's conception of things); the second, however, concerns "anything propos'd to us . . . which causes surprise and admiration, giv[ing] such a satisfaction to the mind, that it indulges itself in those agreeable emotions, and will never be perswaded that its pleasure is entirely without foundation."[21] Without explicitly mentioning wonder here, it is this latter "surprise" or "admiration" that causes pleasure in the soul and sparks the beginning of philosophy. One of these "wondrous" theories of philosophy or science that everyone thinks about at one time is the doctrine of infinite divisibility, especially concerning space and time. How can one even begin thinking about infinity without being under the influence of wonder?

One way of thinking about how wonder initiated philosophical thought, as Adam Smith analyzes in his history of astronomy (published in 1795, but delivered as lectures during the mid-eighteenth century), is to describe how, rhetorically speaking, the structure and content of a text should arouse wonder in the reader. In other words, the text itself should induce surprise, and thus wonder. Adam Smith opens his text on "Astronomy" in the following way:

20. *Treatise*, 177 (1.4.7.14). See also annotation on 481 (second annotation to 1.4.4.3).

21. *Treatise*, 23 (1.2.1.1).

> Wonder, Surprise, and Admiration, are words which, though often confounded, denote, in our language, sentiments that are indeed allied, but that are in some respects different also, and distinct from one another. What is new and singular, excites that sentiment which, in strict propriety, is called Wonder; what is unexpected, Surprise; and what is great or beautiful, Admiration.
>
> We wonder at all extraordinary and uncommon objects, at all the rarer phaenomena of nature, at meteors, comets, eclipses, at singular plants and animals, and at every thing, in short, with which we have before been either little or not at all acquainted; and we still wonder, though forewarned of what we are to see.[22]

A friend of David Hume's, Adam Smith expresses precisely what is absent in the *Treatise*, a categorization of the difference between wonder, surprise, and admiration (but, strangely enough, not curiosity). Whereas towards the beginning of Hume's *Treatise*, he uses the words "surprize and admiration," preparing the way for a discussion of infinite divisibility, Smith begins his history of astronomy with "Wonder, Surprise, and Admiration" before mentioning his topic at hand, namely, "meteors, comets, eclipses, singular plants and animals," etc. However, what Smith and Hume agree on is that there must be some original desire to even want to study philosophy or science. This natural desire is particular to every human being. Some thing, like "space and time," "infinite divisibility," "meteors," or "singular plants and animals" must strike one's imagination and cause wonder, a passion that drives the soul to want to know more about that subject. What is the nature of comets or space and time? How might I understand these further? Once the reader asks these questions, then the reader naturally becomes a philosopher.

At the beginning of his text on "Astronomy," Smith corrects authors such as Milton and Dryden for using the wrong word for what they are describing. Whereas Milton, in his *Paradise Lost*, used the word "admir'd" and Dryden "surprise," they mean the same thing: wonder. Adam Smith describes this further:

> All that I contend for is, that the sentiments excited by what is new, by what is unexpected, and by what is great and beautiful, are really different, however the words made use of to express them may sometimes be confounded. Even the admiration which

22. Adam Smith, *Essays on Philosophical Subjects*, ed. W. P. D. Wightman and J. C. Bryce, vol. III of the Glasgow Edition of the Works and Correspondence of Adam Smith (Indianapolis: Liberty Fund, 1982) §1, 33.

> is excited by beauty, is quite different (as will appear more fully hereafter) from that which is inspired by greatness, though we have but one word to denote them.[23]

In Smith's text introducing astronomy, he describes Milton and Dryden as doing the same thing as Hume, mentioning surprise or admiration when they mean wonder, without which all human beings would be indolent. In Smith's description, surprise is the first "effect of unexpectedness" and in further describing surprise, he mentions Hume's "Of the Connexion or Association of Ideas" (*Treatise*, 1.1.4). Most of the beginning of Smith's lecture here seems to have Hume in mind.

It is key to our understanding of Hume and Smith's contextual ideas of the nature of wonder, surprise, and admiration that all three of these "strike the imagination." Whereas Smith tries to distinguish between these three effects (or sentiments, which, Smith says, are "allied"), they are fundamentally similar in the manner of striking the imagination. In Hume's *Treatise* and in the "Abstract" to that work, he states that the rightness or wrongness of morals, eloquence, and beauty are matters of "taste and sentiment."[24] This means that how one *feels* tells one what is right or wrong, beautiful or ugly, pleasing to hear or displeasing, more than how one *thinks*. Yet, and this cannot be emphasized enough, feelings do not exist without thoughts, nor thoughts without feelings. Furthermore, as we will see, morals and politics "consider men as united in society, and dependent on each other."[25] That is to say, feelings and thoughts need an additional element: other people.

While book 1 of Hume's *Treatise* ends with deepest darkness, or an essentially *skeptical* view of reason, one may nevertheless strive for truth by means of feeling and passion.[26] Since the imagination performs "acts of the understanding" by connecting ideas and impressions (whether naturally or philosophically), one must read book 1 together

23. Ibid., 34.

24. See Hume, *Treatise*, 4, 350, 407.

25. Ibid., 4. For an excellent interpretation of how the three books of Hume's *Treatise* hold together as a 'social philosophy,' see Finlay, *Hume's Social Philosophy*. Whereas Finlay mentions "curiosity, or love of truth" several times in this book, he never states how this curiosity works in relation to the social philosophy (see 66, 72, 74, 86, 109, 154, 171, 173 for the importance of curiosity in Finlay's discussion of Hume).

26. For the interpretation of an essentially Humean view of belief in the contemporary South African author J. M. Coetzee, see Michael Funk Deckard and Ralph Palm, "Irony and Belief in *Elizabeth Costello*," in *J. M. Coetzee and Ethics*, eds. Anton Leist and Peter Singer (New York: Columbia University Press, 2010).

with book 2, "Of the Passions" in order to make the move from impressions to ideas. More fundamental even than differentiating ideas is the underlying natural act of finding resemblances, contiguities, and causes within human nature. One means of finding "surprise and admiration" in the understanding is to try to think the realm of infinite divisibility of space and time as a way of understanding natural philosophy. This endeavor of infinite divisibility may either arouse surprise (i.e., wonder) or despair (i.e., deepest darkness). The line between these two passions is a fine one, and Hume thinks one will be left in despair by relying on the understanding alone. However, the most wonder-inspiring part of the understanding is this desire to even understand causation itself, and yet understanding causation as an epistemological enterprise alone is extremely limited. Moral philosophy is needed to comprehend causation better. As a *transition* then from natural philosophy to moral philosophy, there are further surprises found within the second book of the *Treatise*, "Of the Passions."

OF THE PASSIONS

One can read book 2 of Hume's *Treatise* as a commentary on Descartes's *Passions of the Soul* and Malebranche's *Search for Truth*. But instead of an overtly rationalistic response to passions, Hume will invert the Cartesian method. If the first book of the *Treatise* leaves one with nothing in terms of knowledge, as Descartes's First Meditation had done, so book 2 is intended to build something that will remain standing from the wrecked foundation of the understanding. In place of the understanding, for Hume, one has the most sure thing of all, the passions. Unlike Descartes and Malebranche, however, Hume does not speak of the physiology of passions or examine them "en physicien."[27] Instead, Hume wants to find their natural way of working in one's social life.[28] On the one hand, in the same way that the imagination acts on ideas so the mind and imagination works on the passions. On the other hand, these natural workings

27. See Descartes. *Les Passions de l'âme* (Paris, 1999) 63; Páll S. Árdal, *Passion and Value in Hume's Treatise* (Edinburgh: Edinburgh University Press, 1966) 8; for the comparison of Hume and Malebranche on passion, see Susan James, "Sympathy and Comparison: Two Principles of Human Nature," in *Impressions of Hume*, eds. M. Frasca-Spada and P. J. E. Kail (Oxford: Clarendon, 2005) 107–24. See also the contributions above by Kaposi and Breeur.

28. See Finlay, *Hume's Social Philosophy*, chapters 3–4.

of human nature are not "solipsistic," that is to say, passions cannot occur without some kind of social interaction. Whether speaking of pride, love, fear, or curiosity, Hume believes all passions to be significantly influenced by the imagination *and* social interactions.

As the most central passion for Hume's theory on human nature, pride is an example of an "indirect" passion. Pride and humility both require a complex double relation.[29] The importance of pride for Hume's philosophy points to the fact that one cannot be purely individualist or narcissistic in one's desires. Hume does not encourage the philosopher to achieve the perfect theory at the expense of the philosopher's social surroundings. Of course, despite this fact, pride is a kind of virtue for Hume and thinking highly of oneself is a good thing, but pride is also entirely socially embedded.[30] If your friends or peers did not think highly of you, you could not feel pride. Pride, in a sense, stems from what others think of you, not only what you think of yourself. The point regarding passions is that one cannot even define oneself solipsistically, that is, one requires passions *and* others to define one's very idea of one's self.

In continuity with Hume's discussion of pride is that "of beauty and deformity." For Hume, beauty works in a similar way to that of pride insofar as it also has a double relation. As in pride and humility, if one finds particular pleasure involved in the experience of an object like a house, person, painting, etc., then the result will be something akin to beauty, but if there is pain involved, then the result will be akin to deformity.[31] As Hume writes, "Pleasure and pain, therefore, are not only necessary attendants of beauty and deformity, but constitute their very essence."[32] Since one admires that which is beautiful, there is a necessary connection between admiration and beauty. However, it is impossible to

29. For a further and more detailed discussion of the passion of pride in Hume, see Donald Davidson, "Hume's Cognitive Theory of Pride," *Journal of Philosophy* 73 (1976) 744–57; Annette Baier, "Hume's Analysis of Pride," *Journal of Philosophy* 75 (1978) 27–40. More recently, see Herman De Dijn, "Hume Nonreductionist Philosophical Anthropology," esp. 594ff; Finlay, *Hume's Social Philosophy*, chapter 5.

30. Cf. the relationship of "greatness of soul" and pride discussed by Kate Abramson, "Two Portraits of the Humean Moral Agent," *Pacific Philosophical Quarterly* 83 (2002) 301–34. Thanks to Eric Schliesser for making me aware of this reference.

31. See *Treatise*, 195 (2.1.8.1). Cf. *Dissertation*, 10 (2.17). My discussion "of the passions" (1739) will utilize texts that are, for the most part, repeated in Hume's *Dissertation on the Passions* (1757) and thus I will mention page numbers from both texts, which, as a result, reveals a continuity in Hume's thinking over the course of his life.

32. *Treatise*, 195 (2.1.8.2); *Dissertation*, 10 (2.18).

recognize this beauty without taste, a kind of faculty of discrimination that decides what is worth liking or disliking based on certain criteria. Although beauty cannot be defined for Hume, there is something common to "natural and moral beauty" and that is this "power of producing pleasure." When this power in turn effects the self, it becomes an object of pride: "whatever in ourselves is either useful, beautiful, or surprising, is an object of pride; and it's contrary, of humility. Now 'tis obvious, that every thing useful, beautiful or surprising, agrees in producing a separate pleasure, and agrees in nothing else."[33]

How one understands pleasure in one's soul is central to the passions. Three possibilities for arousing this pleasure are objects which are useful, beautiful, or surprising. These objects will cause pride when: 1) they are naturally and immediately connected to the self; and 2) by means of experience, we are able to remember or imagine this having caused a pleasure before. The object that causes this pleasure cannot be beautiful or surprising in itself. It is here that Hume provides a striking definition of this passion:

> [S]urprize is nothing but a pleasure arising from novelty, it is not, properly speaking, a quality in any object, but merely a passion or impression in the soul. It must, therefore, be from that impression, that pride by a natural transition arises. And it arises so naturally, that there is nothing *in us or belonging to us*, which produces surprise, that does not at the same time excite that other passion. Thus we are vain of the surprising adventures we have met with, the escapes we have made, and dangers we have been expos'd to. Hence the origin of vulgar lying; where men without any interest, and merely out of vanity, heap up a number of extraordinary events, which are either fictions of the brain, or if true, have at least no connexion with themselves.[34]

By means of an example of a "surprising adventure," Hume analyzes the relationship of surprise and pride and the double relation contained therein. If one has taken part in such an adventure (e.g., seeing, hearing, smelling the ocean for the first time, mountain climbing, or travelling to Paris), then one recognizes this adventure as an object of pride. One feels pleasure at having achieved such an "amazing" thing. But if one hears of another's great adventures, only the first relation is achieved, that is, one

33. *Treatise*, 196 (2.1.8.5); *Dissertation*, 10 (2.19).

34. *Treatise*, 196–97 (2.1.8.6); *Dissertation*, 10–11 (2.20).

feels pleasure, but does not feel pride because one cannot attribute this "surprising adventure" to oneself. Thus, one might feel inclined to lie and tell of greater things than what one has actually done in order to picture oneself as a "wonder-inspiring" individual. As Hume continues, "Their fruitful invention supplies them with a variety of adventures; and where that talent is wanting, they appropriate such as belong to others, in order to satisfy their vanity." In order for Hume's analysis to work, the pleasure involved in pride must connect to a surprising, or rare, adventure.

In what sense, then, do all surprising events or objects have a common feature? We find an answer in the sections of book 2 following "of beauty and deformity." Not all surprising objects cause pride in the self, but some such as a

> beautiful fish in the ocean, an animal in a desart [sic], and indeed any thing that neither belongs, nor is related to us, has no manner of influence on our vanity, whatever extraordinary qualities it may be endow'd with, and whatever degree of surprise and admiration it may naturally contain. It must be some way associated with us in order to touch our pride.[35]

Even external objects that have no connection to the self may cause pleasure and this pleasure is the only common feature between objects which are useful, beautiful, and surprising. The degrees of this pleasure can be increased the more an object is connected to the self. For Hume, if it is one's *own* country, county, or town, one feels greater pleasure the closer it relates to oneself. However, Hume does claim that there are also those who find greater pleasure in their ability to relate to distant countries. Due to the fact that "so many" share the same nation, and thus having a particular nationality (e.g., being American) is not special or rare, these people will "always admire the beauty, utility and rarity of what is abroad, above what is at home."[36] Here again, the "surprising" features of an object cause a particular pleasure in the soul.

Can this pleasure of surprising things become excessive? Can one become conceited or prideful to the point of vanity due to one's "amazing adventures"? This danger seems very concrete in Hume's text, when he writes,

35. *Treatise*, 198 (2.1.9.1); *Dissertation*, 11 (2.21).

36. *Treatise*, 200 (2.1.9.8); *Dissertation*, 11–12 (2.24).

> Every thing belonging to a vain man is the best that is any where to be found. His houses, equipage, furniture, cloaths, horses, hounds, excel all others in conceit . . . His wine, if you will believe him, has a finer flavour than any other; his cookery is more exquisite; his table more orderly . . . his fruits ripen earlier and to greater perfection: Such a thing is remarkable for its novelty; such another for its antiquity: This is the work of a famous artist; that belong'd once to such a prince or great man: All objects, in a word, that are useful, beautiful or surprising, or are related to such, may, by means of property, give rise to this passion. These agree in giving pleasure, and nothing else.[37]

One expects a criticism here, on behalf of Hume, or else a critique of pride, vain conceit, haughteous, or self-righteous natures. Yet no critique comes. Hume here only analyzes and describes. He does not moralize. The vain person who finds pride in these things does so due to a particular working of the passions. Thus far in Hume's work, the imagination makes natural connections between objects that strike us as interesting, and the passions also "naturally" find pleasure in objects that are useful, beautiful, or suprising. This is precisely how the account of impressions and ideas in book 1 and the account of passions in book 2 tie together. There is as yet no ethical theory, but one cannot have impressions, ideas, and passions without actions (the subject of book 3).

As with pride, so with love, but instead of oneself being the object of pleasure, the object of love and hatred is another object or person. Pride and humility, though, are unattended by desire and do not excite one to action. What is central, however, begins to be seen when Hume writes, "love and hatred are not compleated in themselves, nor rest in that emotion, which they produce, but carry the mind to something farther."[38] One possible object of this "something farther" is the insatiable human need for truth. Pride or humility is not a sufficient passion to explain this need, nor are any of the other passions Hume describes. As Hume writes, one cannot begin with pride or vanity and move to love and affection, but one can move from love to pride. Therefore, if one gets closer to truth by means of the love of truth, one can become proud. Why is the transition easier from love to pride rather than the other direction? The reason is simple: when one ascribes the "truth" to an object that the

37. *Treatise*, 202 (2.1.10.2); *Dissertation*, 13–14 (2.31). Cf. Abramson, "Two Portraits."

38. *Treatise*, 237 (2.2.6.3); *Dissertation*, 19 (3.6).

self possesses instead of it being beyond or outside the self, that one wrongly becomes proud of something he/she does not *actually* possess. Yet, even for Hume, it appears that the love of truth drives all that for which humans strive.

What Hume says about anger could also be said about the love of truth: "When I am angry, I am actually possest with the passion, and in that emotion have no more a reference to any other object, than when I am thirsty, or sick, or more than five foot high. 'Tis impossible, therefore, that this passion can be oppos'd by, or be contradictory to truth and reason."[39] When I feel thirst, one cannot tell me "it is not reasonable to be thirsty." The same is true for all passions. Nevertheless, "a passion must be accompany'd with some false judgment, in order to its being unreasonable; and even then 'tis not the passion, properly speaking, which is unreasonable, but the judgment."[40]

Concerning love, Hume mentions that when one is in love, one scarcely notices the "little faults and caprice[s]" of the loved one, but they in fact contribute to the passion. It is the same with what may be called "delayed enjoyment." This is a "common artifice of politicians" when they wait to tell you certain news that they know in order to "excite curiosity." They wish to "delay as long as possible" in order to achieve a certain effect. These politicians know that "curiosity will precipitate [one] into the passion they design to raise, and assist the object in its influence on the mind."[41] If politicians can (ab)use the passion of curiosity for a certain end, can one not tap into it for the sake of the good? As in the lover's example, a mixture of passions has great power. As Hume writes,

> Since passions, however independent, are naturally transfus'd into each other, if they are both present at the same time; it follows, that when good or evil is plac'd in such a situation, as to cause any particular emotion, beside its direct passion of desire or aversion, that latter passion must acquire new force and violence. This happens, among other cases, whenever any object excites contrary passions. For 'tis observable that an opposition of passions commonly causes a new emotion in the spirits, and

39. *Treatise*, 266–67 (2.3.3.5); cf. *Dissertation*, 24 (5.1). In this latter text, Hume writes that "*abstract relations* of ideas are the objects of curiosity, not of volition." Although this claim relates closely to the thesis of this paper, I will not discuss it further here since Hume never says anything like this in the *Treatise*.

40. *Treatise*, 267 (2.3.3.6).

41. *Treatise*, 269 (2.3.4.3); *Dissertation*, 26 (6.3).

> produces more disorder, than the concurrence of any two affections of equal force.[42]

If one considers the "opposition of passions" as a principle behind Hume's theory of passions, this is one of the most powerful. The way in which passions gain power over the human being—whether this power be under the aegis of falling in love, going into a rage, desiring revenge, or being full of a love of truth—are various, if not infinite: how such effects mix and combine in each human soul is different with each person and yet uniform. Some humans are more Stoic, some more exuberant in the way they feel passions. Yet, the effects of love can be studied with some precision as causing one to wish to desire and possess the person whom she loves.

Hume's claim about how passions increase their power does not end here. He continues with regards to moral responsibility: "Hence we naturally desire what is forbid, and take a pleasure in performing actions, merely because they are unlawful. The notion of duty, when opposite to the passions, is seldom able to overcome them; and when it fails of that effect, is apt rather to encrease them, by producing an opposition in our motives and principles."[43] But before moving on to what Hume says about morals, which I will do in the final section, it is necessary to show precisely how the love of truth arises in the human soul. As in the 'opposition' of passions spoken of above, so "uncertainty" increases the violence of a passion and produces "an agitation in the mind . . . transfus[ing itself] into the predominant passion." If one is secure in one's love or other passion, then uncertainty will not shake it, but if one fears and does not feel security in one's love, then the fear of uncertainty will become more powerful than the love and take over one's soul. Hume describes this effect further,

> 'Tis certain nothing more powerfully animates any affection, than to conceal some part of its object by throwing it into a kind of shade, which at the same time that it shews enough to prepossess us in favour of the object, leaves still some work for the imagination. Besides that obscurity is always attended with a kind of uncertainty; the effort, which the fancy makes to compleat the idea, rouzes the spirits, and gives an additional force to the passion. As despair and security, tho' contrary to each other, produce

42. *Treatise*, 270 (2.3.4.4); *Dissertation*, 27 (6.5).

43. *Treatise*, 270 (2.3.4.5); *Dissertation*, 27 (6.6).

> the same effects; so absence is observ'd to have contrary effects, and in different circumstances either encreases or diminishes our affections.[44]

Although Hume never uses the word *unconscious* in his work, he could almost be seen as a prefiguration of Freud here in his use of "a kind of shade." The obscurity of the passion increases the power. The power of the imagination over the passion is central to its gaining "additional force." Not only with love, but with anger, fear, joy, and many other passions, the more obscure and uncertain the knowledge, the more strongly one feels.

However, most of Hume's discussion of the passions revolves around a kind of passivity. Nevertheless, the human is an active being. Hume continues,

> When the soul applies itself to the performance of any action, or the conception of any object, to which it is not accustom'd, there is a certain unpliableness in the faculties, and a difficulty of the spirit's moving in their new direction. As this difficulty excites the spirits, 'tis *the source of wonder, surprize, and of all the emotions*, which arise from novelty; and is in itself very agreeable, like every thing, which enlivens the mind to a moderate degree. But tho' surprize be agreeable in itself, yet as it puts the spirits in agitation, it not only augments our agreeable affections, but also our painful, according to the foregoing principle, that every emotion, which precedes or attends a passion, is easily converted into it. Hence every thing, that is new, is most affecting, and gives us either more pleasure or pain, than what, strictly speaking, naturally belongs to it. When it often returns upon us, the novelty wears off; the passions subside; the hurry of the spirits is over; and we survey the objects with greater tranquillity.[45]

One's soul is asleep looking for something to awaken it from its slumber. Each day one looks for something new and different to excite one. Whether this something is worthwhile or not is another question. Are we humans so simple? This passage allows some light to shine through, fundamentally connecting the first two books of the *Treatise*. Taking up the problematic of wonder, discussed by Plato, Aristotle, Cicero, Descartes, Malebranche, and Spinoza, among others, Hume shows here

44. *Treatise*, 270–71 (2.3.4.9–10); *Dissertation*, 27–28 (6.10–11). Cf. the next chapter by Baldine Saint Girons.

45. *Treatise*, 271 (2.3.5.2; emphasis added); *Dissertation*, 28 (6.12).

how philosophy, science, and religion begin. All of the emotions, "which arise from novelty," are the moving "spirits" that initiate the soul's unpliable static nature. One requires something to move one. Something must cause ball a to move ball b, since balls do not move on their own accord, but only because an agent struck the ball, and before that the agent had a desire to strike the billiard ball. Without resorting to an infinite regress, all that occurs must have some principle behind it. Without becoming a skeptic or a metaphysician, Hume names this principle passion. In order to even discover passion itself, there nevertheless must be curiosity or love of truth. Since my goal here is show the underlying connection of ideas, passions, and morals in light of the nature of wonder, the most important section of the *Treatise* for this argument is the final one of book 2, "of curiosity, or the love of truth."

OF CURIOSITY, OR THE LOVE OF TRUTH

Thus far in this chapter, I have described the background of curiosity within an implicit understanding of the effect on the human soul of a particular kind of passion, namely, surprise, admiration, or wonder. This is in turn a natural working of the understanding, especially when something strikes the imagination. While, at an abstract level, surprise, admiration, and wonder may be differentiated from each other, as seen in Adam Smith, it is not so clear in the immediacy of the experience. When a novel object strikes one, the effect of this object is phenomenologically similar: in Descartes's words, what results is "a sudden surprise of the soul."[46]

Since Hume says that it is "the love of truth" that is "the first source of all our inquiries," we wonder how this passion finds "its origin in human nature."[47] Whether we come to truth by means of measurement (i.e. mathematical demonstrations) or by a less scientific route, the former is less pleasurable than the latter. What Hume means here is that the truth of an abstract (e.g., algebraic) fact is much less interesting and thus less pleasurable than the satisfaction of discovering a truth that is not demonstrable in this mathematical way: "The first and most consider-

46. See Dorottya Kaposi's contribution above, and Michael Funk Deckard, "A Sudden Surprise of the Soul: The Passion of Wonder in Descartes and Hobbes," *Heythrop Journal* 49 (2008) 948–63.

47. *Treatise*, 286 (2.3.10.1). There is no comparable discussion in the *Dissertation on the Passions*.

able circumstance requisite to render truth agreeable, is the genius and capacity, which is employ'd in its invention and discovery." It is thus too simple to hear the truth, 2 + 2 = 4, or an equally "analytic" truth. "We never are oblig'd to fix our attention or exert our genius" on such a "matter of fact."[48] Instead of measuring, for example, "the proportions of conic sections," we search for "truths" that may be of importance to us. Philosophers in particular "have consum'd their time, have destroy'd their health, and neglected their fortune, in the search of such truths, as they esteem'd important and useful to the world." Yet most people find the so-called truths of these philosophers completely useless. But if philosophers *knew* that their discovery would be found useless by the vast majority of people, "they wou'd entirely lose all relish for their studies."[49]

There is thus a development from infancy through adulthood regarding how a human finds truth. Even mathematical truths have a particular feeling.[50] Marina Frasca-Spada thinks that it is odd that mathematics occupies 2/3 of the section on curiosity in Hume's book. She writes, "Human nature rules here. But the lovers of truth are strange creatures... The misanthropic philosopher who has sacrificed his whole life to the search after a useful truth is no different from other men dominated by a 'ruling passion.'"[51] In order to keep from sacrificing one's whole life to something seemingly worthless, Hume provides us with this most practical advice: "we must consider, that there are certain desires and inclinations, which go no farther than the imagination, and are rather the faint shadows and images of passions, than any real affections."[52] In other words, there are real passions and there are, like in the world of Plato's cave, opinions which look like real passions, but actually have no substance to them. True passions thus lead to true action, false passions to pure contradiction or ignorance within one's soul.

Since pride by itself does not lead to action, one must look for passions that lead to action. Love of truth is one of these passions. How can one distinguish between a true love of truth and a false love of truth? One of the ways of doing so is to look at what kinds of actions result.

48. *Treatise*, 287 (2.3.10.3).

49. *Treatise*, 287 (2.3.10.4).

50. For a discussion of this phenomenon, see Marina Frasca-Spada, *Space and the Self in Hume's "Treatise"* (Cambridge: Cambridge University Press, 1998) 152–53.

51. Ibid., 155–56.

52. *Treatise*, 287–88 (2.3.10.5).

Earlier in the *Treatise*, Hume states the there is a "whole bent or tendency . . . from the beginning to the end," which determines "the character of any passion." He then continues, "Benevolence or the appetite, which attends love, is a desire of the happiness of the person belov'd, and an aversion to his misery."[53] Hume specifically refers to his discussion of love and benevolence in this section on curiosity. The love, benevolence, or sympathy we have for another person can be tied to the love of truth. Hume's example tells of a person who examines the fortifications of a city, their strength and advantages, the disposition and contrivance, etc. All of these 'abstract' discussions specifically relate, not to a mere "form of the objects," but to the wellbeing of the people inside of the city. The pleasure derived from studying in detail the fortifications of a city "can be no other than a sympathy with the inhabitants, for whose security all this art is employ'd; tho' 'tis possible, that this person, as a stranger or an enemy, may in his heart have no kindness for them, or may even entertain a hatred against them."[54]

The love of truth is thus not a disembodied desire for something remote or abstract from human flourishing and social relationships. There is a particular moral nature behind human curiosity and that is part of the reason why Hume places this chapter at the end of part 2, on the passions, just before part 3, on morals. The transition from passion or passivity to action must pass through desire. Part of the nature of an assent to a certain truth includes its desirability. There are many pieces of truth or fact that are entirely indifferent to us. The nature of news programs, films, radio, or even neighborhood gossip tries to tap into this desire for truth. How does something thus capture our attention? There is no possible truth that one can imagine that is uninvolved with desire. The truth pulls at the human being. It asks to be found. Thus, humans desire this truth and even love it. An argument may be certain and founded on fact, but if we do not feel an attraction to it, we will not care about this truth at all.

What kind of desire in particular is that of "study"? It "consists chiefly in the action of the mind, and the exercise of the genius and understanding in the discovery or comprehension of any truth."[55] Our attention must be directed at some particular object and this object must

53. *Treatise*, 246 (2.2.9.2).

54. *Treatise*, 288 (2.3.10.5).

55. *Treatise*, 288 (2.3.10.6).

be worthy of our interest. Science, philosophy, theology and all of the disciplines that fill the universities have such an object. That is why it is an acceptable value in our culture to attend university. However, just because the subjects taught in universities are all aimed at the same truth, each student may not feel as curious about every subject. "Besides the action of the mind [towards truth], which is the principal foundation of the pleasure, there is likewise requir'd a degree of success in the attainment of the end, or the discovery of that truth we examine."[56] This is Hume's way of saying we must be capable of success at something in order to be curious about it. If one were absolutely unskilled at physics, one will not feel pleasure in pursuing truths in physics. Most of the pleasure we receive in the pursuit of truth is precisely that: the "action and pursuit" of truth is more pleasurable than the attainment of it. The "whole bent or tendency" of this passion lies in the *continuing* pursuit of it. This is precisely what Hume is getting at when he writes, "there cannot be two passions more nearly resembling each other, than those of hunting and philosophy, whatever disproportion may at first sight appear betwixt them."[57] No matter how much one begins philosophizing or hunting with a particular goal in mind (i.e. the truth or the catch), one loses oneself in the action. This is comparable to playing any kind of game in which one enjoys the game so much that one forgets that the goal of the game is to win. In hunting and philosophy it is the same thing. One begins enjoying the search after truth, the pursuit of the game, that one forgets that the object for which one began was the capturing of such truth or food.

Pascal, Montaigne, and Erasmus spoke of this same desire between hunting and philosophy, as David Fate Norton tells us.[58] The mixture of the prize in store, the uncertainty of the catch, and the adrenaline-inspired exercise are like "chymical preparations, where the mixture of two clear and transparent liquids [that] produces a third, which is opaque and colour'd."[59] But now, instead of just searching for mathematical truths, the extension of this same love of truth can be expressed in "mor-

56. *Treatise*, 288 (2.3.10.7).

57. *Treatise*, 288 (2.3.10.8).

58. See the Norton's edition of the *Treatise*, 532–33. See also his critical edition of the *Treatise* (Oxford: Oxford University Press, 2007) vol. 2, 880–82, which has additional annotations to this passage.

59. *Treatise*, 289 (2.3.10.9).

als, politics, natural philosophy, and other studies, where we consider not the abstract relations of ideas, but their *real* connexions and existence."[60] The reality of these topics seem ever more pressing than the "abstract relations of ideas." "But," Hume continues,

> beside the love of knowledge, which displays itself in the sciences, *there is a certain curiosity implanted in human nature*, which is a passion deriv'd from a quite different principle. Some people have an insatiable desire of knowing the actions and circumstances of their neighbours, tho' their interest be no way concern'd in them, and they must entirely depend on others for their information; in which case there is no room for study or application.[61]

Unlike science or study, gossip has a different principle in that it destroys community by undermining social relationships and, as we have seen earlier, Hume's philosophy is based in a social web of relations. These relationships require trust as the cement that holds us together in a community. Sympathy, as Hume shows later in the *Treatise*, is at the basis of these social relationships. Love of truth may easily disintegrate into love of gossip, a semblance of truth. Nevertheless, many desire as much information as they can possibly gather about their neighbors and these ideas strike the fancy with force and vivacity. The hunting metaphor, as true as it is, also expresses a real danger: when the goal of truth (or the catch) is lost, one can become caught up in the pursuit whether or not it is moral. No longer does the truth matter or who suffers. In other words, the nature of trust in the community may deteriorate.

The nature of this search for truth explains the reason for its pursuit, but there is a moral claim behind Hume's theory. However, in January 1739, when the *Treatise* was first published, this is where it ended. The rest of Hume's moral claim had to wait until November 1740 when book 3 of the *Treatise* was published. There, he describes the fact that moral distinctions are derived from a moral sense, instead of reason, and that the basis of morality is shown by one feeling sympathy with one's neighbors and approving or disapproving of their actions. The bond that holds the community together, whether in morals or politics, has a natural basis in human nature. One desires to enjoy the presence of one's friends and family—and this same sympathy applies to neighbors as well. Adam Smith took up Hume's theory in his *Theory of Moral Sentiments* (1759).

60. *Treatise*, 289 (2.3.10.11; emphasis added).

61. Ibid. (emphasis added).

In this article, however, I have been concerned with the relationship of ideas, passions, and action. The reason Hume ends book 2 with a discussion of curiosity or the love of truth points to the danger of wanting to pursue unreal truths or getting lost in the chase of "unreal" or unethical pursuits. Nevertheless, even a "false" passion (e.g., desire to/for gossip) is still central to human nature. Hume does not provide a normative claim about such pursuits, but points to its concrete dangers. He describes more than prescribes in an analogous way to Smith's "impartial spectator." Those truths "of the understanding" that are purely mathematical, abstract, of no purpose and use to humankind, or are neighborhood and community destroying, are "truths" not worth desiring. The *correction* of one's desires and passions leads one as strongly to that which is good and true as the original passions themselves. This leads to a cognitive reading of Hume in which all passions point to the *love of truth*. This passion, otherwise called admiration, curiosity, or wonder, underlies the origin of all philosophical, scientific, and religious pursuits. It has been my intention here to show that "a sudden surprise of the soul" (to use Descartes's language) underlies Hume's *Treatise*, as well as one's own everyday life, and that reading the *Treatise* might help one understand one's own human nature and daily pursuits.[62]

62. For their extremely helpful comments on earlier versions of this text, I wish to thank Lauren Davis, Eva Dadlez, Patricia De Martelaere, Christopher Finlay, Julianne Funk Deckard, James Harris, Maria Mocanu, and Eric Schliesser.

12

A Risk of Testimony

Astonishment and the Sublime

Baldine Saint Girons

In her book, Fiat Lux: A Philosophy of the Sublime *(Paris: Quai Voltaire, 2003, and Librairie philosophique Vrin, 2005, winner of the Morpurgo-Tagliabue International Prize for Aesthetics in 2002), Baldine Saint Girons shows how the sublime refers simultaneously to operating principles, vehicles, and recipients. The principles are active and primary first causes; the vehicles are material and formal signifiers; the recipients are human beings who contribute to the workings of the sublime by bearing witness. The principles are the subjects of ontology and metaphysics, and secondarily of political philosophy and ethics; the vehicles are the subjects of aesthetics, history, and the philosophy of art in particular; the intended recipients have more to do with the philosophy of the subject and critical anthropology.*

The method adopted is that of the systematic study of risks: risks of bewilderment by grandeur, commotion by ugliness, vertigo by obscurity, regression by simplicity; risks of the confrontation with power, of the suffering by passion and of the desire for virtue. But between vehicles and principles, there are the risks of testimony in Chapter V, a passage of which is here translated. While antique enthusiasm takes over the entire being in Longinus's philosophy, and while respect deals with hierarchy and divisions in Kant, Burke's astonishment unsettles the subject. The problem is to understand why astonishment should be chosen to state the major effect of the sublime in the middle of the eighteenth century.

PREFACE: THE DECLINE OF ENTHUSIASM IN MODERN TIMES

WHILE ENTHUSIASM PLAYED A major role in the quarrel between Neoplatonic and Aristotelian philosophers, the determining element of its suspicion was the correspondence between the criticism of heterodox religious feeling and mysticism, on the one hand, and a normative and restrictive conception of artistic creation on the other hand. The role played by the Aristotelian Jules César Scaliger (1484–1558) was, in this perspective, essential, as shown in his polemic with Desiderius Erasmus (1466–1536) and in his systematic critique of Jerome Cardan's (1501–1576) theses, as Marc Fumaroli has established.[1] In his *De subtilitate* (*Of subtlety*, 1550), Scaliger wanted to ignore the concept of *splendor* found in Cardan's autobiography: "I have acquired the use of amplification and lucidity [*splendor*] of understanding partly from practice and partly at the inspiration of my spirit," Cardan wrote. "And so the whole art of writing and of extemporaneous lecturing has in my case been the result of spiritual inspiration and the effort I made to attain lucidity [*splendor*]."[2] These themes were inherited from Plato and Longinus, but Scaliger stigmatized their obscurantism in order to affirm the triumph of rational understanding even in literature. The *Poetices libri septem* of 1561 declared the superiority of the Latin language and genius over the Greek ones and, consequently, the superiority of Virgil over Homer. The splendor of the origins is then inexorably dedicated to dissolve itself in the progress of knowledge.

The Scaligerian influence extends into the 17th century, notably through the writings of the oxonian Meric Casaubon (1599–1671), a familiar of Robert Burton (1577–1640) and of Henry More (1614–1687). Casaubon's *Treatise Concerning Enthusiasm*, published in 1655, was followed a year later by More's *Brief Discourse of Enthusiasm*.

The Roman Leone Allacci (1586–1669), custodian of the Vatican, was still defending the tradition of Cardan and Erasmus in his important dissertation of 1634, *De Erroribus magnorum virorum in dicendo*. Withholding specific judgment on the religious aspects of enthusiasm,

1. See Marc Fumaroli, "Crépuscule de l'enthousiasme au XVIIe siècle," in *Troisième congrès international d'études néo-latines* (Paris: Vrin, 1980) vol. 2, 1279–1305.

2. Jerome Cardan, *De vita propria liber* (1576) posthumously published in 1643, translated by Jean Stoner as *The Book of My Life* (New York: Dover, 1930, reprint 2004) 246–47.

which were, in fact, already condemned, he claimed that the masterworks of literature were written with inspiration. Between the divine *logos*, guarantor of science, and the strictly profane human word, the links seem to grow blurred from this point of history onwards.

But, after Allacci retreated, Casaubon and More dealt ultimate blows to enthusiasm: they excluded enthusiasm not only from religion, but also from art—in order to reject it as a pathology—and they stigmatized it through the confused irony of carnal enthusiasm with the inspiration of the Holy Spirit, in other words, between the infrarational domain of sensibility and imagination, under the dependence of the body and its humors, and the divine supersensible field.

The influence of the Neoplatonic humanism of the Renaissance and that of, more recently, Allacci exercised itself in France thanks to Balzac and his defense of Pierre Corneille (1606–1684). But, by the end of the seventeenth century, this influence had waned due to the *great influence of the moderns*, convinced of the supremacy of reason over imagination and, in parallel, by the *increased offensive of the theologians*—and not only by the Protestant theologians—against all forms of mysticism. How might inspiration, which appeals to a revelation and to an experience that puts nature, dogma, and reason into question, not menace all official forms of discourse concerning religion or science?

Under these conditions, the translation of Longinus by Nicolas Boileau-Despréaux (1636–1711) in 1674, the theory of "je ne sais quoi" in the work of the Jesuits René Rapin (1621–1687) and Dominique Bouhours (1628–1702) or, in another register, the quietist conceptions of the "practice of the presence of God" as "sovereign remedy,"[3] together obstruct the rise of dogmatic rationalism and ecclesiastical authoritarianism in France. In their criticism of religious fanaticism the Encyclopedists were not mistaken in separating good inspiration from bad, recuperating the force of enthusiasm or, moreover, submitting to what Voltaire calls the philosophical spirit or Kant the force of judging.

• • •

3. See François Fénélon (1651–1715), *Spiritual Progress: or Instructions in the Divine Life of the Soul* (New York: M.W. Dodd, 1853) [French original in *Oeuvres* (Paris, 1870), I.276], where he writes, "The sovereign remedy is the habit of dwelling continually in the presence of God. He sustains, consoles, and calms us." (Section 11: "On Temptations").

At the turn of the Great Century (ca. 1700), the concept of enthusiasm was incontestably split: on the one hand, it became quasi-synonymous with superstition, extravagance, fanaticism, and deliriousness; on the other hand, it remained within the field of artistic creation. For example, Roger de Piles (1635–1709) distinguished between the authentic picturesque furor of the simple "dreams of a hot fever," and the "true enthusiasm" of a transport which "surpasses the extremes of accuracy and probability."[4]

But one should notice that de Piles does not endeavor to define the sublime by rehabilitating enthusiasm, which is its source. The sublime seems to him the result of a process, requiring time for penetration, while enthusiasm, on the contrary, characterizes itself by its promptitude. In fact, disengaged by enthusiasm, the sublime tends towards beauty, since Roger de Piles declares: "It seems to me, in a word, that enthusiasm catches us and that we catch the sublime." Certainly, here it is no longer the Longinian sublime; the sublime degraded to become the object of an "infallible secret," emerges again as the required term to qualify the essence of a transport which would be "common to the painter and to the spectator," de Piles states.

Inversely, other writers such as, for example, Jean-Pierre de Crousaz (1663–1750) criticize the sublime in showing the equivocity of the admirable and, in particular, its total absence of beauty and suitability (*convenance*): "As saint Ambrose did not allow Theodosa to enter the church, one sees at first the grand, lively, and extraordinary, and it is difficult to refuse the act of wonder. But if this event still presents more pride in the priest than feebleness in the emperor, one must find nothing there of beauty, because the characters of their rank are not sustained and the proportions are reversed."[5] As the sublime loses its first *dynamis* in de Piles, in Crousaz it also breaks its relations to *kairos* and becomes the simple delusion of grandeur when not rescued by religion.

The suspicion that awakens enthusiasm contaminated then doubly the sublime, by reducing it to a superlative of beauty and instituting a radical break between the sacred and the profane order.

We then should not be surprised that two writers, followers of the genius of Antiquity, took over to elucidate the profound reasons for the

4. *Cours de peinture par principes* [The Principles of Painting], 1708 (ed. by Jacques Thuillier, 1989) 70–75.

5. *Traité du Beau* (Paris, 1715, reprinted in 1985) ch. VIII, 154.

discrediting of enthusiasm in the Christian era. A strange congeniality of thought takes shape between Lord Shaftesbury (1671–1713) and Giambattista Vico (1668–1744). Did they possibly know of each other? The salon of Doria, where Vico was a pillar, possibly united them in the mild atmosphere of Naples where Lord Shaftesbury came to care for his lungs and then died in 1713, at the age of 41.

THE ECLIPSE OF A PAGAN VIRTUE: SHAFTESBURY

Why "that air of enthusiasm, which sits so gracefully with an ancient, should be so spiritless and awkward in a modern?"[6] Such is the question that Shaftesbury poses in his "Letter on enthusiasm" to Lord Somers, published in 1708. In fact, who has not remarked on how many invocations to Apollo and to the Muses seem false and arbitrary in a time where their cult had become null and void? Reviving an unused religion is an impossible thing: in the absence of a living reality which corresponds to it, no ritual will be able to move hearts.

As in Vico, Shaftesbury emphasizes the antique miracle of the intimate union between poetry, religion, and politics. Here is what we should not forget: "But we Christians, who have such ample faith ourselves, will allow nothing to poor heathens. They must be infidels in every sense. We will not allow them to believe so much as their own religion, which we cry is too absurd to have been credited by any besides the mere vulgar."[7]

What is polytheism if not the possibility given to each one to venerate his own god and to profess the ideas of his own faith, without encountering opposition or sanction by a remark that one is ridiculous? "And while some sects, such as the Pythagorean and latter Platonic, joined in with the superstition and enthusiasm of the times, the Epicurean, the Academic and others were allowed to use all the force of wit and raillery against it. And thus matters were happily balanced: reason had fair play; learning and science flourished."[8]

In our day, unhealthy enthusiasm, which invades the sects, becomes involved even in the authorities dealing with repression. Melancholy has become the end of religion, gravity is used as a folding screen for

6. Shaftesbury, *Characteristics of Men, Manners, Opinions, Times*, ed. Lawrence E. Klein (Cambridge: Cambridge University Press, 1999) 5.

7. Ibid., 6.

8. Ibid., 11.

imposture,[9] and reason refuses to submit to the test of mockery. Panic propagates itself without measure, until it suffers, in order to reabsorb it, to denounce the *echoes* that multiply the vociferations of some exaggerations. Has not the god Pan utilized the sonorous rocks and caverns for reflecting the clamors of Bacchus' procession and to sow terror in the enemy? Enthusiasm has its stratagems and employs ruse against ruse itself.

Shaftesbury thus utilizes medical analogies, in the mode of Bacon, in order to demand gentle and innocent remedies which neutralize the enthusiastic imagination, in place of augmenting its roughness under the effect of persecution. "There are certain humours in mankind which of necessity must have vent."[10] It is not necessary to appeal to the "physicians in the body politic" who tamper with these "mental eruptions": they transform these into "mortal gangrene." These councils of prudence, emitted at the same time as the Cevennes prophets, produced good effects; and, doubtless, Shaftesbury's letter successfully lightened the verdict, even on the year of its publication, against the Camisards refugees taken to London.

Shaftesbury thus justifies enthusiasm "in a certain measure" and goes so far as to declare that "enthusiasm is wonderfully powerful and extensive."[11] Distinguishing between a cheerful enthusiasm—polytheist and tolerant of the Ancients—and a somber melancholic and fanatical one—of the Moderns—he resorts to derision as with the sovereign weapon against excesses of a transported imagination. "Now wit can never have its liberty where the freedom of raillery is taken away, for against serious extravagances and splenetic humours there is no other remedy than this."[12] Inasmuch, is he recommending a return to paganism?

The established opposition between enthusiasm conceived as "mistaken sentiment" and an inspiration, which is "a real feeling of the divine Presence" (mentioned at the end of the letter to Somers), may resemble an escape clause. A true hymn to Nature opens the third part of "The Moralists: A Philosophical Rhapsody." It is delivered by Theocles, who asserts "the universal union, the coherence or sympathy of all things":

9. "Gravity is of the very essence of imposture," writes Shaftesbury (ibid., 8). Montesquieu takes up the same idea: "Gravity is the shield of the stupid."

10. Ibid., 9.

11. Ibid., 27.

12. Ibid., 12.

Georges de La Tour, *Job and his wife* (ca. 1650), Épinal, Musée Départemental des Vosges. The entire body of Job would strangely enough seem to be able to enter into this large woman's outfit: the face would rest on the level of the belt and the already contracted limbs, folded up, would coil themselves in a fetal position under the organ pipe-like folds of the apron. There is nothing that is complaisant in this display of misery: the eyes magnetize in a silent face-to-face discussion, beyond insulation and beyond communication. No more enthusiasm, indeed, for a God who inflicts such trials, but endlessly growing doubt and astonishment.

nothing can resist the active universal principle. And Philocles understands Theocles as renouncing "cold reason," in order to continue to deploy his enthusiasm in favor of the universal being. Master and disciple promenade through the galaxy and traverse different regions of the globe, so that they appreciate each other's merits. "Every love and admiration is enthusiasm," the proselyte recognizes, convinced by Theocles of the existence of "this mysterious beauty," in which we have to give "labour and pains" if we ever want to discover it.[13]

But it is in the praise of Job that Shaftesbury finds the most gripping accents. It is not his patience that he praises, in effect, but on the contrary his vigorous tone, when he questions divine Providence. As much as his friends try to justify God, Job is indignant:

> Will ye speak wickedly for God? and talk deceitfully for him?
> Will ye accept his person? will ye contend for God?
> Is it good that he should search you out? or as one man mocketh another, do ye so mock him?[14]

God has no need of man flattering and pleading his cause, as that of vulgar mortals. Nothing is meaner than the renunciation of reason, the affectation of belief and the parade of devotion. Contrary to those twisted beggars who randomly incense those to whom they request generosity, Job dares to expose to God his motives of recrimination. In other words, either God exists and he is good or else he does not exist. The true believer would be one who would not hesitate to sacrifice belief, if the absence of this divine kindness, which makes necessarily the object of its wishes, were shown to him.

I should evoke therefore the painting of Georges de La Tour[15] that is generally interpreted as Job with his wife. The reddish glow of an im-

13. See ibid., 320.

14. Job 13:7–9, King James Version. A more recent translation is the following: Will you speak unjustly on God's behalf? Will you speak deceitfully for Him? Will you be partial toward Him? Will you plead God's cause? Will it go well when He examines you? Will you fool Him as one fools men? (*Tanakh*).

15. On the history of this painting and the German Protestant theme of derision, see François-Georges Pariset, *Georges de la Tour* (Paris: Henri Laurens, 1948) 194ff. Pariset entitled his presentation "Job, a quarrel with his wife, an example of patience." The accent seems to us, however, less carried by derision and the capacity of endurance than on a distress stripped of all complaisance. Job comes to be stricken with an ulcer and takes a shard to cut himself. His wife, hopeless with his state, then said to him: "Dost thou still retain thine integrity? Curse God and die!" Job refuses to obey her

mense robe covers up nearly two thirds of the painting: the silhouette that occupies it is hidden towards the top of the fabric by a gray belt, where the collar and the head are tucked into a strikingly small space. Opposite and below her is the hardened and earthy flesh of Job, sitting on a stool, with raised knees, elbows welded to the top of the thighs, with this unforgettable face that raises towards the woman a scrutinizing glance, where one reads *neither indifference nor complaint.* Job is not a figure of resignation, he incarnates a distress: *he refuses to let his soul sink into itself and, without giving up, continues to question.*

For the Christian, wonder tends in fact to supplant enthusiasm and the meditation on the scandal of suffering, that of the just and innocent God who stretches himself on the Cross for the salvation of the human race, replaces the intoxication of invention. The sublime, however, tends to become a spectacle that plunges into perpetual wonder.

THE SUBLIME OF INVENTION AND THE SUBLIME OF CONTEMPLATION: VICO

Like Shaftesbury, Vico meditates on the unity and coherence of ancient thinking, but insists less on the dangers of religious intolerance than on the misdeeds of the extension of the critical method to all spheres of human activity.

In the time of the Greeks, each philosopher constituted a university of himself alone, the disciplines were artificially cloistered, the professed knowledge rested on a vital choice, the method accords with the personality, and the doctrines had obvious effects upon their manners. Each school implicated itself in the discourse which was named to characterize it. The consciousness of knowing nothing incited Socrates to raise questions, the conviction that the rule of truth was in thought alone led the Stoics in forging *sorites*. Aristotle founded his school on the intellect, but also on the senses, utilizing syllogisms; the Epicureans, who placed all knowledge in the senses, did not worry themselves with working out a logic. "Today, students who may be trained in the art of discourse by an Aristotelian, are taught physics by an Epicurean, metaphysics by a

by blaspheming: "Thou speakest as one of the foolish women speaketh. What? Shall we receive good at the hand of God, and shall we not receive evil?" (Job 2: 9–10, King James Version) But the question remains, that of the absolute scandal of a suffering that one would not be able to justify the punishment for sin. In the painting of La Tour, we interpret a certain form of connivance between the two sidereal beings.

Cartesian . . ."[16] Consequently, the "culture on the whole . . . [which] is really the flower of wisdom" loses in solidity what it gains in extension. And, as we will say later on, the dissociation between knowledge and the personal adhesion to its principles puts enthusiasm out of its field.

In fact, "the condition of the first occupant is always the best" and the Greeks profited from this privileged situation in which, summoned to invent, they did not yet have to take account of the renowned works of their predecessors. In these conditions, Vico demands, "What if I declared that the most outstanding masterpieces of the arts hinder rather than help students in the field?"[17] Consequently, Vico suggests that the success of modern painting could be due to the fact that the ancient masterpieces of painting, Protogenes' "Ialysus" or Apelles' "Venus," have disappeared. "It would seem almost advisable, in order to have great artists, to have the great masterpieces of art destroyed."[18]

However, Vico can scarcely sustain this thesis; all his faculties revolt against this apparent blasphemy. The action would be "an atrocious act of barbarism" and, instead, should be "used for the benefit of lesser minds." Moreover, those with exceptional artistic genius should always "put the masterworks of their art out of their sight, and strive with the greatest to appropriate the secret of nature's grandest creations."[19]

One sees in Vico the Longinian problem: the original perfection of the source leaves the alternative of repetition or opposition. Enthusiasm only becomes truly creative under the auspices of divinized discord, whose name is Eris. But Longinus proved the contrary in a certain way, while he raised literary criticism to the height of philosophy.

The solution proposed by Vico seems at first different than Longinus's, since the former consists in diverting the glance towards what was invented in order to find the originating source *in nature*. However, since truth is *an offspring of our time*, the rivalry cannot but emerge, since it is through the geniuses of the past that those who endeavor to join the geniuses of the present must pass. Vico refers lengthily

16. Giambattista Vico, *On the Study Methods of Our Time*, trans. Elio Gianturco (Ithaca: Cornell University Press, 1990) 77. (This text was originally printed in 1709 as *De nostri temporis studiorum ratione*.)

17. Ibid., 71.

18. Ibid., 72.

19. Ibid.

to Longinus[20]: in order to obtain an eternity of renown, it is necessary that one's works be judged highly and, as Longinus recommends, one must imagine the way in which the greatest geniuses of humanity would welcome his works. A subtle displacement, nonetheless, happens to the work in progress which results in a competition with geniuses of the past in an anticipation of future glory. Just as the temptation of iconoclasm is doubtlessly alive in Vico, so is there also a kind of indignation over the idea that invention is the product of barbarians.

Longinian enthusiasm has provided space for something other than what it itself seeks to define. We will try to show this in two ways: in evoking a text of (Charles Nicholas) Cochin who is obsessed in expressing the sterilization by the great models of the past, but also in rapidly giving conclusions to a dossier concerning the critical fortune of Michelangelo and Raphael between 1662 and 1824:[21] the vicissitudes of the notoriety of these two geniuses, but moreso the ferocity of criticism, plunges us, in effect, in a stupor [. . .].

THE SURPRISE CONTINUED: MONTESQUIEU

No one is more sensible to the grandeur of the sublime than Michelangelo. When Montesquieu evokes the "grand oeuvre" of his *Spirit of the Laws*, what name immediately comes under his pen? Michelangelo's. "When Michelangelo went into the Pantheon for the first time, he says that it should be built in the air. I will imitate him, this great man, in some way, in my own manner. These ancient laws, which lie in the ground, I will expose them to everyone's gaze."[22]

Wonder bursts from his *Essay on Taste* (*Essai sur le goût*, 1757): "No one has ever known art like Michelangelo; no one has displayed it more . . . Art gives rules and taste gives exceptions; taste reveals to us on some occasions that art must submit, and on what occasions it must be submitted to."[23] Montesquieu had already praised Michelangelo's

20. In a passage of the 1730 edition of the *New Science*, which was omitted in the 1744 edition.

21. See Baldine Saint Girons, "Michel-Ange et Raphaël: les enjeux d'une confrontation (1662–1824)," in *Les Fins de la peinture*, ed. René Démoris (Paris: Desjonquères, 1990).

22. *Mes Pensées*, 1720–1755, édition de Daniel Oster, Seuil, 696.

23. Montesquieu, *An Essay on Taste*, from Alexander Gerard's translation of 1758, in Gerard, *Essay on Taste* (London, 1759) 312.

"simplicity," but his most beautiful praise concerns his grandiose manner of saying things without useless pathos, bluntly. "In a piece which represents the Passion of Christ, and is now in the gallery of Florence (*La Pietà de Palestrina* exposed to the gallery of the Academy), he has drawn the Virgin standing in an erect posture, and beholding the cruci-

MICHELANGELO, *Pietà de Palestrina* (1553–1555), Florence, Galleria dell 'Accademia. Twenty-five years after having seen it, Montesquieu remembers the impression which the Pieta of Palestrina left on him. It strikes initially by the absence of projections: the marble is sculpted as a bas-relief. The enormous mass of a body rescued from the fall by a powerful and unfeminine hand occupies almost all the space: the chest and the belly, thick and muscular, tend to ugliness. The Virgin standing erect, as Nicodemus of the Piètá of the Dome, and the face stretched above the neck of her son, looks towards him "without pain, without pity, without regret, without tears," writes Montesquieu. Michelangelo "supposes the virgin is instructed into the great mystery, and by this can withstand with grandeur the spectacle of this death."

fixion of her son without shedding a tear, or discovering the least mark of affliction or pity."

• • •

The contemplation of the works of Michelangelo engenders astonishment, a "surprise" that does not keep one from ceasing to believe, because it constrains the spectator to a true task: "sustaining" the gaze of that which is being shown.

Doubtlessly the ugliness may be the point of departure of the experience described by Montesquieu because it leads from the first impression to divine belief, but the question rises: is such a woman ugly, as we first thought, or is she beautiful because of that which we do not cease discovering in her? "A beautiful nymph," insists Montesquieu, "generally disappoints our expectations, and appears, after some little time, less amiable than at first sight; after having surprised us at first by her charms, she falls greatly off, and surprises us at length by her defects."[24] The greatness manifests the same efficacy as the ugliness. For example, we know that "the cupola of Saint Peter's is of itself an immense edifice," but we ignore where we have to stand, to appreciate his greatness.[25]

But it is the horrible that constitutes the salient point of surprise. If Montesquieu lingers with the example of the *Pietà*, is it not because of the horror of ridiculed maternity expressed there? She who has not known the death of her child is unaware of true suffering; this is what is so admirable about the *Pietà*, which affects the agnostic as much as the believer. Let us look to the *Pietà* in Saint Peter's Cathedral, in which the face and silhouette offer a close relationship to the *Virgin with Child* in Bruges.

Michelangelo's youthful work, exceptionally rendered, the sculpture proclaims so energetically the triumph of beauty, even through suffering and the death, that one accused the artist of having represented the Virgin as too young and Christ unscathed after his martyrdom. "*Vergine Madre, figlia del tuo Figlio*": the verses of Dante articulate the mystery of

24. Ibid., 296: "Car une belle personne fait ordinairement le contraire de ce que nous avions attendu : elle parvient à nous paraître moins aimable ; après nous avoir surpris en bien, elle nous surprend en mal"; see our *Esthétiques du XVIIIe siècle,* 102ff.

25. Ibid., 304. The French original reads: "L'exacte proportion de la fameuse église de saint-Pierre fait qu'elle ne paraît pas d'abord aussi grande qu'elle l'est ; car nous ne savons d'abord où nous prendre pour juger de sa grandeur."

MICHELANGELO, *Pietà* (1497–1500), Saint Peter's Cathedral in Rome.

the splendor of Mary. But the feeling of absolute horror, the dead child in the bosom of his mother, battles against the impression of calm and of beauty. One feels unable, but must assent to the grandeur that seizes the spectator to the point in which pity, becoming useless, leaves only astonishment or respect. It is not necessary to be Catholic: the feeling of divinity inherent in a few beings imposes itself.

The knees of Mary are drawn aside to support the sagging body of her son. Following the example of Giovanni Bellini's Christ, this has no appearance of a stiffened corpse:[26] his supple and living flesh is scarcely touched by death. The Virgin moves back, withdraws, and leans back, in order to receive the body, such that she endeavors to raise his shoulder. A hollow develops between her powerful thighs and between her prominent breasts, undergirded by the vast cord which crosses them and on which the unique signature of Michelangelo is inscribed: MICHAEL ANGELUS BONAROTUS FLORENT FACIEBAT. The closed eyes of Mary are cast upon the hollow between her breasts and, between her

26. On the emergence of the theme of the Compassion or the Piety of Mary at the end of the fourteenth century, see Émile Mâle, *Religious Art in France: the late Middle Ages*, ed. by Harry Bober, transl. by Marthiel Mathews (Princeton: Princeton University Press, 1986) chapter III.6: "The Passion of the Virgin."

arms, the inert flesh of her child. Her gaze descends as much as her son is raised, the head tipped back towards the sky. And, on the right side, the fingers released for testimony, the index finger extended at the end of the tragedy: *Ecce homo.*

Does the horror of death disappear in front of the sublime of consent? The testimony is paradoxically led back to the way of beauty and placed before the threshold of mystery, exactly where what is makes a sign disappears right away.

In contrast to the atrocity just witnessed, Montesquieu also imposes an inflicted atrocity:

> Suetonius recites the crimes of Nero with a certain coolness and tranquility which astonish, and which almost persuade the reader that the enormities he describes excited but faintly, if at all, his indignation and horror. But the historian stops short, changes his style suddenly, and says: "The earth, having suffered such a monster to breathe upon its surface for fourteen, abandoned him at last (*Tale monstrum per quattuordecim annos perpessus terrarum orbi, tandem destituit*)." In this short, but emphatic phrase, the mind is struck with different sensations of *surprise* arising from a variety of circumstances. We are surprised at the historian's sudden change of style, at the discovery of his being affected quite otherwise than we, at first, imagined, and at his describing, in so few words, one of the signal revolutions that happened in the *Roman* empire.[27]

A shock is produced, because Suetonius succeeds in finding a way of expression, which affects the witness, so that, after previous hypocrisy and obliged silence, he felt the revolution of a bestowed freedom.

While insisting, like Suetonius, on the discovery of the unexpected, the renewal of surprise and the revival of thinking, Montesquieu certainly defines the sublime more than the beautiful, if it is true that the beautiful strikes at first by its stability, as an object or an Idea, in the Platonic sense of the term. But, as the soul is "shaken," the "force of judgment" is solicited. Though we have seen the object a long time, "*the mind remains suspended between what it perceives, and what it knows, between the appearance and the reality, and cannot recover from the surprise it feels at the sight.*"[28]

27. *Essay on Taste*, 288–89.

28. Ibid., 304.

Something like pleasure might then attach itself to the constant inadequacy of our immediate perception of an object because we feel that all of our faculties are in a state of alertness. The consciousness of ignorance engenders this sort of *negative knowing* that Socrates, at the dawn of philosophy, reminds us to consider as essential. Substituting the consideration of the beautiful with that of the gracious, the grand, the noble, or the sublime, the genius of Montesquieu thus consists in showing how *philosophy constitutes itself in the eclipse of knowledge, of whom it has for its vocation to recover the traces.*

Left unfinished at Montesquieu's death, the *Essay on Taste* appeared in 1757 in the *Encyclopedie* entry, "Taste" (vol. VII), along with articles by Voltaire and d'Alembert. Edmund Burke (1730–1797) admired Montesquieu's contribution so much that he published a fragment of its translation (the part concerning curiosity and surprise) in *The Annual Register* of 1758. One would readily believe that the sections which open the first two parts of Burke's *Philosophical Enquiry*—one treating novelty, the other astonishment—were directly inspired by the developments of Montesquieu. Such could not, however, have been the case, since the first version of the *Philosophical Enquiry* appeared in 1757.[29]

BURKEAN ASTONISHMENT

The *Enquiry* of 1757 begins with considerations on curiosity. Rather than presenting a metaphysical view, as did Montesquieu, Burke insists on the precocity of this passion and its chronological anteriority to all other passions. "Our soul is made for thinking, that is to say, for perceiving; or one such being *must* have curiosity," declares the first. However, the second states: "The first and simplest emotion which we discover in the human mind is curiosity."

Burke remarks that experiencing life dulls one's curiosity: what passion might, in these conditions, overcome and avoid the "loathing and weariness" that inevitably and infallibly attaches itself to the knowledge of the things of life? Pleasure slips too quickly away from us: "when it

29. In the critical appreciation which precedes the translation, Burke declares that Montesquieu "might eventually appear a little obscure in certain places. If the blame must fall on the translator, it is not necessary to forget that this excellent author, despite his extreme refinement, was not totally exempt from obscurity" (*Literary and Miscellaneous Essays, The Annual Register, or a View of the History, Politics, and Literature, for the year 1758,* 9th ed., London, Dodsley, 1795, 311).

is over, we relapse into indifference." Alone, the pain leaves traces in the mind which strike the witness by their ostensibility.

Consequently, the *Enquiry*'s objective takes shape: to define a pleasure that includes nothing simple or positive, as is the case with novelty, but *rests*, on the contrary, *on a complex relation to pain*. Burke calls it *delight* and explains the difference, from the point of view of the affected subject, between the moment of terror and "a species of passion mixed with terror and surprise": "Let us recollect in what state we have found our minds upon escaping some imminent danger, or on being released from the severity of some cruel pain ... [Our minds were] impressed with a sense of awe, in a sort of tranquility shadowed with horror. The fashion of the countenance and the gesture of the body on such occasions is so correspondent to this state of mind, that any person, a stranger to the cause of the appearance, would rather judge us under some consternation, than in the enjoyment of anything like positive pleasure."

At this point, perhaps the most extraordinary quotation emerges from the *Enquiry*, introducing, in one blow, the representation of horror and the possibility of its attenuation: on the one hand, the crime, or more precisely, the momentary madness, the fate (*atê*) which allows its achievement; on the other hand, the possibility, still only outlined, of an annihilation of its consequences. Can the course of history be partially reversed? Would the crime succeed in no longer involving the crime? Time, suddenly, suspends itself, taking the place of that which dislocates its coherence. The image suspends itself and silence reigns.

> Ὡς δ' ὅτ ἄνδρ' ἄτη πυκινὴ λάβη, ὅς τ' ἐνὶ πάτρη
> φῶτα κατακτείνας ἄλλων ἐξίκετο δῆμον,
> ἀνδρὸς ἐς ἀφνειοῦ, θάμβος δ' ἔχει εἰσοπόωντας.[30]

Burke introduces the translation by Pope:

> As when a wretch, who, conscious of his crime,
> Pursued for murder from his native clime,
> Just gains some frontier, breathless, pale, amazed;
> All gaze, all wonder!

It is necessary to say: "When one is unfortunate, it is known by fate..." because fate (*atê*) is more the cause of the crime than its result, as we know today, thanks most notably to the work of Dodds.[31] The unfor-

30. Homer, *Iliad*, xxiv, 480–82.

31. E. R. Dodds, *The Greeks and the Irrational* (Berkeley: University of California Press, 1959) see ch. 1: "fate is attributed, not to physiological or psychological causes,

tunate person is first of all a victim in the *Iliad*'s world without justice, where shame overrides a culpability which does not yet seem born.

With much accuracy, Burke insists not on the crime, but on the "immanent danger," from which the fugitive just escaped. He affirms that "the sort of mixed passion of terror and surprise, with which [Homer] affects the spectators, paints very strongly *the manner in which we find ourselves affected upon occasions any way similar*." Certainly conscious that the same kind of occasion presents itself rarely, he evokes an easier case to observe: "The tossing of the sea remains after the storm." But the long lasting effect generated by the key example remains: *why does Homer and, after him, Burke, try to sustain by imagination the event which is about to disappear?*

This interpretation of the text of the *Iliad* is delicate. Let us picture the event cinematically in order to grasp its meaning: "the arrested image" of the fugitive rescued from a legitimate revenge. There is nothing more disconcerting than the manner in which it is introduced: King Priam comes to reclaim from his mortal enemy the body of the hero who has killed his best friend. Only accompanied by the *Idaios*, worn down by age and mounted on a carriage that is heavier because of the ransom treasury, the old man comes to the Achaean camp as Achilles' supplicant. Hermes guides him by disguising him from the sight of the mortals and, with a light hand, enables him to cross the enormous barrier of fir trees near Achilles' tent. Peleus' son has just finished his meal, surrounded by his warriors. No one sees Priam enter.

Suddenly the king approached "Achilles, and touche[d] his knees, he kissed the dread and murderous hands that had killed so many of his sons." Then comes the contradictory evocation Burke quoted, of a homicide arriving in a secure place and astounding those who perceive it with astonishment. "So Achilles stared in wonder at Priam. Was he a god? And the others there stared and wondered and looked at each other."[32]

Having miraculously crossed the Achaean lines, does Priam not run the supreme risk in confronting Achilles himself? What does this great king, with courage and stature like an Immortal, have to suffer in

---

but to an exterior and demonic agent." On the opposition between shame culture and guilt culture, utilized notably by Ruth Benedict in *The Saber and the Chrysanthem*, see Dodds, ch. 2 and Thomas Weiskel, *The Romantic Sublime: Studies in the Structure and Psychology of Transcendence*(Baltimore: Johns Hopkins University Press, 1976) ch. 4.

32. Homer, *Iliad,* trans. Stanley Lombardo (Indianapolis: Hackett, 1997) 482 (xxiv.508–18).

comparison with the murderer? Certainly, one can say that this comparison effects astonishment. Moreover, if anguish seizes us, is it that death, whose eventuality seems to have been avoided thanks to the protection of the gods, prowls on all sides: in the murderous hands of Achilles, in the destiny that all mortal men must endure, in the great age of Priam, in the sudden anger of Achilles (who prevented him from obeying the order of Zeus transmitted by Thetis), in his possible contempt for the supplicant, in the "fear" that expresses itself through opposition to Agamemnon and his designs…

"Remember your father": such is the exhortation and the peroration of Priam's admirable discourse, in response to Achilles' "need for sobbing." *The skill of the Trojan consists in obligating Achilles to become like his own father*, such that and so well that Achilles, awakened by filial devotion, laments loudly for being unable to care for the aging Peleus. The astonishment that he has shown regarding Priam transforms itself into respect, so that it is impossible to disentangle what is addressed to the hero and what is addressed to the father: "Ah, the suffering you've had, and the courage. To come here alone to the Greek ships and meet my eye, the man who slaughtered your many fine sons! You have a heart of iron."[33]

Admiration rings at the end of the interview when, having quenched his hunger and thirst, ". . . Priam, son of Dardanus, gazed for a while at Achilles, so big, so much like one of the gods, and Achilles returned his gaze, admiring Priam's face, his words echoing in his mind. When they had their fill of gazing at each other, Priam, old and godlike."[34]

Meanwhile what happened to the ideas of the murder perpetrated under delirium (*atê*) and of the punishment which avoidance became possible? Priam obtained the permission to take the body of Hector with him. Achilles' anger faded away and the war stopped in order to respect twelve days of mourning. But what was the transgression and why was it not punished? The text allows the doubt. According to Weiskel's interpretation, there would be "an unconscious fantasy of parricide"[35] at stake. We cannot but recognize that the father figure will meet the figure

33. Ibid., 483 (xxiv.557–60).

34. Ibid., 486–87 (xxiv.679–84).

35. Weiskel, *Romantic Sublime*, 92. Despite a tendency to overly utilize certain psychoanalytic concepts literally as a lens for interpreting the sublime, Weiskel's work is rich with stimulating insights.

of the one who is "guilty despite himself," in a perturbing way. It arouses feelings in the son that eventually become ambivalent. But only eventually, because Achilles claims not to yield to the enemy father's petition, but instead to the injunction of Zeus: if the petition of the father can be seen as pathetic, then pity imports less than obedience with the king of the gods. "Don't provoke me, old man. It's my own decision to release Hector to you. A messenger came to me from Zeus—my own natural mother, Daughter of the old sea god."[36]

Pity will not be at issue if it is true that this feeling is intended only for one weak and innocent victim. But inasmuch as fear and wonder strike one in the text of Homer, Burke's text is dominated by an anguish that is vanishing, but still going on.

• • •

The vision of the surviving king awakens one to the presence of a *dismayed sympathy*, a pain which is not exempt from a suspect pleasure, not easily avowable, and upon whose presence the text insists: "If this passion [sympathy] was simply painful, we would shun with the greatest care all persons and places that could excite such a passion; as, some who are so far gone in indolence as not to endure any strong impression actually do." How does one deny it? "We delight in seeing things, which so far from doing, our heartiest wishes would be to see redressed" [I.xv] It is certainly not enthusiasm, but the feeling of a brutal request of our faculties, without a predictable intervention of reason.

Where do we then find the sublime? Is it in the fugitive upon whom the gods smile, or is it in the witness who participates imaginatively in his ordeal? The ambiguity is irritating, as it often happens when this concept is evoked. Burke claims that *the sublime is at first a spectacle*, but a spectacle *that we seek to sustain because of our need to master, thanks to imaginative anticipation* in imaginatively anticipating *the horror that threatens us.* Such is the implicit thesis of Burke, as we will try to show.

It is in this sense that the sublime distinguishes itself from the pathetic that seeks to promote pity and has nothing to do with exploiting suffering, which is always a little suspect and nauseating. Only when the signs, which are produced by the unfortunate, are skillfully collected and turned back to the witness, may they give the feeling of the sublime. They may even oblige the lonely victim to retain his egoist claims, at least for

36. Homer, *Iliad*, 484 (xxiv.605–8).

a time, in order to consider suffering and death as simple contingencies. But a turning is nevertheless necessary towards a form of happiness: the happiness of beauty discovered or of *knowing acknowledged, going* with an illusion of *anamnesis*.

If, in effect, the Burkean sublime plays with terror, it is always in order to hold it at the end of a gaze, to transform it, to try at least to master its representation: "When danger or pain press too nearly, they are incapable of delight, and are simply terrible; but *at certain distances, and with certain modifications*, they may be, and they are delightful, as we every day experience."[37]

One might regret that Burke does not explain these "modifications" better, in order to avoid the misinterpretations which his thought has occasioned: hasn't the sublime been used as a sign for the worship of the "shiver" which prevailed in the second half of the eighteenth century? One should not mistake the direction of the famous formula: "Indeed terror is in all cases whatsoever, either more openly or latently the ruling principle of the sublime."[38] Burke wants at first *to avoid only defining the sublime by the effect*, because, as recalled by Michel Delon, "the Moderns cannot accept the sublime if it has not a complete definition,"[39] that is to say, this *real or causal definition*, which they were precisely unable to formulate. While insisting on the limitation which imposes a strictly causal or genetic point of view, Burke undoubtedly tries to take up this challenge: "If it should not appear on the face of the work, I must caution the reader against imagining that I intended a full dissertation on the Sublime and Beautiful. My enquiry went no further than to the origin of these ideas."[40]

The second part of the *Philosophical Enquiry* opens, however, with a definition by the produced effect. Hence the reproach of a vicious circle, when we analyze the title of section 1: "Of the passion caused by the sublime." In place of assigning an origin to the sublime, one transforms it into the producing cause of the effect. Does this not return us to the occult qualities from the Middle Ages? But immediately after this very

37. Burke, *Philosophical Enquiry*, I.vii (emphasis added).

38. Ibid., II.ii.

39. See "Le sublime et l'idée d'énergie: de la théologie au matérialisme," in *Le sublime, Revue d'histoire littéraire de la France* (janv.–fév. 1986) 64.

40. Burke, *Philosophical Enquiry*, preface to 2nd ed.

short section comes the tentative causal explication, followed in 1759 by etymological considerations destined to reinforce it:

> No passion so effectually robs the mind of all its powers of acting and reasoning as fear. For fear being an apprehension of pain or death, it operates in a manner that resembles actual pain . . . Several languages bear a strong testimony to the affinity of these ideas. They frequently use the same word, to signify indifferently the modes of astonishment or admiration and those of terror. θάμβος is in greek, either fear or wonder; δεινός is terrible or respectable; αίδέω, to reverence or to fear. *Vereor* in latin, is what αίδέω is in greek. The Romans used the verb *stupeo,* a term which strongly marks the state of an astonished mind, to express the effect either of simple fear, or of astonishment; the word *attonitus,* (thunderstruck) is equally expressive of the alliance of these ideas.[41]

The difficulty, according to Burke, consists in the fact that *the terrible that causes the surprise is not simple terror, but a defense against it.* Burke twice declares that the effect of the sublime carries *astonishment* to a higher degree and determines its three inferior degrees as being, at one place, admiration, reverence, and respect,[42] and at another place, awe, reverence, and respect.[43] Only the intervention of fear might make one think about an absence of putting terror at a distance. But awe signifies a fear already mixed with respect, which is certainly no longer a panicky terror and constitutes the most elementary degree of reverence.

We will try, then, to understand this scale of the effects of the sublime, in which *astonishment* (l'étonnement) *is placed above admiration, reverence, and respect.*

Admiration constitutes the first degree of the sublime. Descartes's definition is sufficiently complex such that it encapsulates curiosity about the novel, surprise at the extraordinary, and astonishment for the unexpected: "When our first encounter with some object surprises us and we find it novel, or very different from what we formerly knew or from what we supposed it ought to be, this causes us to *wonder and to be astonished* at it. Since all this may happen *before* we know whether or not the object is beneficial to us, I regard wonder as the first of all the

41. Ibid., II.ii.

42. Ibid., II.i.

43. Ibid., IV.vii.

passions."[44] Admiration or wonder is a passion and a *passion par excellence*, because conscience is overwhelmed by Otherness and the anticipations of perception are unable to master the objects which "surprise" the soul and invade its horizon.

Descartes distinguishes between the always useful "force" of admiration and the inevitably toxic excess of astonishment. Admiration leads towards the understanding and the memory of those who, no longer being "stupefied and stupid," are inclined to this passion. Astonishment, in short, "is an excess of admiration and is nothing else but bad." The body remains "immobile like a statue" and the consciousness does not manage to exceed the "first face" of the object.

Why does Burke then "forget" the admiration in the second classification of the effects of the sublime? Perhaps because astonishment includes admiration as well, but, more probably, because the feeling that he describes excludes the representative clarity, which admiration involves. Admiration goes to real objects, but astonishment goes beyond the objects. We do not really find the obscurity of the motives, which characterizes astonishment in reverence and respect.

How, then, might one make a difference between these two last notions? Reverence concerns human beings and means an act of obedience to authority[45]; respect goes to the principles of morality themselves. But such is not the supreme degree of the sublime according to Burke, because the force of the sublime catches us and carries us away, independently of all consideration of practical reason.

As a spectacle, the sublime could not cause the enthusiasm by which the witness identifies himself with the producing cause. As a spectacle, whose principle is indeterminate, it calls this formidable astonishment which we call "dazzling,"[46] that is to say, both the burst of light and the trouble of the vision. That the excess of light might produce obscurity, that ignorance arrives at the recognition of its abyss, that the absolute preeminence of the Other imposes itself on the subject, all that constitutes the experience of the sublime. Surrounded by darkness and

44. René Descartes, *The Passions of the Soul*, article 53, in *The Philosophical Writings of Descartes*, vol. 1, eds. J. Cottingham, R. Stoothoff, D. Murdoch (Cambridge: Cambridge University Press, 1985) 350 (emphasis added).

45. See Samuel Johnson, *A Dictionary of the English Language*, 1st edition, 1755, article "Reverence," 2nd sense.

46. See Saint Girons, *Fiat Lux*, introduction.

deprived of the habitual lights that are our clear ideas, we are subjected to a *threat that paralyses the need to act, without hindering our faculties of contemplation*, because our good fortune enables us to provisionally avoid the powers that surround us. The astonishment is thus the *fear surmounted in a surprising wonder*,[47] as Johnson and Burke remarked at the same time.[48]

But upon what does wonder rely? Upon being rescued from danger? Upon that which sustains the imagination in peril? Upon the capacity to face the representation of the sublime? In any case, Burke is firm on this point, that happiness takes shape at the dawn of the sublime experience rather than being born from an egoist feeling of security.[49]

The principle example of sublime architecture invoked by Burke is *Stonehenge*, which we have elsewhere analyzed in our considerations on grandeur.[50] Here we'll introduce a contemporary of Burke who was obsessed by the ruins of the most ancient edifices, but lived among the debris of the Roman Empire and was endowed with prodigious, inventive artistry. What quality of impression does Piranesi's *Carceri d'invenzione* give us?

Piranesi proves his extreme skill and absolute taste, but we are obliged to overcome a uneasy feeling, as if cornered before the image that we are locked up within ourselves. We are reduced to something insignificant, and it is strange that we almost draw pleasure from this annihilation: such an experience seems an exercise of vital importance for the self, through which the seriousness of the aesthetic experiment is joined to playfulness.

This reminds us of one of the most impressive pages of the psychoanalytic literature, which Freud dedicated, in 1920 in his *Beyond the Pleasure Principle*, to the description of a game of loss invented or repro-

47. When Burke utilizes *wonder*, it serves, like surprise, in the commentary on Homer, analyzed above, but also in relationship to the artificial infinite in the succession of sounds (see *Philosophical Enquiry*, IV.xi). We mention that *Wonder* defines the aesthetic experience according to R.W. Hepburn, notably in his '*Wonder' and other Essays, eight Studies in Aesthetics and neighbouring Fields* (Edinburgh: University Press, 1984).

48. Johnson defines astonishment as "confusion of the mind which changes fear to wonder." For Burke, see *Philosophical Enquiry*, II.ii.

49. We have spoken of this point in our *Fiat Lux*, where we compared the systematic system of Burke's *Philosophical Enquiry*, II.v, to Kant's *Critique of the Power of Judgment*, §28.

50. See our *Fiat Lux*, chapter 1.

PIRANESI, *Carceri d'invenzione* (1760–1761), board XV. The skillful use of the oblique perspective (*scena per angolo*) of the baroque scene, inspired by Giuseppe Bibbiena, permits Piranesi to provide multiple sources of perspectives without fully freeing the imagination. His ladders neither ascend nor descend, the idea of a sky is absent, and any guide is missing in this labyrinthine space. The more the witness contemplates the work of art, the more he feels its oppression taking over. But at the same time there is nothing ugly in this dark and mineral world, although it grows more and more encumbered as Piranesi progresses in his work: an insurmountable solitude reigns there, which is symbolized by the mascarons, in whose lips are closed around rings of servitude.

---

duced by an eighteen-month-old child.[51] The little boy naturally behaves in a particularly gentle way, having already acquired a sense of suitability (*anständig*) and trying not to disturb the others and, notably, not to cry at the departure of his mother. But what is his whim? He would prefer to throw away every little object that he might seize while emitting "with an expression of interest and satisfaction" a sound that signifies "go

51. See *Jenseits des Lustprinzips*, 1920, II, *Gesammelte Werke* (Fischer, 1940) vol. XIII.11–14.

away!" (*Fort!*) No doubt the complete game would include the triumph of the return of the object: "here! (*Da!*) But the most extraordinary thing for us consists in the first act: if any departure of a loved object (even one to which the child is accustomed) is painful, why is there such an enjoyment attached to the simulation of the loved object's departure? Freud hesitates between two conceptions: does the enjoyment rest on the transformation between passivity and activity or on the repetition? Should it be interpreted as the satisfaction of a need for control or as the discovery of the savor of repetition?

In order to mitigate a truly real need, it would be necessary to simulate control over sufferings caused by the loss of the loved object or to reconstitute the object in its imaginary perfection. Has the child not then answered our question through his behavior? Nostalgia of the original splendor and obsession of controlling the uncontrollable join together, producing the astonishment of the sublime enjoyment.

Burke does not go on in his analysis this way. But if he does not find a solution to the problem, his merit lies in posing the problem with the greatest exactitude. How might the greatest enjoyment be born in a certain relationship to pain? To take up again the examples of Michelangelo's *Pietà* or Piranesi's universal prison, their contemplation allows us to understand both the ridiculousness of our pretensions as individuals, and the greatness of our representative faculty, and even, in the best cases, the strength of our inventive power. The difficulty, however, lies in the risk of total disconnection between the worlds of contemplation and action. The illusion of recognition, even the quasi-hallucinatory character proper to a work of art, but also the apparent gratuity of the effort to sustain the gaze, are bringing out a radical suspension of the ego. The game, then, far from preparing the experience, replaced it.

Astonishment is, in effect, neither a principle of creation nor a principle of action. It takes its source in the sensibility of the terrible, in the very complex enjoyment of its re-presentation and in the effort of controlling the play. The complete experience is kept at a distance because its aspects are most tolerable in profile; astonishment opens itself up to the unknowable and the sublime constitutes a certain form of "transcendental experience," using Schelling's expression, but without limiting us to his meaning. It signifies that we become sensible to the enigma of our origin and our destination and, at the same time, to the frightening entanglement of sense and nonsense.

The principal objection comes out: if the sublime arouses wonder, couldn't astonishment conversely cause the sublime? Is not the sublime all that which inspires astonishment? The problem for Burke is, as we have said, to think the aesthetic effect from a philosophy of origin onwards. So his point of departure is a *radical principle of dispossession*, which threatens our physical, psychological, and moral integrity . . .

—Translated by Michael Funk Deckard

13

Two Sources of Wonder in Early Modern Judaism

Roberta Sabbath

Everyman has to find it by himself.
In moments of sensing the ineffable,
we are certain of the value which was worthy
of the world's coming into existence . . .[1]

The ultimate insight is the outcome of moments
when we are stirred beyond words,
of instants of wonder, awe, praise, fear, trembling,
and radical amazement; of awareness of grandeur,
of perceptions we can grasp but are unable to convey,
of discoveries of the unknown, of moments in which we abandon
the pretense of being acquainted with the world,
of knowledge by inacquaintance.[2]

THE JEWISH IDEAS OF wonder constitute a tapestry whose porous and interwoven nature has enriched these notions through the centuries. The threads of this tapestry continue to represent spiritual, rational, and imaginative forces from the earliest traditions of the Jewish cultural experience. The language of philosophy, Christianity, Islam, Eastern beliefs, and aesthetics finds its way into a conversation about wonder and Jewish thought. Thus, the Jewish experience of wonder in early modern

1. Abraham Joshua Heschel, *Man is Not Alone: A Philosophy of Religion* (New York: Harper & Row, 1951) 22.

2. Heschel, *God in Search of Man* (New York: Harper, 1966) 131. Although the citation is in verse form, it appears in the above works as prose. I first heard these words read by David B. Ruderman in "Abraham Heschel—Mystic & Social Activist," *Jewish Intellectual History: 16th to 20th Century* (Chantilly, VA: The Teaching Company, 2002). Heschel is born in 1907 in Poland and dies 1972 in the USA.

Europe has many dimensions; one might even say, it is a palimpsest of ideas that come alive in the minds, imaginations, and the hearts of the people who embrace them, whether passively or actively, innocently or wisely, naively or with self-reflection.

One fundamental concept embraced in this contribution is that the subjective experience is informed by the complex, interwoven nature of the full human experience. The concept of wonder is here assumed to be a daily human experience that comes in many forms, many levels of intensity and immediacy, and is the beneficiary of many names. Within this concept, wonder is considered an essential human experience. The religious believer embraces the divine presence. A philosophical concept of wonder might come with considering what lies beyond the sensible or material human experience. A dancer might experience wonder when performing. Looking at a child might engender a feeling of wonder. And, in a moment of sensible pleasure or emotional and intelligible intensity, wonder can be the surprise of any day.

That having been said, I hope to unravel that tapestry so that the threads can then be understood individually while being appreciated for the complex, lived tapestry that they create. Wonder has been named through the centuries of philosophical, imaginative, and religious thought using many metaphors; its metonymic representations come in many iconic forms; and its allegories provide rich visual narratives that inform, even structure, our cultural lives.

When authors from the sixteenth through the eighteenth centuries write about "wonder" and "awe," about the "ineffable," and about "fear and trembling," they are not talking about an aesthetic experience devoid of theological content but an aesthetic of divine presence. In the practices and experiences mentioned below, the concept of a transcendent God is understood, assumed. Depending on the century, the geography, and the cultural practice, other super-sensible entities also explain phenomena. In the divine mix can be angels, spirits, anthropomorphized forces of good and ill, and dynamic powers, all of which figure into the expression of wonder as found in institutional and non-institutional Jewish practice.

The dispersed Jewish communities of this period share with their non-Jewish neighbors a devotion to religious life. Until the modern period, religion serves to organize society. Yet, the picture is not monolithic. Typical of the nature of European communal life is the porous nature of religious practice. Several major threads of religious and philosophical thought are born in the early centuries of the first millennium with the early Jewish, Christian, and later Muslim patriarchal writings. Their works serve as the well spring from which religious and philosophical practice and discourse of the Late Renaissance and Early Modern period grow.

A SHORT HISTORY OF JEWISH EXPERIENCE

Although Jews live a marginalized life in the medieval period, the occasional integration with host communities produces remarkable results. Three historical moments serve to exemplify the phenomenon of interfaith porousness and the rich traditions that emerge and flourish as a result: first, the Arabic/Muslim Empire of the Iberian Peninsula; second, the Languedoc area of France; and third, in the northwest region of France. These three areas typify the welcoming circumstances that allow Jews and non-Jew alike to flourish. Regions where all strata of society live and participate in urban life generally welcome Jews. This is in spite of laws forbidding Jews to own land and assigning them restricted areas of residence in urban centers (an arrangement formalized as ghetto life by the Renaissance period). Thus the sophisticated cities of the Arabic Empire and northwest and southern France offer urban settings that understand and embrace Jewish contributions.[3]

In northwestern France in the twelfth century, Rashi, hallowed commentator of the Torah, moves his school from Spain to Troyes, France. Typical of the interweaving of cultures possible amongst enlightened political leadership, the wealthy city of trade and commerce not only shows tolerance for the Jews and their pursuit of learning but also serves as the home of Chrétien de Troyes, who, with the patronage of Marie de France, daughter of Eleanor of Aquitaine, produces the first popularized Arthurian legends as a secularized antidote to the failed second crusade. The tradition of taking the idea of wonder out of the institutional religious venue of Church or Temple is thus launched amidst the intense religious world of Christian and Jew. Chrétien de Troyes takes the Christian mythos of individual salvation and adapts the allegorical, religious story to a secular quest narrative that becomes the structure for

3. Mark R. Cohen, *Under Crescent and Cross: The Jews in the Middle Ages* (Princeton: Princeton University Press, 1994)125.

the genre of romance. Yet while the narrative may be secular, the experience of wonder nevertheless continues its intensely personal reality.

In the Languedoc region of southwestern France, another important instance of inter-faith influence gives birth to a new religious/philosophical tradition known as Kabbalah, a Jewish mystical movement that ultimately sweeps throughout the entire Jewish diaspora and Israel. The openness of contact between the Jewish and Christian Cathar community in the region may serve to influence the earliest known ideas of Kabbalah.[4] In the twelfth and thirteen centuries in the Languedoc region, Cathar communities mushroom. The Cathari oppose what they believe to be the corruption of the Roman Catholic Church. The powerful hold of Catharism comprises a serious threat to Church hegemony which considers the Cathars as heretics. By the middle of the thirteenth century as a result of the Albigensian or Cathar Crusade, persecution is ruthless and decisive, eventually resulting in the Church reasserting its dominion. But, before that crusade, Cathari Gnostic and Neoplatonic beliefs spread into the porous Jewish communities of the area. Kabbalah mysticism is specifically dated to this period in Languedoc, France, with the writing of the *Bahir*, the earliest Kabbalah text.[5] The twelfth century Provençal scholars confirm their original discovery by insisting that that the *Bahir* is revealed to them. With its twelfth-century birth in southwestern France, the kabbalah system of beliefs develops and spreads throughout Europe.

Another key of Jewish thought occurs in the twelfth century with the life and work of Moses Maimonides who exemplifies the Golden Age of the Iberian Peninsula Arabic empire in the twelfth century. The greatest of Jewish thinkers in the medieval period, and arguably in post-Talmudic times, Maimonides influences and is influenced by classic Greek, Arabic, and Christian scholarship. For Maimonides, Aristotle is the model of a prophet. Like the Muslim philosopher, Ibn Rushd (Averroes), whom he does not personally know but whose works and those of other Muslim philosophers he knows, Maimonides sees the works of Aristotle as a philosophic tool that improves the practice of theology. Maimonides writes the *Mishneh Torah* and *The Guide to the Perplexed* "which laid the foundations for the entire development of

4. Gershom Scholem, *Kabbalah* (New York: Quadrangle/New York Times Book Co., 1974) 44–45.

5. Ibid., 42–43. See also the contribution above by Elisabeth Blum and Paul Richard Blum on other uses of the Kabbalah in the Renaissance.

Jewish philosophy"[6] He is known as one of the greatest physicians of the medieval period and serves as the court physician under the Grand Vizier, Alfadhil and Sultan Saladin of Egypt. Maimonides influences the European Scholastic tradition including the work of Thomas Aquinas. In spite of his brilliant contributions, the plight of Maimonides and his family is typical of Jewish families throughout the Iberian Peninsula of this period. By Maimonides' time, the clash of empires heralds the end to this Iberian Golden Age. Political disruption, with the loss of control by the caliphate in the Iberian Peninsula, triggers persecutions which begin, at first sporadically but develop more consistently, forcing Maimonides and his family to flee to north Africa and ultimately to Egypt.

Following the rich and relatively peaceful period of the earlier medieval centuries begins a period marked by disease, destruction, and death on the one hand and, on the other hand, dramatic social changes that adumbrates the modern period. In the fourteenth century, Central Europe is hit by the Black Death, which kills between one- and two-thirds of the population. The social structure of society is changed, with a mixed bag of consequences. The Roman Catholic Church receives a serious blow resulting in the persecution of minorities, including the Jews. Urban centers develop. A growing middle class produces a variety of goods and services and thus becomes a threat to the landed aristocracy. Opportunity for travel and trade between regions increases and universities encourage the exchange of ideas between cultures. Persecution of the Jews also increases. Beginning in the fourteenth century, the stereotype of Jew as Christ killer coalesces.[7]

Spear-headed by religious and political forces, the Inquisition intensifies at the end of the fifteenth century, proving catastrophic for Jewish communal life. Queen Isabella and King Ferdinand expel Jews and Muslims from the Iberian Peninsula, thus marking the definitive end of the Golden Age of the Iberian Peninsula. Jews are forced either to convert or die. Many flee. Many, called Marranos, convert to Christianity but continue to practice Judaism in secret.[8] Many of these crypto-Jews, also loosely called conversos, escape at first to Portugal,

6. Isidore Epstein, *Judaism* (Middlesex: Penguin, 1959) 208.

7. Anna Foa, *The Jews of Europe after the Black Death*, trans. Andrea Grover. (Berkley: University of California Press, 2000) 16–22.

8. James Carroll, *Constantine's Sword: The Church and the Jews: A History* (Boston: Houghton Mifflin, 2001) 347.

then Amsterdam, Italy, Germany, and later to France as well as North Africa and the Near East.

From Spain, some Jews flee to Turkey. The Sultan of the Ottoman Empire, Beyazid II (1481–1512), sends ships to transport fifty thousand Jews to Istanbul, famously declaring, "The Catholic monarch Ferdinand was wrongly considered as wise, since he impoverished Spain by the expulsion of the Jews, and enriched Turkey."[9] The sultan recognizes the contribution that his new citizens will make to education, commerce, and culture. These Jews join the already existing, diverse community of Jews who had emigrated from all over Europe. "Within two decades after the expulsions from Iberia, Jewish philosophy flourishes once again alongside an unprecedented creativity in biblical exegesis, *halakhah* (Jewish law), homiletics, and poetry," expressing a determination to preserve memories of a glorious past.[10] During the years of early modernism, Jews enjoy relative prosperity in Turkey. They are the leaders in commerce and trade and serve in diplomatic and other high offices while suffering, in localized events, from persecution, massacre, and deportations.

By the early sixteenth century, the Reformation begins, and the Church is again challenged. Legal restrictions and penalties keep Jews isolated from the general population. High taxes, lack of protection against debtors, accusations of ritual murder and desecration, banishments, extortion, wealth and property "grabbing" by rulers, forced conversions, sudden violence keep the Jewish community theoretically in a perpetual state of imbalance. By the early seventeenth century, the Thirty Years' War, largely a religious war, arguably results in the death of twenty to thirty percent of Europe's population. Amidst all these political and religious vicissitudes, many Jewish communities serve as the scapegoats. The role of Jews as wartime suppliers improves the lot of others.[11] Another defining catastrophe for the Jewish communities in particular occurs in the mid-seventeenth century with the Chmielnicki Massacres (*Pogroms*), 1648–1649, lead by the infamous Cossack, which terrorizes hundreds of Jewish communities in Eastern Europe and brings the total

9. *The Quincentennial Foundation Museum of Turkish Jews* (Istanbul: Jewish Museum of Turkey Press, 2004) 79.

10. Hava Tirosh-Rothschild, "Jewish philosophy on the eve of modernity," in *History of Jewish Philosophy*, eds. Daniel H. Frank and Oliver Leaman (London: Routledge, 1997) 529.

11. Foa, *The Jews of Europe after the Black Death*, 159–60.

devastation to about three hundred of these communities. According to recent estimates, between forty thousand and fifty thousand Jews die—about 20 to 25 percent of the Jewish population. Jews flee westward, into Western Europe, most notably, into Germany.[12]

Whether Jews live apart from the their rural and urban neighbors or in the organized society of the ghetto (instituted in the late fifteenth century by Germany and in the early sixteenth century by Italy), they create, through the centuries of isolation, an environment that encourages close family, communal, educational, and religious ties.[13] The language of Hebrew, languages from the Slavic origins of the community as well as German, all create a diverse cultural life that includes legend, myth, scholarly debate, and inspirational and instructional expression in both oral and written religious and secular traditions. Scholarly and religious life center around the study of the Tanakh (Torah or Five Books of Moses, Prophets, and Wisdom Literature), the Talmud (rabbinic law dating from 200 to 500 CE), and contemporaneous rabbinic judgments contribute to the intellectual acumen among the Jewish community members. In order to participate, a Jew needs to read, to learn a myriad of detailed traditions and practices, and to participate in the communal institutions. Yet, with the glimmers of hope that life outside ghetto walls is possible, most Jews would come to embrace the change.

Finally, by the end of the seventeenth century, with the exception of localized persecutions, Jews begin to become more integrated in the commercial and social life of their host countries. Although ghetto restrictions ban their access to the infrastructure of Christian communal life including educational and social activities, Jews are allowed to serve as peddlers within and between Christian communities. Many Jews travel outside of institutional walls of Judaism's core synagogue life and traditional communal rituals, whether through Spain, France, England, Germany, Eastern Europe, Italy, or the Ottoman Empire. Because Jews move through these areas, thanks to commerce and evictions and worse, their ideas move with them. Layperson and rabbi alike have occasion to travel, whether to earn a living as a peddler, to visit the Holy Land, or to seek out coreligionists.[14] In exceptional cases, philosophical exchanges

12. Ibid., 187–88.

13. Ibid., 138–53.

14. Eli Yassif, *The Hebrew Folktale: History, Genre, Meaning*, trans. Jacqueline S. Teitelbaum (Bloomington: Indiana University Press, 1999) 5.

of Jewish community leaders occur with religious, philosophical, or political leaders of the host State.

A variety of Jewish traditions diffuse throughout Europe during this turbulent period. They provide for several avenues of the personal and communal experience of wonder. The more traditional expression of Jewish practice relies on the prophetic emphasis of rabbinic Judaism with its familiar exhortations of obedience and observance. The more personal attraction to the mystical experience also holds a powerful sway over popular Jewish imagination and practice. Prophecy and rabbinic Judaism address the integrity of the Israelite Society. Mysticism pertains to personal spirituality, the esoteric. Judaism has always included both aspects.

In the seventeenth and eighteenth centuries, each of these vibrant threads within the Jewish tradition takes a unique turn. With the entrance of Sabbatai Zvi on the world stage, esoteric Judaism embraces an apocalyptic dimension and morphs into a messianic frenzy. With the advent of Moses Mendelssohn, a product of rabbinic Judaism embraces the language of Enlightenment philosophy to express his experience of esoteric Judaism. The difference of one century is dramatic. "In the mid-seventeenth century, Jewish communities throughout the world trembled on the brink of a strange future."[15] Zvi enflames the embers of the apocalyptic spirit as Jews try to make sense out of the centuries-old string of catastrophes as they are simultaneously tantalized, yet confused, by the hint of an improved future with enfranchisement in Amsterdam. In eighteenth-century Europe, Mendelssohn rides on a chariot of hope engendered by greater social and economic promise for Jews as ghetto walls begin to fall. The sense of wonder of divine presence moves from the streets into the salon.

MESSIANISM

Deuteronomy portrays Moses as both prophet and mystic.[16] Yet, the Tanakh and the rabbinic Talmudic tradition emphasize the communal

15. Harris Lenowitz, *The Jewish Messiahs: From the Galilee to Crown Heights* (New York: Oxford University Press, 1998) 149.

16. "There has never again arisen in Israel, a prophet like Moses." (Deut 34:10); Moses encounters God "face to face" (Exod 33:11, Num 12:8, and Deut. 34:10); quoted in Mark Verman, *The Book of Contemplation: Medieval Jewish Mystical Sources* (New York: SUNY Press, 1992) 24.

nature of religious practice, resisting the esoteric nature of mysticism as a threat to exoteric Judaism. The core rabbinic tradition also resists the promise that mystical practice will hasten the messianic age. Traditional rabbinic practice focuses on communal integrity and worship.

Yet Gershom Scholem warns that to dismiss Jewish messianism and apocalyptic beliefs is to falsify an age-old aspect of the religion: "It has been one of the strangest errors of the modern *Wissenschaft des Judentums*[17] to deny the continuity of Jewish apocalypticism."[18] Even Maimonides succeeds in "suppressing completely the apocalyptic moment."[19] But esoteric mysticism is well represented in the rabbinic tradition.[20] Eschatological motifs also persist, if among a minority of rabbinic works, from the very beginnings of rabbinic expression.[21]

In the Jewish tradition, apocalyptic eschatology is relatively young. It begins with the Book of Daniel written in the second century BCE, well after the Torah (Five Books of Moses) is completed. End of time eschatology that includes a messianic message is expressed in the Pseudepigrapha, written between 200 BCE and 200 CE, which includes the *Book of Enoch, Jubilees,* the *Syriac Book of Baruch*.[22] This development in Jewish theology can be explained by the advent of Greek Platonic philosophy and its separation of mind and body: "The Greek and the developing Jewish apocalyptic view of the world was both personal and communal in nature and can be indirectly linked with the Socratic and later Stoic concept of duty. The motif of the redemptive messiah suffering to save mankind enters into the Jewish imagination."[23] Mysticism linked to messianism threads its way within the ranks of rabbinic

17. Loose association of nineteenth- and twentieth-century Judaic scholars primarily in Europe.

18. Gershon Scholem, *Sabbatai Sevi* (Princeton: Princeton University Press, 1973) 9.

19. Ibid., 13.

20. Ira Chernus, *Mysticism in Rabbinic Judaism: Studies in the History of Midrash* (Berlin: de Gruyter, 1982) 99.

21. Ibid., 90. *Eschatology* refers to a set of beliefs about the end of days or the termination of familiar life when a day of divine judgment arrives. An apocalyptic moment occurs when believers, or those who have been righteous according to the rituals and acts of obedience, are saved for eternal life.

22. Dan Cohn-Sherbok, *Jewish Messiah* (Edinburgh: T. & T. Clark, 1997) 27–32.

23. Richard A. Freund, "The Apocalypse according to the Rabbis: Divergent Rabbinic Views on the End of Days," in *Millennialism from the Hebrew Bible to the Present*, eds. Leonard J. Greenspoon and Ronald A. Simkins (Omaha: Creighton University Press, 2002) 115–16.

Judaism, a tradition which grows in intensity after the Romans destroy the second temple in 70 CE. A most notable example is Rabbi Akiva, the martyr, who acknowledges his mystical and messianic beliefs. He places his tragic messianic hope in the failed Simeon bar Kochba revolt against the Romans in 132 CE.[24]

Jewish mysticism takes on many forms and has many voices.[25] Merkabah mysticism flourishes early in the rabbinic tradition. Merkabah refers to the chariot that the prophet Elijah is said to have ridden as he ascends to heaven and the chariot upon which the Lord appears to the prophet Ezekiel. Only a select few of the rabbis and sages at the start of the rabbinic tradition express having had an esoteric experience, but they suggest a redemptive effect.[26] The beginning of Jewish mysticism finds its roots in this early spiritual enthusiasm that would be sustained through to the present day.

Kabbalah mysticism explodes out of twelfth-century southern Europe and would develop broadly along three basic paths: theoretical (description of the cosmos), meditative (experience of divine presence), and practical (ways to evoke the supernatural).[27] The theoretical aspect of Kabbalah mysticism incorporates beliefs that describe a cosmic structure which resembles a Neoplatonic view of the world. Under the neo-Platonic view, the world is viewed as an emanation of the attributes of God. Under the Kabbalist view, the world is comprised of the divine emanations, including the world and human life, which are testament to the greatness of God's creations.[28] In the sixteenth century, Rabbi Isaac Luria of Safed, Israel, develops the kabbalah for his messianic vision. Luriac Kabbalah is not militant or ascetic but is "oriented toward the Apocalypse, adhering to a mystic worldview that seeks the repair of the world in a cosmic sense."[29] Kabbalah understands the universe as broken because of God's initial withdrawal (*tzimtzum*). The messiah's success depends on the efforts of each Jew to repair the universe (*tikkun*).[30]

24. Ibid., 122.

25. Cohn-Sherbok, *Jewish Messiah*, 85.

26. Chernus, *Mysticism in Rabbinic Judaism*, 76.

27. Aryeh Kaplan, *Meditation and Kabbalah* (York Beach, ME: Samuel Weiser, Inc., 1982).

28. Scholem, *Kabbalah*, 98.

29. Lenowitz, *Jewish Messiahs*, 125.

30. Ibid., 128.

Among the kabbalah tenets that Luria develops is an eschatology that sees the disasters that befall Jews not as punishment but as cosmological birth pangs announcing the messiah.[31] Luria joins the esoteric experience with strict observance of Jewish law in order to create a world order that would hasten the arrival of the messiah.

The popular nature of kabbalist practice and belief by the sixteenth century contributes to Zvi's ability to intensify already wide-spread esoteric practices. Gershon Scholem writes that, in the heart of Europe, kabbalah is available to everyone:

> One conclusion emerges clearly. Polish kabbalism in its golden age (that is, in the early seventeenth century), consists of Lurianic ideas grafted onto a stock of new and original conceptions evolved by the Polish kabbalists themselves. The popular and almost folkloristic character of many of these undoubtedly contributed to their wide reception and distinguishes them sharply from the closely knit systems of Cordovero, Luria, and Sarug.[32]

In practical ways, kabbalah moves from an esoteric practice, whose mystical experience is only available to a select group of the initiate, to the streets and thereby into the hearts and minds of the uninitiated, available to anyone who wishes to believe in its promises of ecstatic closeness to God. At the same time, kabbalah moves from purely mystical aspirations to a set of tenets that includes the apocalyptic and utopian tradition against which Maimonides had so sternly warned.

The stage for the arrival of the next messianic hopeful is set, by the early seventeenth century, with the expectation of the arrival of another messiah. Despair feeds the fire of hope for a messianic time. Sabbatai Zvi (1626–1676), born in Smyrna (Izmir today), Turkey, responds to the call. By the time he is twenty two years old, in 1648, Zvi decides he is the messiah.[33] After his early years of traditional yeshiva education, Zvi discovers the kabbalist traditions as developed a century earlier by Rabbi Isaac Luria. Zvi becomes a self-proclaimed messiah incorporating the messages of asceticism of the Luriac Kabbalist tradition. His message of salvation spreads like wildfire throughout Europe and the Ottoman Empire. It is fueled by the desperate straits of many

31. Ibid., 126.

32. Scholem, *Sabbatai Sevi*, 85.

33. With parents of Greek ancestry, Zvi was born in Smyrna (today's Izmir, a city on the Aegean coast of Turkey). See Scholem, *Sabbatai Sevi*, 136.

of the Jews in these areas. The frenzy of Sabbatianism knows no limits. Communities of all kinds are taken by the power of this movement. The desperate communities of Northern Morocco, the up-and-coming communities of the Middle Eastern Ottoman Empire, the solid communities of Constantinople, Salonika (Greece), Leghorn (Italy), Amsterdam, and Hamburg all experience the messianic force. Rich and poor, rabbis and parishioners, peddlers and businessmen, bequeath their earthly possessions to the messianic movement. Whole communities institute ascetic practices, penitential prayer, and additional study of the holy books.[34]

The following excerpts are taken from the testimony of Rabbi Leib ben Ozer, written not long after Zvi died which is ten years after his conversion to Islam. According to Harris Lenowitz, ben Ozer's testimony is believable because he is neither a follower nor a critique of Zvi. Also ben Ozer documents his reports with close attention to detail, even mentioning when he lacks confirming evidence. In one report, ben Ozer details the account made by the Zvi's loyal follower and interpreter, Nathan the prophet, who first identifies Zvi as the messiah and who accompanies him, spreading his messianic message, throughout Europe. Nathan explains how the Holy Spirit from God enabled Zvi to attain redemption for his generation and that the Shekhina had arisen out of exile. He also reports that the messiah, Sabbatai Zvi, had suffered exile and torture and in so doing had atoned for all of Israel. He continues, describing the second resurrection which would include all the faithful: "The resurrection of the dead will take place in the Land of Israel for the righteous buried there, and the evil will be cast out of the land and will not rise until the general resurrection, after forty years, when all the dead outside the Land of Israel will rise."

Ben Ozer's report is characteristic of Jewish eschatology and messianic aspirations. Thus, the culmination of the apocalyptic judgment will result in a utopic vision of life on this earth. In another report, ben Ozer expresses his disbelief that so many became enthralled by what turned out to be the false messianic claims of Zvi. Yet, just as he shares his disbelief, he attests to one of many miraculous events. Ben Ozer reports, in the month of Tevet 1665, in places as removed as Izmir, Constantinople, Adrianopole, and Salonika, thousands of people, including little children, prophesy in the holy tongue, in the language of the Zohar [kabbalah

34. Scholem, *Sabbatai Sevi*, 553.

scripture], without ever having known a word of either, and speak, as someone struck with epilepsy, secrets of the kabbalah and proclaim:

> The reign of Shabtai Zvi, our lord, our king, our messiah, has been revealed in Heaven and Earth and he has received the crown of the kingship of Heaven . . . And wherever you went you heard nothing but that Mr. So-and-so had become a prophet and that Miss So-and-so had become a prophetess; and here there was a company of prophets, some prophesying in one way and others in another way, but the sum of the matter was always that Shabtai Zvi was the messiah and our righteous redeemer. Now, a man might think that all this was trickery and some people created themselves prophets. Not so![35]

At the height of this frenzy, Zvi is eventually discredited. He is excommunicated by the Polish rabbis, flees to Turkey, is arrested by Ottoman authorities for causing disruption and, in 1666, when given the option to die or to convert to Islam, Zvi converts.

Even with the apostasy of Zvi, his messianic movement has considerable influence on the state of Jewish religious institutions. Change had been in the air for a long time. Because kabbalah is theocentric and anthropocentric, Jews and Jewish scholars begin to look differently at nature and the self, anticipating the transition to the modern era.[36] And, because kabbalah frenzy links communities of Jews to this shared belief, its existence threatens the very nature of the relationship between the rabbi and his flock. Also, by the end of the medieval period with the invention of the printing press, literacy spreads amongst the lay population and democratized learning in the ghetto.[37] The end of the seventeenth century sees the power of the rabbinic control challenged and diminished. The winds of change not only blow from within. Outside the ghetto walls, the opportunity for Jewish involvement with host communities in commerce, education, and the Enlightenment philosophical project

35. Lenowitz, *Jewish Messiahs*, 152–57. Lenowitz writes, "The passages here are taken from the account written by Leib ben Ozer, in Yiddish, and are taken here from Sh. Zucker and R. Plesser, eds., Zalman Shazar's *The Story of Shabbetay Zevi by R. Leib ben R. Ozer (Amsterdam, 1711–1718)* from the original Yiddish ms. with Hebrew translation, introduction, and notes (Jerusalem 1978).

36. Tirosh-Rothschild, "Jewish Philosophy," 528.

37. Nathan Ausubel, *A Treasury of Jewish Folklore: Stories, Traditions, Legends, Humor, Wisdom and Folk Songs of the Jewish People* (New York: Crown Publishers, 1979) xxiii. See also Tirosh-Rothschild, "Jewish Philosophy," 531.

is possible. With the end of the persecutions of the late Renaissance and the seventeenth century, the eighteenth century dawns on a promising future for Jews in Europe. The ghetto is still a part of reality, but the ghetto walls have become more porous.

PHILOSOPHY AND AESTHETICS

In Berlin, Moses Mendelssohn, a practicing Jew, represents the hope for a different kind of future for Jews. Mendelssohn stands within his rich Jewish milieu with its synagogues, mystical and messianic traditions and history, revealed religion, commandments, communal life; with its rabbis, its love of education, and its ghettos. Mendelssohn steps out of the ghetto and enjoys the fresh air. But unlike many of his Enlightenment colleagues in philosophy, he never abandons his religious beliefs. He just doesn't talk about them until he writes *Jerusalem* in defense of his fellow Jews and against the anti-Semitism that persists in keeping the Jews separated from society and vulnerable to its whims.

For Mendelssohn, the experience of wonder is not confined to Jews but rather is universal. He embraces the possibility of the universal, an Enlightenment axiom, through the common experience of wonder. Like other Enlightenment thinkers, Mendelssohn, in his works on aesthetics, generously opines on the nature of the sublime. The rhetoric of the aesthetic of the sublime finds a welcome home among those Enlightenment thinkers who, on the one hand, embrace the optimistic and even utopic vision of the Enlightenment and its rationalist philosophies while, on the other hand, remain averse to the encroachment of a secular Enlightenment world and a drive towards universal religion with an implicit threat against particular religious practices.

In his comprehensive discussion of Mendelssohn, *L'esthétique de Moses Mendelssohn*, Jean-Paul Meier argues of this Berlin Jew that "ses modestes travaux auront constitué une brève tentative d'esthétique pure"; that is to say, Mendelssohn sustains his discussion of the arts and the psychological and emotional responses they inspire without an over weighted emphasis on morality, teleology, faith, or education.[38] Meier places Mendelssohn at the crossroads of philosophic and aesthetic discourse in eighteenth-century Europe. In the salons of Berlin in the

38. Jean-Paul Meier, *L'esthétique de Moses Mendelssohn (1729–1786)*, 2 vols. (Paris: Diffusion, Librairie Honore Champion, 1978) 2:881–82. This is the publication of his dissertation.

early part of the eighteenth century, Leibniz dominates the discourse. Mendelssohn consistently retains his orientation toward Leibniz whose rationalism combined with belief in the monotheistic God and his systematic philosophical approach to proving that this world is the best of all possible worlds constructed by a beneficent and loving God underpins Mendelssohn's belief that Judaism is a religion based on reason, one requiring no leaps of faith. In his "Dialogues," Mendelssohn embraces Leibniz's cosmological view of the world, of God, and of the human condition.[39] Leibniz is the genius whose philosophy acknowledges both the Enlightenment ideas of reason and the God as Creator whose reasoned omniscience and omnipotence selects the wisest of all creations.[40]

By the end of the eighteenth century, the rationalism of Leibniz and Christian Wolff is replaced, in philosophical circles, by the transcendental idealism of Kant, rejecting the universal belief in reason as enabling the discovery of universal truths.[41] Because Kantian idealism theorizes no personal, monotheistic God, creator, ex nihilo, of the world and all of its glories, Mendelssohn's orientation is surpassed. Also, disagreeing with the progressive teleology of Kant, Mendelssohn understands history as an ebb and flow of benefits for humanity. His emphasis, therefore, would always be on the individual ability to progress through education, spirituality, and social contact.[42]

Mendelssohn falls in the line of philosophers who find that the concept of the language used to describe the sublime is an essential to expressing the experience of the divine and that of the greater cosmic order. In his description of the sublime, its experience, and the works of art it inspires, Mendelssohn stands on the shoulders of the classical writers like [Pseudo-]Longinus as well as his contemporary writers, including the Englishmen Robert Lowth, the Irishman Edmund Burke, and several German theoreticians, including Christian Wolff and Alexander Gottlieb Baumgarten. Mendelssohn disagrees with a limitation of the

39. Alexander Altmann, "Moses Mendelssohn on Leibniz and Spinoza," in *Studies in Religious Philosophy and Mysticism* (Ithaca: Cornell University Press, 1969) 246–74.

40. Moses Mendelssohn, *Philosophical Writings*, trans. and ed. Daniel O. Dahlstrom (Cambridge: Cambridge University Press, 1997) 119. Gotthold Ephraim Lessing (1729–1781) and Moses Mendelssohn sustain a lifelong relationship as friends, social peers, and philosophical colleagues.

41. See the contribution below by Patrick Frierson on Kant.

42. See, for instance, Moses Mendelssohn, *Jerusalem and Other Jewish Writings*, trans. and ed. by Alfred Jospe (New York: Schocken, 1969) 90.

sublime according to [Pseudo] Longinus whose aesthetics only deal with what Mendelssohn would call "the sublime of the second type," which focuses on the power of the work itself. But Mendelssohn does agree with [Pseudo-]Longinus' description of the ideal style for the sublime, *noble simplicity*, as when Longinus writes: "So, too, the lawgiver of the Jews [i.e., Moses], no ordinary man, having formed a worthy conception of divine power and given expression to it, writes at the very beginning in the opening words of his *Laws*: 'God said'—what?—'let there be light,' and there was light. 'Let there be earth,' and there was earth."[43]

Longinus also notes that "figurative writing has a natural grandeur and that metaphors make for sublimity . . . However, it is obvious without my stating it, that the use of metaphor, like all other beauties of style, always tempts writers to excess."—another point that earns Mendelssohn's agreement.[44]

Bishop Robert Lowth, in his lectures at Oxford between 1741 and 1750, also points to the Hebrew language and more specifically to the poetry in the Old Testament as exemplars of noble simplicity that convey the immediacy and power often lost by flowery and figurative. Lowth argues:

"Nothing can be conceived simpler than the ordinary Hebrew language: all in it is bare, straightforward, sane, simple; the words are neither far-fetched nor carefully chosen . . . the spirit dashes on unchecked, having no leisure or will to attend to minute and frigid details; its conceptions are often not clothed or adorned by language, but laid open and bare; a veil is drawn aside so that we look straight into every condition, and movement of the mind, the sudden impulse, the onward rush, the manifold turnings."[45] The fact that an English bishop also acknowledges the sublimity of the language of the Old Testament, Mendelssohn's beloved Torah, must have brought great satisfaction to the Berlin Jew.

With the publication of his treatise on the beautiful and the sublime, Edmund Burke brings, according to Mendelssohn, a freshness and

43. Longinus, *On the Sublime*, translated by W. H. Fyfe, revised by Donald Russell. The Loeb Classical Library (Cambridge: Harvard University Press, 1995) ch. IX.9.

44. Longinus, *On the Sublime*, ch. XXXII.7.

45. See Robert Lowth, *Lectures on the Sacred Poetry of the Hebrews, translated from the Latin . . . by G. Gregory*, 2 vols. (London, 1787). See also Longinus, *On the Sublime*, ed. A. O. Prickard (Oxford: Clarendon, 1906) 115; *The Sublime: A Reader in Eighteenth-Century Aesthetic Theory*, eds. Andrew Ashfield and Peter de Bolla (Cambridge: Cambridge University Press) 106–10.

immediacy to aesthetic discourse. Burke describes the terror inspired by the sublime. "Infinity has a tendency to fill the mind with that sort of delightful horror, which is the most genuine effect, and truest test of the sublime."[46] Acknowledging his debt to Burke, Mendelssohn incorporates the negative reactions to the sublime, including feelings of impotence, bitterness, and terror, into his treatises. He acknowledges the human propensity to enjoy watching disasters, executions, and representations of other horrible events. Our displeasure with the object source of the sublime reaction is founded on our feeling of limitations, our inability to understand or grasp infinity. Our pleasure may exalt in being spared the disaster, in love of danger, curiosity about a source of terror, or feelings of empathy for those who suffer. These works enthrall and seduce us to observe and to study and learn the most vile images.[47] With the work of Burke, Mendelssohn finds the dimension of awe that is missing from the discourse of his contemporary German aestheticians such as Baumgarten, who find in the sublime the tranquility of the soul.[48]

At the same time that he includes the aspect of terror in his description of the sublime, Mendelssohn retains Christian Wolff's position, following the Leibnizian position regarding divine beneficence, that there is still an admiration for that power, that perfection that reaches beyond the human experience and ability to understand.[49]

In his "On the Sublime and Naive in the Fine Sciences," Mendelssohn divides the sublime into two types. The first type is known as "objective," generated from the work itself; the second type, "subjective," as it is directed in the way the object is depicted.[50] According to his definition, the sublime of the first type is not dealt with by Longinus in his discussion of the sublime. For Mendelssohn, the experience of the sublime of the first type is a mystical experience. Wonder is the irrational state of recognizing that which is beyond explanation by reason. Mendelssohn writes: "God is called 'the most sublime being.' A truth is said to be 'sublime' if

46. Edmund Burke, *A Philosophical Enquiry into the Origin of our Ideas of the Sublime and Beautiful*, edited by Adam Phillips (Oxford: Oxford University Press, 1990) part I.7 [this text was originally published in 1757]. For more on this text by Burke, see Baldine Saint Girons' contribution above.

47. Mendelssohn, *Philosophical Writings*, 132, 142.

48. Meier, *L'esthétique de Moses Mendelssohn 1729–1786*, 2:829.

49. Ibid., 856.

50. See Henry E. Allison, *Kant's Theory of Taste: A Reading of the Critique of Aesthetic Judgment* (Cambridge: Cambridge University Press, 2001) 401.

it concerns a quite perfect or complete entity such as God, the universe, the human soul, and if it is of immense use to the human race or its discovery would require a great genius."[51] It is even that which is beyond any sensible perception. Wonder is reserved for the experience of the ineffable, the experience of love, the joy and the trembling in the face of a power that is beyond the explanation of words. The sublime is a complex feeling, a mixture of the pleasant and unpleasant, the seductive and the terrifying. It is experienced suddenly, with a shock, and often takes away our breath.

Words are at once beautiful because of their poetry, novel because of the surprising juxtaposition of thought, and sublime because the result is to awaken the reader to reflection. As Mendelssohn writes, "[T]he genuinely sublime occupies the powers of the soul in such a manner that all adjoining conceptions which are somehow attached have to disappear."[52] In his description of the sublime of the first type, Mendelssohn anticipates the sensibilities of Wordsworth in his *Lyrical Ballads* and Romanticism's beguiling wish to express a "spontaneous overflow of powerful feelings" in terms of the everyday,[53] exuding, in an apparent desire for an emotional transparency which will "take us by surprise sweep our soul up with them, elevating it, as it were, above itself."[54]

When the works are those of the Creator, then the awe, experienced according to the sublime of the first type, responds to the object itself. When the work is that of an artist imitating divine work, a subgroup of the sublime of the first type is produced. Artistic genius communicates sublime sensibilities or the heroic, causing awe and admiration, suggesting that which a divine work produces: "The human artist . . . gathers together in a single viewpoint what nature has diffusely strewn among various objects, forming for himself a whole from this and taking the trouble to represent it just as nature would have represented it if the beauty of this limited object had been its sole purpose."[55] For works that

51. Mendelssohn, *Philosophical Writings*, 195.

52. Mendelssohn refers to sublime passages, such as Homer's *Odyssey*, bk. XI.V, line 563; Sophocles, *Oedipus Rex*; Shakespeare's *Hamlet* (quoted in Mendelssohn, *Philosophical Writings*, 206–9).

53. William Wordsworth, *Preface to Lyrical Ballads (1800). William Wordsworth: Selected Poems and Prefaces*, ed. Jack Stillinger (Boston: Houghton Mifflin: 1965) 448.

54. Mendelssohn, *Philosophical Writings*, 198.

55. Ibid., 176.

may still inspire awe and admiration but that lack something elevated, Mendelssohn recognizes a second type of the sublime which responds more to the art of representation rather than in what is represented.[56] He explains that this second type consists of the sublime as it is produced by the artistic object itself and its embellishments and the artistic genius.[57] This is the form of the sublime covered most eloquently by Longinus.[58] The (art)work may seem like a tour de force or have some exceptionally novel dimension that inspires a slight degree of awe and admiration but without the elevated dimension that includes the suggestion of divine goodness which characterizes the sublime of the first type.

Regarding the subgroup of the first type of the sublime, that group of works produced by the artist-genius, inspiring the sublime of awe and admiration of the most elevated type, Mendelssohn is specific about the style required. The concrete, transparent style of the Old Testament, language without too much embellishment, intense brevity all contribute to its intensity and power.[59] All focus on the actions of its patriarchs. According to Vico, "*logos*, or word, meant "deed" to the Hebrews."[60] According to Mendelssohn, "Men must be driven to action but merely stimulated to contemplation."[61] The Old Testament style is primarily metonymic, about the deeds or action of men and women.[62] Robert Alter explains that the radical shift from the epic/mythical form of earlier cultural expression to the prose/metonymic form of Tanakh compilers is an apparent self-conscious strategy designed to mark the theological shift.[63] The narratives of the patriarchs, the genealogies, the codes of laws, the speeches all are expressed in concrete and unembellished style described

56. Ibid., 210.

57. Ibid., 216.

58. Ibid., 217.

59. Ibid., 199.

60. Giambattista Vico, *New Science*, revised edition, trans. Thomas Goddard Bergin and Max Harold Fisch (Ithaca: Cornell University Press, 1968) sec. 2, chap. 1, para. 401 (the first edition of this text was originally published in 1725). See the contribution above by Baldine Saint Girons and her discussion of Vico in particular.

61. Mendelssohn, *Jerusalem*, 90.

62. Metonymy has been variously explained as a single characteristic representing a more complex entity ("White House" for the "presidency" where "White House" holds not an organic relationship, but a relationship closely associated with the relationship to "presidency.").

63. Robert Alter, *The Art of Biblical Narrative* (New York: Basic, 1981) 27.

by Lowth. Except for the creation and flood mythologies at the start, most of the Torah presents as a history, a testament that has believability, suggesting real people, issues, events, and legal codes.

It is the very historical style of the Torah upon which Mendelssohn grounds his argument that Judaism is the most rational of religions. According to Mendelssohn, the only Revelation made by God to the Jews is the Revelation of the Law, not of any faith.[64] Mendelssohn acknowledges that miracles are reported to support the gift of the laws, but he also maintains the triviality of miracles.[65] While miracles may carry temporary weight to prove to the skeptics about a divine act, miracles cannot be repeated over the centuries to impress further cynics. "Miracles and extraordinary signs, according to Judaism, are no proofs for or against the eternal verities [that can be demonstrated by reason]."[66] Miracles are to be looked at with a certain healthy suspicion, even if, on occasion they occurred.[67] For this latter point, he argues that our acceptance of any historical truth relies solely on the authority of those who have passed down the information and on their reasonable nature. According to Edward Breuer, "It was precisely [the] notion of historical knowledge that allowed Mendelssohn to blur any real distinction between sacred and profane history."[68] Most importantly, history is to be looked upon as a matter of concrete experience brought to the present day by authoritative reports. History is a reasonable report of events.[69] Unlike the teleology of Kant whose view of history includes progress, human history does not progress but rather only makes some progress and suffers some faltering.[70] History makes and loses progress, as the persecutions and successes of the Jews had shown. The individual has control over personal progress with the help of the laws as revealed to man by God.[71] For this reason, individuals have a responsibility to improve their abilities

64. Mendelssohn, *Jerusalem*, 61.

65. Ibid., 65.

66. Ibid., 70.

67. Mendelssohn's view of miracles here resembles that of Spinoza. See the contribution above by Gábor Boros on Spinoza.

68. Edward Breuer, "Of Miracles and Events Past: Mendelssohn on History," *Jewish History* 9 (1995) 27–52; 33.

69. Alexander Altmann, *Moses Mendelssohn: A Biographical Study* (Birmingham: University of Alabama Press, 1973) 537.

70. Mendelssohn, *Jerusalem*, 104.

71. Ibid., 92.

to experience the pleasures of life and, to neglect those pleasures—both emotional and sentient—is to make ourselves miserable and to neglect the intent of the Creator.[72]

Mendelssohn provides numerous examples of the subgroup of the first type from Greek, Roman, and European classics. He applauds Virgil's guarding the silence of Dido when addressed by Aeneas in the Elysian Fields; Shakespeare's eloquence which is the greater for being sparse, as when Macbeth discovers the slaughter of his family, he says simply, "He has no more children"; or when Hamlet despairs, "To be or not to be."[73]

But it is the dramatic work of Gotthold Ephraim Lessing, Mendelssohn's life long friend and colleague, which exemplifies the concept of the subgroup of the first type in a most personal and intense fashion. Both men look to theater as the great educator and exemplar of morality. Theater, in the age of German Enlightenment, is based on the rational concept of the universe where evil is punished and good is rewarded, or, if the good suffer, it is for moral improvement.[74] Lessing works against the current tide of German theater and "maintains virtue in its old position of dominant interest . . . and to make virtue attractive and irresistible . . . the brave and magnanimous way these characters meet and accept their misfortune."[75] For Lessing and Mendelssohn, tragedy is supposed to be the organ of instruction for the Enlightenment. "It is a pleasant spectacle for the gods, an ancient philosopher says, to see a virtuous soul wrestle with fate, to see him surrender everything to it but his virtue."[76] In the following statement, Lessing suggests the Leibnizian concept that God is a personal God and that God had a choice to create worlds and chose to create the one in which we live. Lessing states that "action and character] should not disturb our optimistic theory, because we can say to ourselves that if we knew all the facts as God does, we could see the justice that was operative; ". . . thus it will have its good reasons in the eternal infinite connection of all things. . . . [T]he whole of this mortal

72. Mendelssohn, *Philosophical Writings*, 140.

73. Ibid., 206–209.

74. See Robert R. Heitner, *German Tragedy in the Age of Enlightenment: A Study in the Development of Original Tragedies, 1724–1768* (Berkeley: University of California Press, 1963).

75. Ibid., 176, 178–79. The Lessing German tradition comes after the German Baroque theater, about which Walter Benjamin writes in *The Origin of German Tragic Drama,* trans. John Osborne (New York: Verso, 1998).

76. Mendelssohn, *Philosophical Writings*, 199.

creator should be a silhouette of the Whole of the eternal Creator."[77] Both men see the sublime in the form of the heroic figure who demonstrates the sublime because of courage in the face of misfortune. The figure of Job is a model tragic figure who sustains the wrath of God but who is made better for the challenge.

Like Mendelssohn, Lessing critiques the Enlightenment project of envisioning a universal, natural religion, based on reason as the only avenue to achieve its universal goals. In response to this Enlightenment vision, Lessing writes two plays suggested by his friendship with Moses Mendelssohn, *The Jews: A Comedy in One Act* (1749) and *Nathan the Wise: A Dramatic Poem in Five Acts* (1778–1779).[78] In *Nathan the Wise*, Lessing suggests that positive religions are important. On the one hand, Lessing acknowledges that joining religions is ideal, achieving universality as much as possible. On the other hand, he insists that the historical moment has not yet arrived. The particularity of each religion is still important. The best that a positive religion can do is to comply with natural, reasoned religion as much as possible.[79] Lessing, like Mendelssohn, sees the limitations of the Enlightenment as they see the limitations of rationality and the historic terror that can occur in the name of rationality. Yet they both embrace the optimism of Enlightenment and the hopes for a world improved by the application of reason to the problems of humanity.

For Mendelssohn, Lessing's play, *Nathan the Wise*, represents an example of the sublime, inspiring the sense of wonder[80] and reflects the high esteem and thirty year friendship of the two men.[81] According to Mendelssohn, whose praise was unrestrained, the play inspires the intuitive experience of the ineffable, of the infinite, of God. The play portrays that thoroughly believable tragic hero, the hero who is neither so good nor so evil, but completely sympathetic, who struggles bravely and

77. Heitner, *German Tragedy*, 91, 242.

78. Gotthold Ephraim Lessing, *Nathan the Wise*, in *Nathan the Wise, Minna Von Barnhelm, and Other Plays and Writings*, ed. Peter Demetz, foreword by Hannah Arendt (New York: Continuum, 1991) 137, 173. See also *Two Jewish Plays: The Jews: A Comedy; Nathan the Wise*, trans. Noel Clark (London: Oberon, 2002).

79. Gerard Vallée, *Soundings in G. E. Lessing's Philosophy of Religion* (Lanham, MD: University Press of America, 2000) 53–68.

80. Altmann, *Moses Mendelssohn*, 578.

81. Ibid., 298.

with great dignity to overcome the injustice that fate throws his way.[82] Wondrous sublimity has an innocence that allows for the connection between the infinite and the reader. Because of the language strategy of parsimony, excess words, even phrases, are absent. Oxymoron, with its unlikely juxtaposition of ideas and references, is another characteristic tool that creates the sublime response, the sense of the wondrous surprise, even in the everyday event. Additional dramatic strategies require that the reader's intellectual and emotional activity be invested into the work. Thus *Nathan the Wise* inspires not so much admiration of the artistic work, as would be the case of a work of art of the second type of sublime that Lessing describes. On the contrary, *Nathan the Wise* demonstrates a subgroup of the sublime of the first type because the audience becomes attuned to the ineffable, to the infinite, to the possible in the grandest, largest sense.

The wonder of *Nathan the Wise* is the wonder that Mendelssohn describes in *Jerusalem*. It is a world where people, no matter the race, creed, and color, live together in peace and mutual respect of difference. As Mendelssohn writes, "Let every man who does not disturb the public welfare, who obeys the law, acts righteously toward you and his fellow men be allowed to speak as he thinks, to pray to God after his own fashion or after the fashion of his fathers, and to see eternal salvation where he thinks he may find it."[83]

In *Nathan the Wise*, in spite of the murder of his wife and sons by the crusaders, Nathan raises a Christian child as a compelling necessity to retain his humanity. In thanking the Christian Templar who saves the life of his adopted daughter, he cries out "This is a man!" And when the Templar responds, "We must be friends!" the wonder of surprise and the connection to something beyond our sensible natures explodes through.[84] The universal experience of wonder is not confined to a religious ritual of any one creed but to the universal human experience that crosses boundaries and provides a venue for friendship of the most wondrous kind. Such is wonder for Moses Mendelssohn.

The ideas of wonder are documented from the earliest of biblical times, particularly in the prophet Isaiah, whose ecstatic vision of the throne continues to inform the mystical expression to this day. From

82. F. J. Lamport, *Lessing and the Drama* (Oxford: Clarendon, 1981) 152.

83. Mendelssohn, *Jerusalem*, 110.

84. Lessing, *Two Jewish Plays*, 102.

Merkabah mysticism of the second century, with its stories of ascent to the heavenly realm, to sixteenth-century Kabbalah, with its comprehensive cosmological explanation and, in some traditions, ecstatic meditation, to eighteenth-century Chasidiam, with its emphasis on community, celebration of life, and strict observance of Jewish law, a mystical tradition has arguably always been represented alongside the more traditional rabbinic Judaism. The Chasidic communities that followed Zvi's failed apocalypticism emphasize the improvement of life in the here and now in preparation for a future messianic arrival. As with the thread of mysticism, the expression of the mystical sense of wonder through philosophic discourse, in the manner of Moses Mendelssohn, continues. For Jews living in the sixteenth through the eighteenth centuries, just as for their ancestors as well as for the generations that would follow, the experience of wonder lives inside and outside synagogue walls, up and down the social hierarchy, and in public and private spaces. These are avenues to experience the spiritual link to a sense of divine accessible to all.

14

Kant and the End of Wonder

Patrick Frierson

IN THE CONCLUSION TO his *Critique of Practical Reason* (1788), Kant famously writes: "Two things fill the mind with ever new and increasing wonder [*Bewunderung*][1] and reverence, the more often and more steadily one reflects on them: *the starry heavens above me and the moral law within me*. I do not need to search for them and merely conjecture them as though they were veiled in obscurity or in the transcendent region beyond my horizon; I see them before me and connect them immediately with the consciousness of my existence" (*Critique of Practical Reason*, 5: 161[2]).

In accordance with this revealing passage from the end of Kant's *Critique of Practical Reason*, we can see virtually the whole of Kant's philosophy as flowing from his wonder at either the starry heavens above or the moral law within. Not only does his philosophy *begin* in wonder, however, Kant also *ends* his discussions of the starry heavens and the moral law with a kind of wonder. Unlike philosophers who begin in

1. *Bewunderung*, which can be translated as "admiration" or "wonder," is a type of *Verwunderung* (cf. 5:269, 365), which is the closest one comes to the Latin *admiratio* (cf. 7: 261; 15:51) or the English "wonder" in Kant's writings. There is no strict German equivalent to the English term wonder. (Kant's use of *Wunder* focuses on religious wonders, cf. e.g., 6:84–88.)

2. All references to Kant's works are to the volume and page number in the German "Academy Edition." Translations are taken from the Cambridge Edition of the Works of Immanuel Kant. Each volume in the Cambridge Edition includes Academy Edition pagination in the margins. The *Critique of Practical Reason* (along with the *Groundwork of the Metaphysics of Morals*) is included in Cambridge's volume titled *Practical Philosophy* (see next footnote). Both the *Critique of Pure Reason* and the *Critique of Judgment* are in Cambridge volumes with those titles.

wonder and end in complete understanding, Kant leaves his philosophy incomplete in such a way that wonder persists.

However, Kant still insists that wonder must lead to inquiry, and he seeks to focus this wonder-inspired inquiry properly. As he explains in a passage that occurs shortly after the one quoted above, Kant claims, "Though wonder . . . can indeed excite to inquiry, [it] does not supply the want of it" and asks "What, then, is to be done in order to enter upon inquiry in a way that is useful and befitting the sublimity of the object?" (5:162). This question is particularly urgent because it is so easy to enter into this inquiry in a *bad* way, in a way that leads to superstition or dogmatism or skepticism. As Kant puts it, "Consideration of the world began from the noblest spectacle . . . and it ended in astrology. Morals began with the noblest property of human nature . . . and it ended in . . . superstition" (5:162).

For Kant, wonder brings with it opportunities but also serious dangers. Thus Kant seeks to guide wonder-inspired inquiry and thereby bring wonder to its proper end in two key respects. First, Kant shows the ways in which wonder must be transformed and even replaced, by understanding in the case of the starry skies and by respect in the case of the moral law. Second and more importantly, however, Kant shows the "end" of wonder in the sense of its purpose or goal. Kant explains what role enduring wonder plays for understanding, acting within, and enjoying our world. Finally, however, Kant maintains an important place for an *un*ending sort of wonder in the aesthetic appreciation of both nature (the starry skies) and one's own transcendence (the moral law within).

• • •

Kant's philosophy can be divided into three main parts, each of which is tied to wonder. First, there is the metaphysics and epistemology that Kant lays out in his *Critique of Pure Reason*, the book for which he was most famous during his own day. This book focuses on offering a complete and systematic account of everything that can be known a priori about the world of experience, the world of the starry skies. However, the second half of the book deals with a "dialectic" between reason and the understanding, a dialectic that is a sort of rigorous unfolding of wonder at the ways in which the world outstrips our need to understand it. By making the structure of this dialectic rigorous and clear, Kant in a sense brings wonder to an "end," but because the dialectic does not go away, he really provides an "end" (in the sense of a goal) *for* a wonder that continues.

Second, there is Kant's moral philosophy, laid out primarily in his *Groundwork of the Metaphysics of Morals*, *Critique of Practical Reason*, and *Metaphysics of Morals*.[3] Here, Kant turns to the moral law within. The *Groundwork* explains this moral law in terms of a "categorical imperative" that dictates whether an action[4] is right or wrong, the *Critique* unpacks the metaphysical presuppositions and implications of the demands of morality on human beings, and the *Metaphysics of Morals* gives detailed accounts of various moral duties. All of this might seem, again, to bring wonder at the moral law to an end. But in fact, all of Kant's attempts to make the moral law clear and rigorous are merely ways to *preserve* wonder at it, but a wonder that gets taken up into "respect" and it thereby effective in motivating action and bringing about virtue. In that way, wonder at the moral law is given an end (goal): the perfecting of the human will.

Third, there is Kant's aesthetics, laid out primarily in the first part of his *Critique of Judgment*.[5] Here Kant analyzes two distinct aesthetic feelings: a feeling for the beautiful and a feeling for the sublime. Both involve wonder. The former focuses wonder at beauty in nature, not primarily of the starry skies sort, but of those natural objects (bird's songs, pretty flowers, etc.) that can provoke constant aesthetic contemplation. The latter, while often inspired by natural grandeur (probably including the starry skies), is fundamentally a sense of wonder at one's own moral nature. In both cases, again, Kant gives detailed accounts of how such objects bring about wonder in human beings, and in that sense, he makes the experience of wonder less mysterious. But in both cases, he also shows how wonder is proper, how it can even be an end in itself.

3. All three of these works are included in the volume *Practical Philosophy* (*The Cambridge Edition of the Works of Immanuel Kant*), translated and edited by Mary J. Gregor, with a general introduction by Allen Wood (Cambridge: Cambridge University Press, 1996).

4. Strictly speaking, the categorical imperative applies to *maxims*, not actions. A maxim is a personal rule for action and generally has the structure, "In situations like S, I will performs actions like A in order to achieve goals like G."

5. Wonder also plays an important role in the rest of the *Critique of Judgment*, which deals with Kant's philosophy of biology and teleological account of nature, but the role of wonder in that account is beyond the scope of the present discussion.

WONDER AT THE STARRY SKIES: THE *CRITIQUE OF PURE REASON*

Kant published the *Critique of Pure Reason* in 1781, after a "Silent Decade" during which he published virtually nothing. The work was and still is his magnum opus, the work that defined him as a philosopher. The specific focus of the book is "getting to the bottom of that faculty we call the understanding, at the same time . . . determin[ing] the rules and boundaries of its use" (A xvi).[6] Thereby, Kant's "inquiry . . . is supposed to demonstrate and make comprehensible the objective validity of [the understanding's] concepts a priori," that is, to answer the question, "What and how much can understanding and reason cognize free of all experience?" (A xvi–xvii). In other words, Kant aims to save but also limit metaphysics as an a priori science of the nature of reality.

Saving the possibility of a systematic and a priori metaphysics of the nature of reality might seem like neither a natural response to wonder nor a good way to show the ultimate value of wonder. For Kant, however, metaphysics is inextricably linked with wonder. Human beings make use of various principles for understanding our world, but "with these principles, [we] ris[e] . . . ever higher, to more remote conditions" (*Critique of Pure Reason,* A vii). As we rise higher and higher and reach the heights of the starry skies above, human beings feel wonder at the majesty of these heavens, but we also invariably seek to understand them, and, in the end, human reason "sees itself necessitated to take refuge in principles that overstep all possible use in experience" (A viii). And thus metaphysics—the systematic study of these highest principles—is born. But for Kant, this sort of metaphysics is far from innocent. Out of an apparently reasoned response to wonder, reason "falls into obscurity and contradictions" (A viii). Thus is born a history of metaphysics within which philosophers seem to propose ever greater numbers of ever more ridiculous and conflicting theories of the nature of reality.

For Kant, then, wonder must be brought to its proper end in a system that completely lays out all of the fundamental metaphysical principles that can be known a priori. More importantly, however, this "complete" metaphysics will, in the end, be fundamentally *in*complete

6. Kant's *Critique of Pure Reason* was published in two different editions. In referencing this work, I follow the customary practice of referring to page numbers in the first edition with the letter A and page numbers in the second edition with the letter B. This particular quotation is from the preface to the first edition, hence it has only an "A" page number.

in that rather than reducing wonder at the starry skies to lifeless and static knowledge, Kant will show that there is an *irreducible* wonder that is *active* and drives human beings on to a never-ceasing exploration of our world.

Kant's project, then, has two parts. First, he aims to lay out the basic principles of metaphysics that can be known a priori. Second, he shows the *limits* of these principles. The fundamental move that makes both parts of this project possible is a shift in perspective: "Up to now it has been assumed that all our cognition must conform to the objects; but all attempts to find out something about them *a priori* through concepts that would extend our cognition have, on this presupposition, come to nothing" (B xvi).

Kant's fundamental idea here is to shift our perspective about what we think "knowledge" involves. One "knows" something when one's justified beliefs about the world actually correspond to the way the world is. Typically, when one thinks of having accurate knowledge about the world, one assumes that this means that one's beliefs about the world have been *molded by* the world. But on this account of knowledge, Kant claims that one can never have *a priori* knowledge about the world, since such knowledge is—by definition—not based on experience of the world. Thus Kant proposes to think of knowledge differently. Knowledge still involves having a correspondence between one's beliefs and the world, but Kant considers the possibility that the world is molded by our beliefs, rather than vice versa. Of course, this sounds crazy, and Kant builds lots of caveats into this claim to make it less crazy. It's not that the *details*—what Kant calls the "matter"—of the world are determined by our specific beliefs. Rather, the *structure*—what Kant calls the "form"—of the world is determined by the structure of our cognition. Just as someone wearing rose-colored glasses could know *a priori* that whatever he sees will be rose-colored because his glasses make his world rose-colored, our ways of perceiving and thinking about the world have a certain structure, and we can know that any world that we could possibly experience must conform to those structures.

Making this move has two effects. First, it becomes possible to have a priori knowledge of the world, as long as the "world" is defined as a world *of possible human experience*. On the basis of this fundamental move, Kant proceeds to analyze the structure of human's ability to sense objects and to think about them. The former capacity—which Kant calls "sensibility" and which gives rise to what he calls "intuitions"—has a spatiotemporal structure. Every object of possible experience will be spatiotemporal, since it must be perceived through our sensibility, which makes its objects spatiotemporal. The latter capacity—which Kant calls the "understanding," and which gives rise to concepts—has a structure that is captured in a set of twelve a priori concepts including such things as "substance" and "causality." Every possible experience must fit into these a priori concepts. For example, every change in the world of possible experience must be a change *of a substance* and must *have a cause*. Putting together Kant's a priori forms of intuition (space and time) and his a priori concepts, Kant gets a whole a priori metaphysics of the experienceable world.

However, the move that made this metaphysics possible has a second effect. As Kant explains,

> From this deduction of the possibility of our faculty of cognizing *a priori* in the first part of metaphysics, there emerges a very strange result, and one that appears very disadvantageous to the whole purpose with which the second part of metaphysics concerns itself, namely that with this faculty we can never get beyond the boundaries of possible experience, which is nevertheless precisely the most essential occupation of this science. (Bxix)

If metaphysics is made possible by thinking of knowledge as being a matter of the world conforming the structures that make it possible for us to experience it, then we cannot make any claims about what the world might be like *independent* of those structures. That is, we can't say anything about what a world that we could not experience would be like. Our a priori claims don't extend that far. And that means that much of traditional metaphysics—which focused on claims about God, or the immortality of the soul, or freedom—is out of bounds. Kant's complete metaphysics of possible experience also sets an absolute limit on metaphysics: we can't go beyond possible experience.

But Kant does not think that the story simply ends there. The desire to ask questions that are out of bounds did not come from nowhere. Kant begins his *Critique of Pure Reason* with the following words "Human reason has the peculiar fate in one species of its cognition that it is burdened with questions which it cannot dismiss, since they are given to it as problems by the nature of reason itself, but which it also cannot answer, since they transcend every capacity of human reason" (A vii).

Kant's *Critique of Pure Reason*, it turns out, is a philosophical treatment of an aspect of the human intellect that is inextricably linked with wonder. While Kant sometimes uses the term "reason" in the way in which we are most familiar, as an all-purpose term for human thinking, he also often distinguishes between "reason" and "the understanding." The understanding, for Kant, is that form of human thinking that categorizes the world in accordance with concepts, making it understanding for us human beings. And much of the *Critique of Pure Reason* is an extended analysis of the *understanding*, showing how the *a priori* concepts of the understanding structure the world that we experience and thereby make metaphysical claims about that world possible. But in the second half of the *Critique of Pure Reason*, Kant turns to a "dialectic" between this understanding and reason in a narrower sense. And in that dialectic, it becomes clear that reason is Kant's term for the intellectual work that wonder does in investigating the world.

The second half of the *Critique of Pure Reason* focuses on what Kant calls a "transcendental dialectic" (A293/B349) or a "transcendental illusion" (A295/B352). Kant spends this part of his *Critique* offering detailed analysis of various illusions that arise from conflicts between the understanding—which has a priori concepts that structure the world of possible experience—and reason. The "reason" on which Kant focuses here is an intellectual ability, like sensibility and the understanding, and it too has an a priori concept. But whereas sensibility is what makes it possible to perceive objects and the understanding is what makes it possible to think about these objects, reason is where "all our cognition . . . ends" (A298/B355); it is what one *ultimately* seeks in cognitions. Kant explains,

> So the transcendental concept of reason is none other than that of the *totality of conditions* to a given conditioned thing. Now since the *unconditioned* alone makes possible the totality of conditions . . ., a pure concept of reason in general can be explained through the concept of an unconditioned. (A322/B379)

Put more simply, reason is the human's insatiable desire to know *why* something is the way that it is. But what reason wants is not simply *an* explanation, but the *ultimate* explanation. That is, reason seeks the kind of answer to the question, why? that doesn't require any follow-up. The "unconditioned" is just this sort of an answer, an answer that is not conditioned on any further answer.

In that sense, reason is the human intellect's natural response to wonder. Wonder is a sort of recognition of something beyond one's understanding, and reason is the desire to know that springs from this wonder. For Kant, all of this is natural and good. The problem, however, is that this desire to know the "unconditioned" runs up against the forms of sensibility and understanding that make it possible for us to experience. Kant gives many examples of this in his *Critique*, touching on all of the major debates in the history of metaphysics. For the purpose of this chapter, though, two examples will suffice.

Kant explains, for example, how metaphysicians get trapped by the question of whether or not the world is made up of atoms[7] that cannot be subdivided into smaller parts (see A434–43/B463–71). On the one hand, reason seeks the "unconditioned," and that leads human beings to want to find a smallest possible atom, something that can bring our restless wonder to an end. On the other hand, the very structure of *space*—one of the forms of our sensibility that determines the structure of the world—is such than we can conceive of any particular space as being divided. No matter how small we think of an atom being, if we think of that atom as existing in the world of experience and thereby taking up space, we have to think of it as being—at least in principle—divisible. Reason's demand for an unconditioned smallest part runs up against the nature of that space which determines the structure of the world.

For another example, Kant discusses the possibility of freedom, both human freedom and the freedom of a "first mover" such as God (see A444–51/B472–79). On the one hand, reason requires an unconditioned, some cause that is not itself caused by anything else, since otherwise there can be no *ultimate* explanation of anything. And freedom is just this sort of uncaused cause. On the other hand, one of the concepts through which the world is structured is causality, and Kant's metaphysics shows that everything in the world of possible experience must have a cause. Thus there cannot be any freedom. Reason's demand for an unconditioned first cause runs up against the concept of causality that determines the structure of the world.

7. For Kant, as for most philosophers up until the twentieth century, the term *atom* just referred to the smallest constituent parts of the universe. Thus, what we might today call subatomic particles, would not really be subatomic in Kant's sense. Whatever we consider to be the *most basic* particles in the universe—quarks, leptons, superstrings—would be what Kant refers to by the term *atom*.

In both of these (and many other) cases, reason's efforts to deal with wonder at the magnificence of the world run into unavoidable and impassable barriers. The collision between reason's demands and the structures of experience causes a "transcendental dialectic," an illusory sense that one can get metaphysical insights into the world beyond the possibilities of experience. From this illusion springs an endless series of metaphysical debates in which each side argues for its view based on either the demands of reason or the requirements of the empirical world. Fortunately, Kant does not merely *diagnose* this transcendental illusion. He also provides a framework for understanding how illusion can be put in its place, and he shows the positive value in the impossibility of metaphysically satisfying reason's response to wonder.

Kant provides a framework for putting this illusion in its place by showing that the limitations imposed by spatiotemporality and a priori concepts are limitations on the world *of possible experience*. Such limitations tell us nothing about what Kant calls the "thing in itself." That is, these structures do not determine what a world would be like independent of the possibility of its being experienced. Kant calls this his "transcendental idealism," and its central claim is that while the formal structures of space and time and the a priori categories are completely real *in the world of experience*, they are "ideal"—that is, they don't necessary apply—with respect to things in themselves. And that at least leaves open room for the unconditioned among things in themselves, even if there can never be an unconditioned among objects of possible experience. Kant is careful here not to claim that he has any *knowledge* of such things in themselves. As he puts it with respect to freedom, "I have not been trying to establish the *reality* of freedom . . . [and] have not even tried to prove the *possibility* of freedom . . . [but rather] that nature at least *does not conflict* with causality through freedom [at the level of things in themselves]—that was the one single thing we could accomplish" (A558/B586).

Transcendental idealism cannot itself establish that there *is* or even *could be* an unconditioned free cause, but it can show that Kant's metaphysical proofs of causality in "nature" (the world of experience) does not preclude the possibility of freedom for things in themselves.

This transcendental idealism sets up a structure for reconciling the demands of reason and the understanding, but in itself it does nothing to show what *legitimate* purpose reason could have in our experience of the world. To do that, Kant develops the notion of "regulative" principles and ideals. As Kant explains,

> The principle of pure reason we are thinking of retains its genuine validity only in a corrected significance: not indeed as an *axiom* for thinking the totality [or the unconditioned] in the object as real, but as a *problem* for the understanding, thus for the subject in initiating and continuing, in accordance with the completeness of the idea, the regress in the series of conditions for a given conditioned. (A508/B536)

The a priori forms of sensibility and the understanding function like axioms to define what is *real in objects*. Kant thus calls them "constitutive" principles because they help *constitute* what the objects actually can be. By contrast, the "regulative" principles of reason merely *regulate* a human being's intellectual *activity*. They define what one is to *seek for* in objects, without promising that one can find it.

To go back to the example of freedom, we have seen that the search for the unconditioned in a series of causes gives rise to the transcendental illusion of freedom. If we were to take the unconditioned as a *constitutive* part of the real objective world, then we would have to believe that there is some cause that does not itself have a cause; that is, we would have to believe in freedom. As we saw previously, this belief gets us into trouble because we cannot think of any event (and hence not of any beginning to act of a cause) as being without a cause in turn. However, as a *regulative* principle, the idea of an unconditioned first cause is extremely valuable. If we seek merely for *a* cause of an effect, our understanding will not be extended very far. But if we seek for an *unconditioned* cause of a particular effect, then for each cause that we find, we will move back to a further underlying cause, and then to a yet further cause, and so on. We will never find an *actual* unconditioned cause in nature, but our understanding's search, *regulated* by the *idea* of an unconditioned cause, will lead us to greater and greater knowledge of our world.

In the end, though, even this account of regulative principles is not the whole story for Kant. Kant constantly reiterates that reason's response to wonder, while it provides an end (goal) for wonder, does not actually bring wonder to an end (completion). As we continue to search for the unconditioned, we learn more and more about the world in which we live. But unlike the kind of termination of wonder that might come with actually *finding* the unconditioned, this discovery of more and more aspects of an always conditioned world only inspires even *more* wonder.

> The present world discloses to us such an immeasurable showplace of manifoldness, order, purposiveness, and beauty, whether one pursues these in the infinity of space or in the unlimited division of it, that in accordance with even the knowledge about it that our weak understanding can acquire, all speech concerning so many and such unfathomable wonders must lose its power to express, all numbers their power to measure, and even our thoughts lack boundaries, so that our judgment upon the whole must resolve itself into a speechless, but nonetheless eloquent, astonishment. (A622/B650)

For Kant, metaphysics is born in wonder, facilitated in wonder, and completed—but never finished—by wonder.

WONDER AT THE MORAL LAW: KANT'S MORAL PHILOSOPHY

In the last section, we saw that Kant's *Critique of Pure Reason* laid out the structures of sensibility and understanding that provide conditions of the possibility of experience. Kant thereby established both the legitimacy and the limits of metaphysics. And in setting limits to metaphysics, Kant emphasized the importance of wonder at the starry skies above. Reason responds to these starry skies with a search for the unconditioned, a search that has a valuable role to play in increasing knowledge of and thereby wonder at the natural world. But reason can never *find* the unconditioned in any possible experience, and thus one's wonder at the starry skies continues unabated. The last section also opened up another important way in which Kant deals with the unconditioned. Not only does the unconditioned play a regulative role in the search for knowledge, it is also a constant reminder that the world known through metaphysics is always only the world of possible experience. We have no theoretical knowledge of "things in themselves." And that means, for example, that while we can never know anything that is freedom, we also cannot rule out the possibility of free things in themselves. In the *Critique of Pure Reason*, Kant explains *this* important role of his transcendental idealism by claiming, "I had to deny *knowledge* in order to make room for *faith*" (B xxx). For Kant, however, this "faith" is not a religious leap of faith, nor an arbitrary choice to believe one thing rather than another. Instead, Kantian "faith" is a reasonable response to another kind of wonder: wonder at the moral law within. So it is to that wonder that we now turn.

The key components of Kant's moral philosophy are laid out in his *Groundwork of the Metaphysics of Morals*. There Kant advances three key claims, all essential for understanding his conception of the moral law and how that moral law is connected with wonder. First, Kant begins the *Groundwork* with the famous claim that "It is impossible to think of anything at all in the world, or indeed even beyond it, that could be considered good without limitation except a *good will*" (4:393). Second, Kant claims that a good will is a will that acts in accordance with the moral law because it is the moral law and for no other reason. That is, a good will acts from duty. Famously, Kant considers someone who "without any other motive of vanity of self-interest . . . find[s] an inner satisfaction in spreading joy around them" and who therefore does good to others solely for the satisfaction he gets from seeing them happy. Kant argues that such actions, though "amiable," have "no true moral worth" because they are not done for the sake of duty alone (4:398). Third, Kant argues that for human beings, this moral law takes the form of a "categorical imperative" that has a very specific form (or a set of specific forms). In particular, a good will is one that "acts only in accordance with that maxim through which [one] can at the same time will that it become a universal law" (4:421). All three of these claims are deeply connected to wonder, and all three end up being prominent parts of Kant's eventual moral philosophy.

The first claim, that nothing can be considered good without limitation other than a good will, connects the *Groundwork* with Kant's theoretical philosophy, in that the idea of being "good without limitation" evokes the standard of the unconditioned. To be good without qualification is to be *unconditionally* good. Morality, in the form of the good will, thus provides an end to the series of conditions that, in the empirical world, is endless. In the *Groundwork*,[8] the good will is not "unconditioned" in quite the same way as the unconditioned of the first *Critique*. The good will is unconditionally *good*, not unconditioned as

8. In the third part of the *Groundwork*, and again in the *Critique of Practical Reason*, Kant connects this unconditionally good will with an unconditioned free cause. At the end of this section, I discuss the way in which Kant does this in the *Critique of Practical Reason*.

such. Still, it provides a kind of stopping point, a place at which one no longer needs to ask the question "why?" And this stopping point, here as in the *Critique of Pure Reason*, is connected with wonder.

Kant's second claim grows out of his first. The first section of the *Groundwork* focuses on articulating what sort of will could be a *good* will, if a good will is understood to be good without limitation. The first key move is to "explicate the concept of a will that is to be esteemed in itself and that is good apart from any further purpose" (4:397). Precisely because the will is good in itself, its goodness cannot depend upon anything external to it. Thus the *results* of the will cannot make it good. But then the goodness of such a will cannot be derived from any purposes that such a will accomplishes, since this would put the focus of the will outside of itself. Instead, the good will must be a will that acts solely *from duty*[9] and not for any particular ends. And that means that a will the actions of which are ultimately justified by *any* ends cannot be an essentially good will. Kant gives several examples of this. A person who acts honestly, but for the sake of profit, does not have a good will. And someone who chooses not to commit suicide, but only because they want to continue living their generally happy life, does not have a good will. Only when one does what is right *because* it is right can one be said to have a good will.

But the implication of this abstraction from ends, for Kant, is that even one who helps others in order to see others experience joy does not exhibit a good will, since her will is directed not towards the goodness of the action but towards the effect it will bring about. Of course, Kant recognizes that human actions have purposes, and he recognizes that pursuing the joy of others is a purpose that a good will can have. His point is just that if the *ultimate* reason that one acts is to promote the particular end, then one is not acting out of duty, and hence one lacks a good will. To better understand Kant's point here, it helps to distinguish between the *goal* at which one is directed when one acts and the *reason* that one acts for that goal.[10] One can imagine two people who both pursue the joy of others, but for one, the ultimate reason for that pursuit is found in

9. Strictly speaking, Kant shifts from considering the good will to considering the concept of duty, which is that of a good will "under certain subjective limitations" (4:397). While important in many respects, this distinction is not important for the present purposes.

10. This distinction is taken from Christine Korsgaard's discussion of the *Groundwork* in *Creating the Kingdom of Ends* (Cambridge: Cambridge University Press, 1997).

the satisfaction that she takes in seeing others joyful, while for the other, the ultimate *reason* is found in the fact that making others joyful is one's duty. If we focus on the structures of will of the two people, the first has a will that aims for good goals when those goals are satisfying, while the second aims for good goals when the pursuit of those goals is morally good. In this context, it should be clear that the first will is not good *without limitation*. Rather, the first will is good only insofar as it happens to find satisfaction in good things. The second, by contrast, is a will that *always* pursues the good precisely because it pursues things *out of duty*.

Once Kant makes clear that the good will is a will that acts only out of duty, he is faced with a problem: How can a will be motivated to do something solely for the sake of duty? Kant's surprisingly simple answer to this question appeals, fundamentally, to *wonder*. In the *Critique of Practical Reason* (a sequel to the *Groundwork*), Kant explains how one can

> show in an example the mark by which pure virtue is tested and, representing it as set before, say, a ten-year-old boy for his appraisal, see whether he must necessarily judge so of himself. . . . One tells him the story of an honest man whom someone wants to induce to join the calumniators of an innocent but otherwise powerless person . . . Then my young listener will be raised step by step from mere approval to *wonder* [*Bewunderung*], from that to amazement [*erstaunen*], and finally to the greatest veneration and a lively wish that he himself could be such a man (though certainly not in such circumstances) . . . All the *wonder*, and even the endeavour to resemble this character, here rest wholly on the purity of the moral principle . . . Thus morality must have more power over the human heart the more purely it is represented. (5:155–57; emphasis added)

Kant's point here is that the moral law, when exhibited in all of its purity, is *immediately motivating through wonder*. Even a ten year old boy will experience wonder when he is led—in this case through a poignant example—to confront the moral law within. And this wonder will give rise to a "lively wish" and even an "endeavor" to be a person who always follows that moral law.

In the *Groundwork* and again in detail in the *Critique of Practical Reason*, Kant connects this wonder at the moral law within with a practical attitude that he calls "respect." As he explains in the *Groundwork*, "duty is the necessity of an action from respect for law . . . Only what is connected with my will merely as ground and never as effect, what does not serve my inclination but outweighs it or at least excludes it altogether from calculations in making a choice . . . can be an object of respect" (4:400). Respect is the feeling that follows from wonder at the majesty of the moral law. It is the feeling of being motivated by that law to act in accordance with it. Respect is distinct from other forms of wonder or admiration, however, in that it is rooted in an unconditional wonder. Kant explains how one might admire or even be amazed by the "courage and strength" of another person without feeling true *respect* for that person. As he puts it (quoting Fontenelle), "I bow before an eminent man, but my spirit does not bow" (5:76). But the wonder that one feels at the moral law manifested in another person, even if that other person lacks any other excellences, is a wonder that gives rise to true veneration, a respect that motivates one to "endeavor to resemble this character" (5:156). It is a wonder that can be felt only at the unconditionally (and hence morally) good.

For Kant, then, morality in all of its purity inspires wonder. And that helps explain why it is so important for Kant to lay out a *Groundwork of the Metaphysics of Morals* that frees morality from the influence of anything empirical. In his moral philosophy, Kant famously insists that morality cannot be based on anything empirical. And in particular, he argues vociferously against what he calls "eudaimonism"[11] in morals, that is, any attempt to show that morality either springs from or ultimately serves our own self-interest. Many of the reasons for this insistence are purely philosophical. If morality is to be truly universal, for example, it cannot be based on any contingent empirical facts about human nature. But most fundamentally, Kant's concern with an empirical basis for morality is related to *wonder*. Whether one derives ethics from human nature or shows how what is ethical is ultimately what will make human beings happy or even ties ethics to divine rewards, in all of these cases, one tries to provide a reason for what is unconditionally good. And in the process, one makes morality more understandable and more appealing to the senses, but also less pure and ultimately *less wonderful*. Morality is all its purity is a source for wonder, and wonder at the moral law gives rise to a respect that is truly unconditional.

11. "Eudaimonism" comes from the Greek word *eudaimonia*, which is typically translated as "happiness." The point is that any attempt to reduce morals to happiness ends up degrading morals.

Having laid out the unconditional value of the good will and the importance of wonder at and respect for the moral law as the good will's sole motive for acting, Kant must turn to explaining what sort of moral law could actually warrant this kind of respect. And just as Kant's second claim—that the good will must act out of respect for duty—followed from thinking about what would warrant calling the good will unconditional, so his third claim—his specific formulation(s) of the moral law—follows from thinking about what kind of moral law could be worthy of the kind of wonder that gives rise to respect. Kant asks, then, "What kind of law can that be, the representation of which must determine the will, even without regard for the effect expected from it, in order for the will to be called good absolutely and without limitation? Since I have deprived the will of every impulse that could arise for it from obeying some law, nothing is left but the conformity of actions as such with universal law, which alone is to serve the will as its principle, that is, *I ought never to act except in such as way that I could also will that my maxim should become a universal law*" (4:402).

The argument here is straightforward. In order to be a proper object of respect, the moral law, whatever it is, must appeal to nothing other than itself. Thus no particular goals or inclinations or characteristics of oneself or the world can be the reason that a particular action accords with the moral law. The moral law itself must be the only reason. But then the content of the moral law cannot include anything other than the fact that, whatever one does, one must do the sort of thing that could be a law. But if something is suitable for being a law for one person then it must be suitable for being a law for another, since every particular fact about oneself has been excluded from being relevant. Thus the moral law commands nothing other than that, whatever one does, one acts in such a way that one's actions can be make "universal law."

Though Kant distinguishes this moral law from the classic Golden Rule (4:429n), the basic idea is quite similar. One ought not make an exception of oneself. One ought not do the sorts of things that depend on other people *not* doing those same sorts of things. Kant discusses how to apply this law in several cases, the clearest of which is the case of false promising. If one considers making a false promise, it might be quite uncertain what the effects of that promise would be, but one can immediately know that a false promise cannot be done out of duty, since if the making of false promises because a "universal law," then there would not even be such a thing as promises. When one thinks of false promising as a universal law, it contradicts itself and is therefore impermissible.

Kant goes on to articulate several different versions of this moral law. The most important (in addition to the "universal law" formulation in the quote above) is a version that focuses on the unconditional value of other people: "So act that you use humanity, whether in your own person or in the person of another, always at the same time as an end, and never merely as a means" (4:429). The basic idea here is to always value people for their own sakes. Kant gets this version by thinking about what end one could have for one's actions if they are truly to be unconditionally good. The only possible legitimate end would be an unconditionally good end. Following from the first version of the moral law, Kant then thinks about what ends one must always want to be universally considered to be worthy of respect. Since any choice of any ends depends upon one's capacity to choose those ends, one can never coherently deny the value of one's capacity to choose, and (by virtue of the universal law formulation) one must therefore respect *every* capacity for choice. Hence human beings, as free choosers, must always be respected.

Kant's practical philosophy, then, moves from wonder at the moral law to the importance of *respect* as the proper motive for obeying that law, and from the unconditional value of morality implied in respect to several specific formulations of the nature of morality. Kant begins this articulation of the moral law with his fundamental principles in the *Groundwork*, and he goes much further in the actual *Metaphysics of Morals*, which details our duties in areas as diverse as property rights, sexual ethics, compassion for the poor, gossip, and taking care of one's physical health. In a sense, the rigorous formulations of the moral law and its practical applications to all the details of life make it comprehensible, and they might seem to reduce the wonder that one feels at the moral law within. But for Kant, his account of the moral law (unlike accounts of eudaimonists) actually focuses attention on precisely those aspects of the moral law within that make it worthy of wonder. Moreover, by giving a clear and distinct formulation of that moral law, Kant facilitates the transformation of wonder into respect. The end (goal) of wonder at the moral law within is not a passive admiration, but a respect that brings the moral law into play in all of one's actions.

Kant's moral philosophy thus brings wonder to an end in two senses. On the one hand, "wonder" is transformed into respect, and thus one's wonder at the moral law is no longer the purely contemplative feeling that one might have towards the starry heavens. On the other hand, Kant shows the proper end-goal of wonder at the moral law. By fleshing out the moral law's specific demands, Kant shows that wonder at the moral law ought to direct us to specific practical activities and ultimately to a "good will" that is unconditionally good.

There is one final way in which Kant's moral philosophy intersects with his account of wonder. I started this section by pointing out that Kant saw his practical philosophy as completing his metaphysics and epistemology. While Kant's account of wonder at the starry skies ended up "denying knowledge" about the ultimate unconditioned grounds of the empirical world, his practical philosophy provides a basis for a "faith" that consists in making justified metaphysical claims about such unconditioned grounds. In particular, Kant's moral philosophy is based on the idea that we ought to act out of respect for a moral law that does not appeal to any empirically grounded desires. But if one is *obligated* to act in a way that is not determined by empirical grounds, then it must be *possible* to act in that way. And that implies that human beings are transcendentally free since we can act in a way that undetermined by any empirical motives; we are uncaused causes of events in the world. As Kant puts it, one "judges, therefore, that he *can* do something because he is aware that he *ought* to do it and cognizes freedom within him, which, without the moral law, would have remained unknown to him" (5:30). The "faith" to which Kant alludes in his *Critique of Pure Reason*, a faith that the denial of metaphysical knowledge made possible, is filled in by the *Critique of Practical Reason*. The search for the unconditioned that had its beginning in wonder at the starry skies finds its end in the freedom that is a condition of the possibility of the moral law.

WONDER, BEAUTY, AND THE SUBLIME

In the previous two sections, we have seen how wonder at the starry skies and the moral law finds its "end" in Kant's philosophy. Wonder does not completely go away, but in both cases, Kant gives wonder a specification and a purpose. In the *Critique of Pure Reason*, wonder at the starry skies is supplemented with a rigorous metaphysics of the conditions of possibility of experience and with a set of regulative ideals that drive human beings to deeper and deeper understandings of nature. But because our knowledge is always limited to the realm of experience, metaphysics never gives access to the unconditioned, and so wonder at the starry skies persists. In Kant's moral philosophy as laid out in the *Groundwork of the Metaphysics of Morals*, wonder at the moral law is transformed into a *respect* that motivates moral action purely for the sake of duty, and in the process one finds both an unconditioned *good* and the unconditioned freedom that one sought in metaphysics. In Kant's metaphysics and moral philosophy, wonder is transformed into something that motivates *activity*, whether as a regulative ideal of theoretical reason or as respect for moral law.

Kant's aesthetics turn again to wonder, but in the context of aesthetics, wonder is *not* taken up into any sort of practical activity. The work in which Kant offers his most developed aesthetic theory is called the *Critique of Judgment* because Kant sees aesthetics as linked with humans' ability to make judgments about particular things. In this work, Kant revisits both the regulative principles of the *Critique of Pure Reason* and the moral law of the *Groundwork*. The former ends up being closely linked with Kant's account of the beautiful, and the latter with the sublime. Correspondingly, the beautiful is based on a pleasurable wonder at nature that does *not* give rise to any deeper understanding. And the sublime is based on a wonder at one's own moral nature that does not (directly) give rise to any practical activity. Throughout the *Critique of Judgment*, Kant emphasizes that these kinds of wonder "continually recur."[12]

• • •

Starting with the beautiful, then, Kant raises a fundamental puzzle about judgments that a thing is beautiful. Kant points out that "the judgment of taste is aesthetic," by which he means, fundamentally, that we claim that something is beautiful just because it *seems* beautiful, because it gives us a distinctive sort of pleasure that we associate with beauty. For Kant, the beautiful cannot be explained by virtue of subsuming it under any concepts. (So, for example, the concept of the golden rectangle, or of exact resemblance to its object, do not define what beauty is.) Rather, something is beautiful when and only when it feels beautiful to someone of good taste. And there's the problem, because we also think that if

12. *Critique of the Power of Judgment*, edited by Paul Guyer, translated by Paul Guyer and Eric Matthews (Cambridge: Cambridge University Press, 2000) 5:365, cf. 5:301, 5:245. All quotations in this section will be from this edition.

something is beautiful, then it really *is* beautiful, and anyone who does not think that it is beautiful has bad taste.[13] What makes this even more complicated—but what also seems basically right—is that Kant is unwilling to use either epistemological categories or ethical ones to make sense of this bad taste. That is, someone who does not find a particularly beautiful sunset beautiful is not misapplying concepts; they might perfectly well understand that it is a sunset, that it has a particular hue of orange, and so on. And they are not *morally* wrong; there is nothing *evil* about not enjoying a beautiful sunset. But they are still missing something essential. They aren't *feeling* the right things. So Kant's puzzle is how to make sense of the fact that the beautiful is both *subjective*, in that it is based on my own personal feeling, and *universal*, in the sense that everyone *should* feel the same way.

One clue to answering this comes from a similarity between the wonder that gives rise to regulative principles in the *Critique of Pure Reason* and the feeling of pleasure at a sunset. In the *Critique of Judgment,* Kant reiterates that human beings seek a finality and completeness to our understanding of nature, one that is not constitutive for nature itself but that governs how we must think about it. But Kant adds that insofar as we successfully bring nature into accordance with increasingly general principles, we experience a *pleasure* in this increased understanding. As Kant explains,

> [The] discovery [of the order of nature] is a task for the understanding, which is aimed at an end that is necessary for it, namely to introduce into it unity of principles . . . The attainment of every aim is combined with the feeling of pleasure; and, if the condition of the former is an *a priori* representation. . . then the feeling of pleasure is also determined by a ground that is *a priori* and valid for everyone. (5:187)

Kant's point here is that the search for the systematicity in nature brings with it periodic successes, such as when Newton discovered that the

13. Admittedly, this universalism in aesthetics is not as common today as in Kant's day. We might be inclined to say that "beauty is in the eye of the beholder" and mean by this that there is no difference between good taste and bad or that no one can be *wrong* about whether something is beautiful. But I suspect that much of the reason that we hold this view is due to the fact that we can't see how a judgment that is based solely on a feeling can possibly be universal. If *that* is the reason that one is opposed to universalism in aesthetics, then Kant's argument is precisely what is necessary to vindicate a universalist aesthetics today.

gravity that affected the planets and the gravity that caused apples to fall were two forms of the same force. Because this search for systematicity is something that is a natural response of a universal wonder at the world, the pleasure that we feel in these successes should be a similarly universal pleasure. Since everyone must seek to find systematicity in nature, everyone must be pleased by really understanding such systematicity.

These successes are *not* examples of *aesthetic* pleasure, however, since, for one thing, they bring wonder and its related pleasure to an end. One experiences pleasure at the process of gaining greater understanding of the unity of nature, but not in the understanding itself. *Discovering* gravity, or even learning about it for the first time (in the right frame of mind), is a pleasurable delight that flows from the satisfaction of an impulse to which wonder gives rise. But just knowing about gravity is not pleasurable. The pleasure is in the moment of coming to understanding, not in the understanding itself. For Kant, however, truly aesthetic pleasures are different in that they are *ongoing*. Some objects that we encounter in the world are such that "pleasure is connected with the mere apprehension of the form of an object" (5:189), but the pleasure here, like the pleasure of discovering new natural laws, is a pleasure rooted in the nature of our cognitive capacities. As Kant says, "the pleasure can express nothing but its suitability to the cognitive faculties that are in the reflecting power of judgment, insofar as they are in play, and thus merely a subjective formal purposiveness of the object." (5:190) The idea here is that the same activity of cognition that takes place in bringing the apple and the planets under the general concept of gravity takes place in the contemplation of the sunset. But whereas the cognition of gravity ends up reaching an end, the cognition of the sunset does not. Now the cognition of gravity is nothing more than the outworking of wonder transformed into a regulative principle that seeks unity in nature. The cognition of the sunset involves that same wonder, working through the same cognitive capacities and with a similar directness, but now in what Kant calls a "free play." The play is free because we can continue to think about the sunset from all sorts of different directions and never reach any "definite cognition" of the nature of the sunset. But unlike a merely confusing representation, where we just "don't get it," something that is beautiful is constantly *conducive to contemplation*, without ever being finally contemplat*ed*.

As Kant goes on to say,

> That object the form of which . . . in mere reflection on it (without any intention of acquiring a concept from it) is judged as the ground of a pleasure in the representation of such an object—with its representation this pleasure is also judged to be necessarily combined, consequently not merely for the subject who apprehends this form but for everyone who judges at all. The object is then called beautiful; and the faculty for judging through such a pleasure (consequently also with universal validity) is called taste. (ibid.)

Since the cognitive capacities with which we think about the object are the general cognitive capacities of all human beings, anyone who spends the time and attention to properly contemplate something beautiful experiences the pleasure that comes from the free play of these cognitive capacities. The experience of the beautiful is a kind of enduring and contemplative wonder. Like the wonder that gives rise to regulative principles and an increasing understanding of nature, the wonder that underlies the beautiful is directed towards the world of nature—the starry skies, beautiful sunsets, and pretty flowers—and is rooted in humans' capacities for understanding that world. But unlike the intellectual activity of those regulative principles, wonder in the beautiful does not seek anything beyond itself. It is not part of a quest for greater understanding but simply a pleasure in the contemplation of a beautiful form.

• • •

Kant's discussion of the sublime falls under his more general discussion of aesthetic judgment, so much of what Kant says about the beautiful also applies to the sublime. In particular, Kant emphasizes,

> The beautiful coincides with the sublime in that both please for themselves. And further in that both presuppose neither a judgment of sense nor a logically determining judgment, but a judgment of reflection: consequently the satisfaction does not depend on a sensation, like that in the agreeable, nor on a determinate concept, like the satisfaction in the good . . . Hence both sorts of judgments are also *singular*, and yet judgments that profess to be universally valid in regard to every subject, although they lay claim merely to the feeling of pleasure and not to any cognition of the object. (5:244)

Like judgments about the beautiful, judgments about the sublime are at once subjective in that they are based on mere feeling and universal because this feeling is the result of the interaction of universal human cognitive capacities.

In the case of the sublime, however, the feeling is not *wholly* pleasurable, and the cognitive faculties involved do not play together in the friendly manner in which they do in the case of the beautiful. Where the beautiful reflects the way in which humans' striving to understand the world conceptually is well suited to the presentation of particular objects, the sublime arises when one's relationship to the world is frustrated in one way or another.

> Natural beauty (the self-sufficient kind) carries with it a purposiveness in its form, through which the object seems as it were to be predetermined for our power of judgment, and thus constitutes an object of satisfaction in itself, whereas that which . . . excites in us the feeling of the sublime, may to be sure appear . . . to be contrapurposive for our power of judgment, unsuitable for our faculty of presentation, and as it were doing violence to our imagination, but is nevertheless judged all the more sublime for that. (5:245)

Kant primarily describes two forms of this frustration. He starts with the case where one's attempt to understand the world in terms of a systematic whole can be frustrated by the sheer immensity or complexity of something. Kant describes, for example, the feeling of the sublime at the contemplation of St. Peter's in Rome: "The very same thing can also suffice to explain the bewilderment or sort of embarrassment that is said to seize the spectator on first entering St. Peter's in Rome. For here there is a feeling of the inadequacy of his imagination for presenting the ideas of a whole, in which the imagination reaches its maximum and, in the effort to extend it, sinks back into itself, but is thereby transported into an emotionally moving satisfaction" (5:252). Similarly, when contemplating the starry skies or grand mountains or the Grand Canyon, one can have an experience of not being able to "take it all in." And this inability to take it all in, while it involves a certain kind of displeasure, can also be strangely satisfying. In Kant's terms, it is sublime.

For Kant, the sublimity in contemplating the starry skies or the Grand Canyon or even St. Peter's is not something that is present *in* those starry skies. It is not *actually* the case that one literally cannot take it all in. Kant explains that we can come up with ways of measuring these things that *do* let us take it all in. By taking the distance between the self and the sun as a standard or by developing units like "light-years," we can measure the distances between stars and come to a detailed understanding of them. Even with manmade wonders like St. Peter's, Kant points out that to experience the effect of being blown away, "we must avoid coming too near just as much as remaining too far away." We *can*, it turns out, "take it all in" in all of these particular cases. But what the experience of the sublime shows us is that we have a capacity to seek a complete knowledge of things that goes beyond what we actually experience of them. When we think about *what* precisely we aren't "taking in" in our contemplation of the Grand Canyon, we realize that it *cannot* be simply the sense data. We *are* always taking that in. But there is a demand for completeness, an ideal of reason, that is frustrated in moments of awe-struck wonder at these grand objects. And the recognition that there is in us a faculty of reason that outstrips all literal experience and takes us beyond the merely natural world; providing an insight into a world beyond nature itself. And *that* insight strikes us with deep wonder. As Kant insists, "we express ourselves on the whole incorrectly if we call some *object of nature* sublime. . . . We can say no more than that the object serves for the presentation of a sublimity that can be found in the mind; for what is properly sublime cannot be contained in any sensible form, but concerns only ideas of reason" (5:245).

This first sort of sublimity thus comes when we find it impossible to fully take in the objects of our perception, but a second sort of sublimity is evident when we are confronted by "the irresistibility of [the] power [of nature, which] certainly makes us, considered as natural beings, recognize our physical powerlessness" (5:261):

> Bold, overhanging, as it were threatening cliffs, thunder clouds towering up into the heavens, bringing with them flashes of lightning and crashes of thunder, volcanoes with their all-destroying violence, hurricanes with the devastation they leave behind, the boundless ocean set into a rage, a lofty waterfall on a mighty river, etc., make our capacity to resist into an insignificant trifle in comparison with their power. (ibid.)

Just as the previous case brought a kind of frustration of intellect but at the same time a deep sense of pleasure, so here one feels an intense fear at these sublime forces in nature, but at the same time a distinct sense of pleasure. And again, Kant explains this source of pleasure as a recognition of something about *oneself* that takes one beyond the realm of nature. In this case, however, what is revealed is "a capacity for judging ourselves as independent of [nature] and a superiority over nature on which is grounded a self-preservation of quite another kind than that which can be threatened and endangered by nature outside us" (ibid.). In other words, the recognition of the fragility of one's life brings one face to face with one's nature *as a free being* who transcends nature and at the same time has a higher *moral* vocation that goes beyond one's mere mortal life.

In the beautiful and the sublime, Kant shows that the "end" of wonder orchestrated in his epistemology and ethics is compatible with an ongoing and pleasurable experience of wonder at the world in which we live. And especially in the feeling of the sublime, one returns precisely to the two sources of wonder with which Kant began. Despite all of the progress of the understanding in following through the regulative principles of reason to come to a greater understanding of the world, the contemplation of the starry skies still provokes a feeling of sublime majesty that proves that reason's wonder outstrips anything that the world can satisfy. And the contemplation of the awesome power of nature over one's mortal life brings attention back to the most important source of one's own worth and dignity, the wonderful moral law within, a moral law that places human beings in a realm of unconditional worth beyond any mortal danger.

15

Ways of Wondering

Beyond the Barbarism of Reflection

William Desmond

WONDER AND THE BARBARISM OF REFLECTION

NEAR TO THE BEGINNINGS of modernity Giambattista Vico famously speaks of what he calls the barbarism of reflection (*barbarie della riflessione*). While the connection with wonder is not explicitly made by him, this barbarism is intimately related to the loss of wonder that recurrently befalls humanity. The barbarism of reflection comes at the end of a cycle of unfolding: human beings originate as humans from more feral conditions when struck by the lightening bolt of Jove, develop their powers from the barbarism of the senses, arriving eventually at what seems like the consummate self-conscious, reflective form of life. But this consummation brings an unanticipated end in the form of the barbarism of reflection. I connect this barbarism with a fall from wonder, and the malformation of its endowing gift. The barbarism of reflection arises when *our very powers of knowing* have seemingly perfected themselves, but in so doing they have managed to clog the deeper ontological and epistemic springs out of which original wonder flows. The light coming at this end reverts to a darkness, deeper than the darkness at the beginning, for this darkness thinks that now at last it is enlightened.

The barbarism of reflection might be called a counterfeit double of true enlightenment, and like all counterfeits takes itself to be the true original. Why is it this counterfeit double? Because it has malformed the endowment of original wonder, while claiming for itself the crown of wonder's true fulfilment. Vico suggests that the barbarism of reflection is a condition casting its light, or shadow, over an entire epoch, and if that epoch is ours, naturally we must ask if our enlightenment is under its shadow. For this barbarism entails a condition of knowing that knows nothing of the truly worthy, an enlightenment devoid of religious reverence and wisdom.

The striking passage in the *New Science* (1106) where Vico famously describes this barbarism is powerful in its eloquence, and worth quoting at length:

> But if the peoples are rotting in that ultimate civil disease . . . providence for their extreme ill has its extreme remedy at hand. For such peoples, like so many beasts, have fallen into the custom of each man thinking only of his own private interests and have reached the extreme of delicacy, or better of pride, in which like wild animals they bristle and lash out at the slightest displeasure. Thus no matter how great the throng and press of their bodies, they live like wild beasts in a deep solitude of spirit and will, scarcely any two being able to agree since each follows his own pleasure or caprice. By reason of all this, providence decrees that, through obstinate factions and desperate civil wars, they shall turn their cities into forests and the forests into dens and lairs of men. In this way, through long centuries of barbarism, rust will consume the misbegotten subtleties of malicious wits that have turned them into beasts made more inhuman by the barbarism of reflection than the first men had been made by the barbarism of sense. For the latter displayed a generous savagery, against which one could defend oneself or take flight or be on one's guard; but the former, with a base savagery, under soft words and embraces, plots against the life and fortunes of friends and intimates. Hence peoples who have reached this point of premeditated malice, when they receive this last remedy of providence and are thereby stunned and brutalized, are sensible no longer of comforts, delicacies, pleasures, and pomp, but only of the sheer necessities of life. And the few survivors in the midst of an abundance of the things necessary for life naturally become sociable and, returning to the primitive simplicity of the first world of peoples, are again religious, truthful, and faithful. Thus providence brings back among them the piety, faith, and truth which are the natural foundations of justice as well as the graces and beauties of the eternal order of God.[1]

1. *The New Science of Giambattista Vico*, abridged translation of the third edition (1744), by Thomas Goddard Bergin and Max Harold Fisch (Ithaca and London: Cornell University Press, 1970) 381–82.

This passage strikes one as saying something very true, in a number of senses: true to the way human knowing sometimes develops; true to the counterfeits of wisdom that some forms of knowing create; true to the corruption of the soul that follows when its secret reserves of wonder are exhausted; true to a fulfilment that is really a dissolution of this wonder. I take Vico's remarks as emblematic of the loss of wonder that often shadows the perfection of the project of determinate cognition. I am thinking of the instrumentalizing approach to being such as we find in the project of science-technology, or the administrative-bureaucratic mind that holds it can regulate human affairs with abstract universals but that lacks the finesse of concrete practical wisdom. The collusion of scientistic and aesthetic will to power drives persons into themselves as monadic caves, making them over as machines of self-serving desire. The outcome is loss of wonder, loss of piety, and the clogging of the porosity that places us outside our selves in ethical community with our fellows and in reverence before the divine.

Worth noting is that the barbarism of reflection is the outcome of a certain kind of *knowing*. Vico says (*New Science,* axiom 66): "Men first feel necessity, then look for utility, next attend to comfort, still later amuse themselves with pleasure, then grow dissolute in luxury, and finally go mad and waste their substance." This axiom comes back to haunt us as the project of science–technology gains more and more power and seeks more and more to perfect itself. The madness to which he refers, the substance we waste, relates to our being as wondering creatures, the meaning of how we live our wonder. I will say that it is not primarily wonder that drives on to the barbarism but rather a hyperbolic curiosity, subserved by and serving a sometimes secret, sometimes open, will to power. If there is wonder in the beginning, what comes after is not more wonder but the dispelling of wonder. And with this dispelling, the subjection of what is other to our measure, not our subjection to the wonder that exceeds our measure.

There is a wonder before the barbarism of reflection, there is barbarism in the middle when wonder is reconfigured as voracious curiosity that spends itself in ceaseless accumulation of determinate cognition. Is there the hope of wonder renewed after the barbarism of reflection runs its course? So long as life continues, one has to say yes to wonder. This is not a matter of reviving our capacity for wonder. In a way, we do not have a capacity for wonder; rather we are capacitated by wonder—capacitated to wise mindfulness. Since this capacitation is not determined through ourselves alone, *we alone* cannot revive it. Wondering is not a power over which we exercise self-determination; it witnesses to a given porosity of being that endows us with the promise of mindfulness. If there is to be a revival of the capacity, it is in coming home again to this porosity—and its capacitating of our powers. Ingredient in this homecoming is our capacity to know incapacitation.

Wonder is not a univocal concept. It is not first a concept at all, but a happening, and as a happening it is plurivocal. For reasons that will become clear, it is important to distinguish three different modalities: astonishment, perplexity and curiosity. These are internally related to each other, but they reveal a different stress in the unfolding of our wondering. If we do not properly understand these different stresses, we can mistakenly think all wonder is subsumable into the curiosity that makes of all being an object of determinate cognition. This subsumption might consume curiosity, but it is the death of wonder. This death seeds the arising of the barbarism of reflection.

WONDER AS ASTONISHMENT

Turning to the first modality, it is important to stress the primal character of wonder as astonishment. There is a wonder preceding reflection that takes the form of a certain ontological astonishment. Wonder before the being there of being and beings is precipitated in this astonishment. In a certain sense, all human mindfulness is seeded in this astonishment. A caution: the word "wonder" strikes one today as a bit too subjectivized—it is seen as the "gosh" feeling, the "wow" experience to which we give vent before the surprising and the strange. I do not deny the gosh and wow of wonder but there is an ontological bite to original wonder, perhaps captured better in English with the word "astonishment." Consider some ordinary ways of thinking about wonder: What precipitates astonishment, often implying admiration, we call a wonder, a marvel, a prodigy, something marvellous. The wonder may be even seen as a miracle, something extraordinary or even supernatural. Likewise, a wonder can be taken as a communication or a sign, such as an omen or portent. Those who perform wonders are wonderworkers, *thaumaturges*. We do not have to be with Alice in Wonderland to be struck into wonder. If there is something ordinary about wonder as astonishment, there is also in the ordinary something communicating the extraordinary.

We say, for instance "no wonder" and thus we acknowledge that what happened was no surprise, was as one would have expected. If "No wonder" is a sign of continuity, wonder marks the happening of a discontinuity. Thus too if the wonder can be a sign communicating discontinuity, the sign can also strike one dumb—one is dumbfounded, for something spectacular and stupefying is intimated about the wonder. In astonishment there is *the stress of the emphatic*: the unexpected is not anticipated to happen and yet it happens. When we say "The wonder of it is . . ." and refer to a happening, we are suggesting something beyond expectation—the surprising has communicated the emphatic. Being struck by astonishment has something of the blow of otherness in it. Extreme astonishment can seem even to deprive one of sensation. The blow of otherness stuns us, seems to stupefy us, as if inducing a kind of black-out. Many of the characteristics of astonishment—bewilderment, shock, consternation, deprivation of self-possession, benumbing, being "stricken" by amazement—astound one even unto a kind of ontological stupor. The happening is idiotic, with a singularity prior to all reflective thinking and deliberate action.

If this idiocy of astonishment sounds negative, in fact, the rupture of surprise, in striking into us, takes us beyond ourselves. This is not just a subjective feeling. It is more like the seeding and first fertilization of the promise of "subjectivity" by an enigmatic communication to sleeping mindfulness from out of the otherness of being. We go from our minds to the things; our mindfulness wakes to itself by being woken up by the communication of being in its emphatic otherness. Already before we more reflectively come to ourselves, there is the more primal opening in astonishment—an opening of which I speak in terms of a certain porosity of being. There is no fixed boundary between there and here, between outside and inside, there is a passage from what is into the awakening of mindfulness as, before its own self-determination, opened to what communicates to it from beyond itself. We do not open ourselves; being opened, we are as an opening. Astonishment awakens the porosity of mindfulness to being, in the communication of being to mindfulness, before mind comes to itself in more determinate form(s). This is an important point in relation to the difference between wonder in the modalities of astonishment and curiosity.

It is hard to think this more original porosity of astonishment, for all thinking already presupposes it as having happened. All determinate knowing proceeds from it, but it is not yet determinate knowing; nevertheless some sense of it can be communicated. I behold the majestic tree and murmur: "This is astonishing!" I am not projecting my feeling; I am being awakened by the tree, and am awakening to myself, in a more primal porosity, where the striking otherness of this blossoming presence has found its way into the intimate recesses of my now roused and receiving attention. This astonishment is not a vector of intentionality that goes from subject to object; it is a porosity prior to intentionality, and hence refers us back to a patience of being more primal than any cognitive endeavour to be. It is a mindful *passio essendi* prior to and presupposed by every *conatus essendi* of the mind desiring to understand this or that. First we do not desire to understand. Rather we are awoken or become awake in a not yet determinate minding that is not full with itself but filled with an openness to what is beyond itself—filled with openness, if that is permissible to say, for such a porosity looks like nothing determinate and hence seems almost nothing, even entirely empty. Being filled with openness and yet being empty: yet this is what makes possible all our determinate relations to determinate beings and processes, whether these relations be knowing ones or unknowing.

A certain reversal is signalled here in terms of how we often think: the communication is from what is other to us first, and then from ourselves towards that otherness as other. The first initiative does not lie with us, and yet something is initiated. As initiating, we are always seconds. Consider the notion of "beholding." We tend to think of beholding as a movement from us to something other to us. Beholding something seems to put the perceiver in a position of active superiority to the being beheld as other. My beholding seems to confer on me the pre-eminence: the other beheld may be marvelous but my beholding seems to be the privileged glory. This kind of beholding, I would say, is too full of itself, and hence lacks the fertile emptiness that is filled with openness. Not being properly nothing, it beholds nothing. I would say that the first astonishment makes us ask if there is rather a "*beholding from*" being-other.

There are lines from William Wordsworth's "Lines Written above Tintern Abbey" (102–6) which suggest the point. Wordsworth says:

> Therefore am I still
> A lover of the meadows and the woods
> And mountains; and of all that we behold
> From this green earth; of all the mighty world
> Of eye and ear . . .

Note the wording: "behold from." First astonishment is not unlike such a "beholding from." "Beholding from": What is there as other communicates to us, and we are called beyond ourselves. There is *a reversal* of directionality relative to the more normal reflective understanding of being "conscious of . . ." "Beholding from" seems to be first a "consciousness from . . ." Perhaps this is already indicated in the word "consciousness" with its "with" (*cum*): *con-scientia* as a "knowing with." Such a "knowing with" originates in a relating, coming as much from the other known, as from the knowing itself. "*Cum*": there is the coming towards us of what is other, there is the giving of the other, and astonishment is overtaken by the gift of an original porosity. The porosity is not the knowing of this or that but is the awakened opening that enables further the knowing of this or that, the "knowing with" that is communicated from the otherness of being as given. There is not at all to deny the possibility of a two-way mediation in beholding: both a "beholding of" and a "beholding from." But this two-way mediation, this intermediation is possibilized by the opening of a more original porosity between what is other and ourselves. In "beholding from," the otherness of being opens itself from itself in its otherness, offers itself for minding, minding which therefore is only a "beholding of" because it is first a "beholding from."

Another sign of what is at issue with first astonishment: consider how when hearing the word "beholding" we are tempted to think of a *taking hold* of something. One thinks of the German word for concept, *Begriff*: *begriefen* (verb), as "conceiving," seems to suggest a taking hold of, a gripping, a grasping, a seizing. I would say that in "beholding from" there is something more like a *being taken hold of*. In first astonishment what is other seizes us, and we then go towards it out of our seizure. Being gripped puts us in the hold of the other-being, in the sense that it enthralls us, puts us in its thrall. Our receptivity to it is both ruptured and enrapt. One is captivated, as if the sight of the other descends on us, showers itself upon us, and we fall in love with it. First astonishment thus implies a kind of love. Wordsworth speaks very aptly when he describes himself as still a "lover . . . of all that we behold from this green earth." Beholding is a kind of "being beholden" to the green earth for the marvel of its givenness. We feel the call of a certain fidelity to the earth in "being beholden," but this is not imprisoning but *releasing*—releasing of mindfulness towards what is offered to us in given being as other to us. We answer for our "being beholden" in seeking to be truthful and to be true to being.

It is a well-rehearsed theme that philosophy begins in wonder and Aristotle is often cited: "All men desire to know" (*Metaphysics*, 982b11ff.). Aristotle sees the connection of marveling and astonishment when he reminds us of the affiliation of myth and metaphysics, and also the delight in the senses. Nevertheless, the desire to know is understood essentially as a *drive to determinate intelligibility*, which on being attained dissolves the initial wonder launching the quest. The end of Aristotle's wonder is a determinate *logos* of a determinate somewhat, a *tode ti*. This end is the dissolution of wonder, not its deepening or refreshing. Significantly, Aristotle invokes *geometry* to illustrate the teleological thrust of the desire to know (*Metaphysics*, 983a13ff.). I take geometry here as representative of determinate cognition whose *eureka* solves the problem but also surpasses the wonder.

I think the issue is better put in Plato's *Theaetetus* (155d3–4) where *thaumazein* is named as the *pathos* of the philosopher. *Pathos*: there is a patience, a primal receptivity. This is not the self-activating knowing such as we have come to expect from Kant and his successors in German idealism, as well as in varieties of the constructivist epistemology we find in different contemporary inheritors of this Kantian stress. There is a pathos more primal than activity, a patience of the soul before any self-activity. One could say: there is no going beyond ourselves, no activation of our self-surpassing powers of transcending, without this more primal patience. I stress this since in modernity patience has often been relegated to a servile passivity supposedly beneath the high dignity of human power as self-activating, as self-determining. The truth is that no one can self-activate themselves into wonder. It comes or it does not come. We are *struck* into wonder. "Being struck" is beyond our self-determination. We cannot "project" ourselves into "being struck." It comes to us from beyond ourselves. It does not come in the spiky oppugnancy of a hostile estrangement, though the hateful can strike one. It comes in the communication of an intimate strangeness that makes us porous to what before us is enigmatic and mysterious.

Helpful here might be a brief comparison between first astonishment and curiosity (to be more fully treated later). Curiosity is more to be correlated with a determinate cognition of a determinate somewhat (*tode ti*) or "object." By contrast, in astonishment it is not that an "object" as other simply seizes us, making us passive while it is actively dominating. What is received, as we undergo it in "beholding from," cannot be thus objectified. What seizes us is the offer of being beyond all objectification, and the call of truthfulness to being. This is not first either subjective or objective, but transsubjective and transobjective. "Trans": we witness a *crossing between* "subject" and "object" and an intermedium of their interplay which is more primordial than any determinable intermediation between the two. The happening of this "being-between," in the occurrence of "beholding from," reveals a porosity beyond subjectification and objectification and we are beholden to what eventuates in this between, making us answerable to its truth in our own being truthful. In the intimate strangeness of the porosity an excess of being flows, and overflows towards one. This is astonishing, not because initially we make no sense of it, but simply because the surprise of being's being there at all is there at all.

Ontological surprise is there at what is there at all and as "too much" for us. We could not anticipate it, so surprising is the pure gift of what is granted in the porosity. The surprise is not only before the being there of being, as transobjective, as more than any determinate being. It is also the surprise *qua* surprise at surprise itself. And this, so to say, double surprise, redoubled surprise is transubjective—the being-awakened of minding at all, to being as other to it and to its own very self as minding. The first astonishment is a mindful joy in this redoubled surprise of being. One might say there is a primal "*being pleased*" with being, prior to all cognitive thematizations and objective determinations. We come to ourselves more mindfully as concrete singularizations of this ontological surprise. There is an ontological pleasure in our being at all. We love the "to be" in our own singular "to be." There is nothing "egoistic" in this self-love. Our pleasure with simply being greets, and is greeted by, the elemental agape of being itself, and in this astonishment draws its breath, so to say, or has its breath taken away in an unpremeditated love.

If there is something *childlike* about such a beginning, this does not mean it is merely *childish*. The childlike opening is our finding ourselves astonished already in the porosity of being. We do not produce astonishment; astonishment *opens us* in the first instance, and there is joy in the light. The child *lives* this primal and elemental opening; hence wonder is often noted as more characteristic of earlier stages of life. Thus too, as has been also noted, children have a spontaneous tendency to ask the "big questions." First astonishment is more intimate with the primal porosity that constitutes the human being as metaphysically opened from the outset. The later developments of curiosity and sophisticated scientific knowing are seeded in the primal porosity but what its grant enables we too quickly take for granted. Then, alas, this maternal porosity can be long forgotten when the project of science comes more fully on the scene. The child points to the night sky and murmurs, "Look, the moon!" The astonishing has won its way into its heart. Later, the astonished child is recessed, even driven underground, in the curious project of (say) space exploration which lifts off the earth on the technical constructs of determinate cognition. The child is not only father to the man, but the man is the shield of time that shelters, or denies, the idiotic child it was originally born as. If the child dies, the shield shelters nothing, and the man dies too—a self-guarding hollowness, and not the elemental porosity. The callous of a self-circling *conatus essendi* covers over the idiotic pathos of the exposed child. The barbarism of reflection is the mania of reason that erupts when the ontological tenderness of the reverent child has been calloused.

This more primal porosity of first given minding is at the origin of all modalities of mind, but as intimate with the giving of the first opening, it can be passed over, covered over. It enables the passage of mindfulness but the endowed passing can be passed over, since we come to ourselves in this passing. First a happening it is only subsequently gathered to itself in an express self-relation. In this being gathered to itself, there is the risk of a contraction of what the first opening communicates. The gathering concretizes us as determinate, and as thus ontologically concentrated we can contract the opening of the porosity to just what *we* will grant as given. I will come back to this when dealing with perplexity and curiosity, and with modes of minding that are determinate and self-determining. Nevertheless, as coming to mind in astonishment the porosity happens, we do not produce it, we do not determine it, it communicates from beyond our self-determination. Prior to the more determinate and determining selving of mindfulness the porosity that is neither of self or other happens as the between space in which, and out of which, a variety of determinate and self-determining forms of minding come to be. These latter are derived, not original. What is more original is the between of porosity.

Some ways of indicating how this happens might be correlated with what I call the hyperboles of being: the idiocy of being, the aesthetics of happening, the erotics of selving, and the agapeics of communication.[2] As already indicated, first astonishment opens up the marvel of being at all: the "that it is." I call this the idiocy of being. Idiocy seems to have the implication of the unintelligible, but the question here is rather: While some surds might be absurd, is there a surd that is not absurd? Is the givenness of being such a surd? Idiocy also has a certain implication of intimacy. Does not idiotic astonishment open us to a feel for the *intimate strangeness* of being—an intimacy that yet intimates something beyond estrangement? We often undergo this intimation in astonishment. It is true that after the first astonishment the impulse comes to search for the intelligibility of the idiotic. But the intelligibility sought is not exhaustive of the idiotic, as the surplus givenness of the "that it is." Beyond even a complete summation of determinate characteristics, this "that it is" always abides in the mystery of its surd givenness. The mystery is not dispelled, no matter how extensive our determinate cognition. There is a wonder that comes to this mystery again and again, as hyperbolic to all determination.

We see something of this also in what I call the aesthetics of happening. In opening his *Metaphysics* Aristotle interestingly says: A sign (*sēmeion*) of our desire to know is our love of the senses. His words are significant: *hē tōn aisthēseōn agapēsis* (*Metaphysics*, 890a23). There is an *agape* of the aesthetic. We are not talking about sensing as the intake of the empiricist's sense impression. The aesthetics of happening is a feast (agape) for the senses. It is shot through with marvellous beauty and communicates the excess of something sublime. But both beauty and the sublime are not proportionately regarded by determinate cognition such as the finite sciences give us—they are hyperbolic to determinative science. Not that they are merely subjective—again a characteristic response of modern thought—they are both transobjective and transsubjective. There is an overdeterminacy about them, again in mirror of the agapeic surplus communicated in the happening of original astonishment.

Not only the idiocy of being, and the aesthetics of happening, but the "too-muchness" can be communicated in the erotics of selving also. First astonishment might rock us back on ourselves but it also releases

2. See, for instance, William Desmond, *God and the Between* (Oxford: Blackwell, 2008) chapter 6.

us towards what is there in a great energy of going out, of self-surpassing towards being in its otherness. The "too-muchness" of the energy of self-surpassing—this is first something that overtakes us. We are already beyond ourselves before we come to ourselves, surprised by the energy of self-surpassing that surprises even the selving. If I can refer to the account of the double parentage of eros given by Diotima-Socrates in Plato's *Symposium*, I would say we often think of eros too much in terms of the *penia*, and not enough in terms of the *poros*. In the tale told there of origins, beside *penia* the other parent of eros is *poros*: *poros* who at the feast of the gods is drunk, and with whom, sleeping, somehow *penia* couples, and out of their union eros is born. There is a divine porosity, drunk with divine festivity, at the secret origin of eros. I see in the sleeping festivity of the drunken divinity the promise of something more agapeic, in any case something not reducible to the *penia* of the other parent. There is a secret surplus at work, too much for any account that would privilege only the lack of our desire. There is more. I will return to the erotics of selving with perplexity as the second modality of wonder.

Finally, I would say this surplus of the "too-muchness" is most signalled in the agapeics of communication: in community there is a paradoxical conjunction of surplus and poverty—something may be lacking but it is not that lack alone is communicated—communication communicates beyond lack. This is why I would say there is something of the promise of the agapeic in the first astonishment. There is a communication beyond lack, an intimate community beyond estrangement, in first astonishment. This is not initially known as such, nevertheless, the first wondering is rather akin to a kind of love, to a kind of falling in love. This is a happening that enters intimately into us before our autonomy can pre-empt exposure or impose closure. Likewise, it exceeds fixation on this determinate thing or that. Even more: if we are astonished at this quite singular thing, the porosity seeps beyond fixation on any finitized locus of wonder. It is the intimate communication of the "more" in the this, here and now that seeps beyond finite boundaries, that calls us further in an enigmatic permeability of one thing to another, and of ourselves into the promise of a love with all things other.

WONDER AS PERPLEXITY

I now turn to perplexity as a second modality of wondering. We are apt to think of perplexity as signifying our being troubled with doubt or uncertainty, our being puzzled. The word "plexus" in perplexity suggests a plaiting, a twining, an entanglement. We find the sense of something involved, com-plex, interwoven, something intricate and difficult to unravel, perhaps so knotted we wonder where to start with trying to untangle it. Plagued by perplexity, as we sometimes put it, our thoughts seem to be tormented with some vexing matter we cannot comprehend. Not only is it difficult to understand, but we may find ourselves thinking: we do not know what to think. There is nothing of calm serenity in this modality of wonder—there is often anxiety, bewilderment, distress, trouble and perturbation.

Perplexity arises out of first astonishment. How so? An important element of the agapeics of first astonishment is the way original wonder does not so overtake us as to squash us as selving but comes to release us into our more evident being for ourselves, into selving as promising of itself, and perhaps of more than itself. We are granted to come to be ourselves, freed also into our own self-becoming in the between. This is part of what I call the erotics of selving. The original "too-muchness" is not indeterminate, not determinate, but exceeds all determination. In the first instance it is overdeterminate, but as such endows the promise of self-determining. If astonishment holds the promise of the agapeic, there is awakened in it the erotics of self-surpassing. In being thus awakened, we are as selving, and come to be ourselves. Out of this more primal awakening both perplexity and determinate curiosity are born as further modalities of wondering.

Being awakened in the primal astonishment, the "too-muchness" of given being can seem to *oppress* us.[3] Given to be as ourselves, the intimation intrudes that we cannot be its full measure. Though we are not the measure of the "too-muchness" we yet want to know it in full measure. In this disjunction troubled perplexity arises: we do not know, we would know, we know we do not know. We are stressed in the baffling difference between what we know is too much for us, and our intimately known de-

3. Burke connects astonishment with horror: "astonishment is that state of the soul, in which all its motions are suspended, with some degree of horror" (*Enquiry*, part 2, section 1). He is not wrong, but he is not entirely right either, since in terms of the analysis I offer here, there is in what he says of astonishment a kind of mingling of wonder as agapeic astonishment and as erotic perplexity. The sense of horror becomes more overt. I think, *on the turn* of wonder from first astonishment to perplexity. Some of these nuances will be evident from my analysis, itself less psychological and more ontological-metaphysical than Burke's.

sire to know just that "too-muchness." Perplexity is born in the baffling difference wherein our mindfulness is torn between its desire to know and its intimate knowing that it does not know what is too much for it. To live with this baffling difference is not easy, and there is the inevitable urge to diminish its stress in seeking a knowing which reduces the "too-muchness" to proportions that allow us to appropriate its difference. We are then faced with the urge to develop the desire to know as our way of subjecting the given "too-muchness" to our measure, that is, to the proportionate measure of ourselves as knowers. Often perplexity takes off in this direction, but not always, and the entire situation is always more equivocal, since wonder as perplexity is recurrently haunted by faces of otherness that are just so as *disproportionate* to the determinate measure of our determinative cognition.

Perplexity comes to live the porosity of mindfulness differently to first astonishment. Coming to ourselves as ourselves, we also come to more explicit mindfulness of the *otherness* of the rest of being, just as other to us. Being is not over against us, but it is not identical with what we ourselves are. This otherness can be intimated even in the intimacy of our own being—it is not only an issue of an external otherness. The "too-muchness" can come over us or surge up in us, even unto the threat of drowning us by its excess. Instead of the green earth, the element of the porosity of perplexity is here more like that of water. As we say, we are "put at sea" and ask, what is this otherness of the "too-muchness"? The porosity seems to have no boundaries in its open space of communication. For this reason, if we are only overwhelmed, we risk not being able to relate to it, risk not to know at all what is sent or flows towards us. Inevitably in perplexity we stake ourselves as questioning what it is. We come to ourselves as questioning, and finally too as questions to ourselves. Perplexity is still wrapped around with the overdeterminacy of being, and it can sometimes be unbearable to be caught in the cloud of unknowing. Inside not seeing, we seek to see, cloaked in too much of light or too much of darkness. Wordsworth speaks of "the burden of the mystery"—it is with perplexity that this burden troubles us, almost without cease.

What we call everyday life is the intermedium of this mystery, though often what we deem ordinary is a dimmed-down version of this mystery. We make the mysterious intermedium of being into something familiar in this sense: no longer taken as granted but taken for granted.

The intimate strangeness of being becomes too familiar and diminishes in strangeness—and perhaps in ontological intimacy also. But the mystery of the intermedium is not made the less perplexing because it is made the more familiar. The "too-muchness" becomes the too familiar. The strangeness of the "too-muchness" keeps coming back, keeps resurrecting itself beyond familiarization. There is much of this overdeterminate perplexity at work in philosophy, in great creations of art,[4] indeed in the struggling religious soul that finds itself on a dark way along which it seems destined to be the bearer of the cross of the mystery.

Perplexity is a modality of wondering that brings us more into the *equivocity of being*: the play of light and darkness, the chiaroscuro of things and ourselves; the dark light of unformed things and things forming, of ourselves formless and seeking form and being returned to formlessness, of all things enigmatic and intimating, of ourselves the most baffling of beings, at once shouting absurdly and absurdly singing. Perplexity is not the reverse of astonishment but our waking to the troubling equivocity of the "too-muchness," given in the astonishment. The equivocity is shown on both the sides of self-being and other-being. One might say the equivocity of the perplexing "too-muchness" is both transobjective and transsubjective. There is too much to the thereness of what is there; there is too much to the intimacy of being waking up to itself as our selving. Other-beings and selvings come from formlessness beyond form, are themselves as forming and coming to form, and finally point beyond themselves and all finite form. The troubling equivocity can fill us with great foreboding in face of the mystery of life, and every human being knows something of its disconcertment and dismay. It can drive us to distraction; it can drive us mad. It is never too much to say that it is always and ever too much for us.

Perplexity awakens a *seeking* for what is true in all significant art, in all intellectually honest philosophy, in all spiritually serious religion. Mostly, however, the seeking has no fancy names, as ordinary persons in

4. One thinks of Keats's "negative capability": "when a man is capable of being in uncertainties, mysteries, doubts without any irritable reaching after fact and reason." "Negative capability" is somewhere *between* first astonishment and perplexity—suspended in first wonder and touched with the troubled perplexity of doubts and uncertainties; and not yet overtaken by determinating curiosity, which brings obsession with univocal fact and reason. I would say that the perplexity of the poet calls secretly on a wonder before art and more primal, a *prior* porosity opening a metaxological mindfulness that is not hostile to reason or fact but is not univocalizing of them.

accustomed community, mostly out of the limelight, seek to thread the way of truth (with a bow to Parmenides). I see the wondering of perplexity to be more erotic than agapeic in the following respect. We are aware that we lack the measure of what faces us—we wake to our non-knowing in knowing something of our participation in the "too-muchness." There is porosity, there is always porosity—but there is the *penia* (poverty) told of in the myth of Eros's parentage. *Penia*: the sense of something we need, the sense we do not now have it. It is as if we have fallen out of the astonishing rapture of porous participation in the intermedium of being, and instead of finding ourselves as knowing in that outside, we find we are ignorant—but ignorant in a manner that oppressively knows it does not know. The oppression of perplexity now is, so to say, felt in the bones, not yet made the theme of a more explicit philosophical reflection or religious practice. Perplexity is the living of a lack of knowing—a lacking living itself as knowing it is lacking.

The equivocity of this perplexity is in the *doubleness* of being *both* the dismaying destitution of not-knowing *and* the ignorance of a voracious desire to know. Perplexity is second-born from original astonishment, but we wake up to ourselves even before and beyond the second-born desire to know. As a modality of wondering, something about perplexity is more primitive than what we normally call the desire to know. For we have already passed through treasures and dispossessions to get to the quotidian awakening of what we more ordinarily call the desire to know. This more primitive perplexity takes shape in the archaeology of the selving that comes to be out of the original ontological porosity. Of course, as transobjective and transsubjective, this perplexity is not just a matter of selving alone with itself. As coming to awakening out of the porosity, it is already an equivocal way of "being with" what is other than selving—a "being with" that is ingredient in waking selving both to itself and what is other to itself. Perplexity as wondering, like the primal astonishment, is a way of being between the "too-muchness" of other-being and selving coming to wakefulness of both itself and what is other.

In wondering as perplexity, given the equivocal play of light and darkness, we are closer to something like Plato's condition of the Cave. In perplexity, however, we are not in the Cave as prisoners who do not know they are prisoners. These latter do not know perplexity as an awakening. Perhaps these prisoners, that is, we ourselves in this condition, do have a dull presentiment that not all is as it seems.

There is presentiment in perplexity but the dullness has already been tenderized into the pain of not being able to take for granted what now more and more enigmatically presents itself as being opened for questioning. To be perplexed is to realize that one is held in check by something too much for one's own power. The chiaroscuro of being shows the troubling face of the equivocal "too-muchness" which holds us in a kind of thrall. To be enthralled is to be under a spell, but some thralls stop us, stupefy us. The "too-muchness" bewilders, befuddles, bemuses, bewitches us. Perplexity can be nonplussed by the equivocity. Nonplussed, we may appear to be stupid, but there is a salutary stupefaction in the wondering of perplexity. In moments of more ontological porous mindfulness that break into perplexity, we know that there is light, and that there is an access of light in perplexity truthfully undergone. That light might be the Siamese twin of the darkness, and yet the twinned darkness does not make it any the less the light. It is *we ourselves* who are twinned: participants in the perplexity which both burdens and enlightens, double-headed between the burden of the mystery and the godsend of light that gives ontological uplift.

There is also the following deep equivocity. This we can see with an extremity of perplexity when it takes on the shape of *horror*: ontological horror before the being there of being in its excess to our rational measure. In the Cave we can turn *downwards* as well as upwards. Perplexity can come over us in the feeling of *being blocked from ascending into the light*. We would find light, but we find ourselves darkened—darkened in the very seeking for light itself. Not the measure of the light, we are also not the measure of this darkness. We cannot go up; perplexed we find ourselves falling. We may not want to fall, but we still find ourselves falling.[5]

5. Think of the *aporiai* of thought as showing a lack of *poros*: we are unable to find a way across, are at an impasse. In the *Theaetetus* Plato again and again stresses the philosopher's suffering of the aporetic. The question is related to perplexity. In the end perplexity is not dissolved, but it can be the anticipation of a new occasion of trying to understand. It can also be addressed by myth or likely stories. Univocal theories are not enough. The way of philosophical perplexity is wayless. There is a noplace that is the place of thought (Socrates is described as *atopos*, see *Symposium*, 215a2, 221c2–d6). This noplace witnesses to the porosity of the soul. Concerning perplexity one also thinks of Kant and metaphysics: there are questions we cannot avoid raising but cannot also answer; we must raise them, but we cannot put them to rest in a univocal science or theory. Perplexity here is not like pure reason. It reminds one of trying to rest but being unable to find a comfortable position; one keeps casting around for a better

Different responses to this are possible. One might respond in still seeking to hold true to the presentiment that there is the true above us. Or we might respond in the manner of a Schopenhauer or a Nietzsche or a Dostoevsky: we dig *down* instead of ascending on high. Notice that in this going down there must still be some *fidelity* to the sense of something true, if we are to affirm such a descent as worthy of pursuit. Perplexity can be *re-doubled* in such a descent, and the re-doubling has to do with our worry about the absence of the true, in our seeking for the true. Paradoxically and perplexingly, the worry about the lack of the true is in the light of the seeking of the true. Up or down, the perplexing light keep coming back to haunt us—even as a ghost there of what, it seems, should not be there at all.

I want to suggest that perplexity tinged by horror at the possibility that there is nothing true is very widespread in the ethos of post-Enlightenment modernity. *Below* Plato's Cave, we can come to wonder if there is any ground *under* the underground. Perhaps there is a second underground. Perhaps there is an abyss below rather than the sun above. We come to be perplexed between the abyss and the sun. When perplexity is turned back downwards, into the sunless (under)ground under the underground, we can find the kind of mixture of eros and horror that we encounter in a thinker like Schopenhauer; or the mixture of horror and erotic ecstasy that we find in Nietzsche—perplexity tinged by horror paradoxically mutates into wonderful affirmation of even the horror. Schopenhauer: because the dark primal origin is prior to reason—a voracious Will, an *eros turannos*; art may put this into abeyance, and so alleviate us; ascetic religion may uproot it and so seek to save us from it. Nietzsche: because horror before the dark origin is very like facing Medusa—the Gorgon's head will turn all flesh to stone if looked on too directly—the horror can be deflected from this petrification, Nietzsche hopes, into erotic ecstasies in the Dionysian rapture that also comes up

position but finally the perplexity does not get dispelled. With some thinkers, it can be the opposite: they try to get away from perplexity by a strategy: "on the one hand, this," "on the other hand, that." Indeed, is there not much of zigzag in Kant? Perplexity can remind one of a fever where we restlessly turns this way, that way. Of course, this can generate the idea that thinking is itself a kind of sickness, reflection a curse, as happens with the underground man of Dostoevsky, and here and there with Nietzsche. The barbarism of reflection makes reflection itself the barbarism of the mind. This is perplexity sickened with itself, not the first astonishment, nor the posthumous wonder, I will discuss at the end.

from the dark depths. But as Heraclitus says: "Dionysus is also Hades" (frag. 15). Nietzsche, we know, wanted to stand at the meridian under the sun and sing his fidelity to the earth, but how ultimate can such standing in the sun be, if under the ground under us the Dionysian origin is a Hades that turns every tongue to stone? I do not ask the question in criticism of Nietzsche. It is the horrifying question all true perplexity must face: Is death the petrifying truth of life, or is there to life something good in excess of all death, and to be affirmed as such? Is death the true, or is the true beyond death? This is the perplexing question about which tragic art makes us wonder. It is an ultimate perplexity deeply intimate to all the great religions. The wonder of true philosophizing brings us to it too.

I note a connection with Vico's barbarism of reflection. I mean that the danger of madness from such extreme horrors of perplexity might well *drive more moderate rational men away* from all such perplexities. Moderating rationality must protect us from horror, protect us from Hades. Sanity will seem to reside in keeping one's distance from the horror. This is very understandable and rational philosophers have often been banner carriers for this salutary sanity. But in pursuit of this sanity there can come to be an unperplexed apotheosis of univocal knowing, such as we find, for instance, in the ideology of more scientistic world views. The rallying cry then is: Tame perplexity with manageable curiosity and methodical rationalization of all epistemic and metaphysical excess. But for those souls touched by the extremer perplexity, this salutary sanity can hide its own shadow and indeed drive more thoughtful humans to madness. I do not mean now the madness of the barbarism of reflection, but the madness of knowing that any such scientistic apotheosis of univocal knowing is truly mad. Scientistic madness secretes its own counterfeit double as that rational enlightenment that absolutizes itself.

I think of Dostoevsky's underground man and his bitterness against the regnant "geometricization" of the world. This latter is the barbarism of reflection in the form of modern rationalism. Against the certainty of 2 plus 2 equals 4 the underground man rages. His rage is madness to *l'esprit de géométrie* but perhaps *l'esprit de géométrie* has already generated the form of enlightened madness that keeps an epoch in thrall. The underground man rages in a fallen world but it is a world fallen partly because of the absolutization of univocal knowing, one that evacuates every space of the extremes of perplexity, or any secret hiding place where wonder at the mystery of being might fructify. It is unfortunately also true that the perplexity of the underground man has itself become so embittered and spiteful that it seems to have evacuated its own intimacy with the primal astonishment. Without this intimacy perplexity itself easily becomes just horror and hate, just as *eros uranios* becomes *eros turranos*, an eros that is no longer love at all but secretly inspired by the spirit of hatred. The underground man is himself so fallen in this fallen world he seems only able to dig down further into dark Hades, neither seeking to climb up out of the Cave, nor awaiting any resurrection that might release him from hell. He curses the darkness, but his curse is itself darkness and only darkens darkness. It desecrates the world of fallen univocity but in that desecration what restoration of the world is there, what saving knowing?

Perplexity comes back again and again, and again as double-faced: we can be opened, and open ourselves, and opened either to moving up or moving down, and sometimes both to moving up and moving down. Often we are enabled to move up only because we have moved down. We sometimes have to spend our season in hell before the light of heaven breaks on a new dawn. Even despite the scientistic will to absolutize *l'esprit de géométrie*, the power of perplexity to outlive geometry, as *posthumous perplexity*, so to say, is evident in that religion and art continue to grow on the soil of the dark equivocity. The darkness of the earth communicates the burden of the mystery. This side of death, the perplexity is ineradicable finally, and the soil of art, religion and philosophy will always put forth new sprouts. We can, of course, treat that soil as a problem and try to turn its fertility for art, religion and philosophy into a disposable resource for more and more determinate cognition of this, that or the other. Encounter with enigma happens in a between space—alternatively enchanting and tormenting. Because it is tormenting, we are inclined not to let it be; we want to univocalize its fertile equivocity. Once again then begins the process which culminates in the barbarism of reflection. Rather than dwell with the perplexity, we become curious about this thing and that, and thus we seek to surpass, indeed even to do away with, both astonishment and perplexity.

WONDER AS CURIOSITY

Turning now to curiosity as a third modality of wondering, we cannot but be curious, given our inextirpable desire to know the world around us and ourselves. The devil is in the details, or God is, we say; and often we think of the curious person, in his or her desire to know, as giving careful attention just to the details of things. Such attention, we think, can sometimes be carried to excess; it can be addressed to unworthy objects; we inquire into things, but too minutely. There is healthy curiosity; there seems also to be an undue or too intrusive inquisitiveness in which we are curious about what does not properly concern us. Curiosity, in a good sense, finds things interesting and surprising; its desire to know is open to the novel and strange; in turning to what is curious in things, inquiry fastens on their *interesting determinacy*, often with the twist of the odd. Novelty is important for the curious mind: the queer, the peculiar, whatever arouses closer attention. We also talk of a curious argument – one marked by ingeniousness or excessive nicety or subtlety. Those who are collectors of curiosities search in out of the way places for things or people out of the ordinary. Interestingly, inquisitiveness, whether in approved or unapproved senses, can lead to *inquisitions*, in which novelty itself is suspect. The inquisitor is particular about details because the details revealed are unapproved. There is a desire to know what one has no right to know; *prying curiosity* intrudes on what properly does not concern it.

This double-edged character means that, qua wonder, curiosity is not a pure porosity to what is true. What we are in the idiotic recesses of our being infiltrates our manners of being curious. There can be something closer to the purer porosity, the reception of astonishment, the awakening of perplexity. There can also surge up a will to know marked by a *conatus essendi* that wills to overtake, subordinate, if not extirpate the porosity and patience that are more intimate to the idiotic, ontological heart of our being. I stress this doubleness again, since one might claim that in our time this second possibility has taken on such an all-pervasive life that it seems to have a life of its own, and not really to come to be out of the more original porosity at the origins of wonder as astonishment and perplexity. The fuller plurivocity of wonder can only be recovered if we keep these differences in mind, within curiosity itself and between curiosity, astonishment and perplexity.

Curiosity can be said to emerge from perplexity, both continuing what is at work in it, and also redirecting it in significantly differently ways. Out of astonishment perplexity sees the light, as the indeterminate seeking is born out of the gift of overdeterminate wonder; while out of perplexity can descend curiosity, as the determinate questioning that tries to overcome and put to rest the discomfort of indeterminate seeking. Curiosity is mediately related to astonishment in that it is shaped by the fact that we are not the measure of what perplexes us. What disturbs perplexed irritates curiosity—not to be the measure of what exceeds our grasp. There is an undergoing about being perplexed, just as there is about astonishment, a patience. In perplexity we are overtaken by troubled thought about what faces us—we are lacking in the knowing of it and know our lacking. But we also know our lacking desire to be more than lack, and so we can turn differently to all that is before us. We can forget our "being between" because we stand on one side facing the "too-muchness" and wonder if we can make ourselves the measure of the other side. Though not finally the measure, now we come more into the ordinary light of day where we think of ourselves as subjects over against objects. If originally we are not the measure of what perplexes us, now we think we are or might be that measure. Thinking we can determine it in our measure of thought, curiosity arises as minding that carries the seed of wanting to be the measure. This wanting lays the foundation stone of the world as the crystal palace—with a bow to Dostoevsky again.

I recall here how Vico talks also about a *barbarism of the senses* as well as a barbarism of reflection. See this as Vico's naming something of the "too-muchness" at the beginning, experienced as our being overwhelmed by the aesthetics of happening. Lost in this "too-muchness," we are barbarians in our own participation in the aesthetics of happening because asleep in the saturated immediacy of what is sensuously given. The artist often helps us articulate this "too-muchness" of the aesthetics of happening, without the reduction of its equivocity to some set of univocal propositions. Vico reminded us, against the conceit of scholars, that the first humans were poets. The philosopher who remains faithful to *thaumazein*, wonder as astonishment, also seeks to articulate it, and indeed to lay out as best as possibly the contours of its intelligibility. The pathos does not vanish in the effort to understand, but rather finds itself inhabiting the "too-muchness" with a light-accessed mindfulness, and this, even granting the extremities of perplexity. There can also be, however, a will to univocity that becomes impatient with this first wonder in its overfull brimming, as well as the troubled thought of perplexity. It wills to reduce the equivocity to theoretical and pragmatic univocity. Wonder as curiosity is crucial to this process.

If perplexity is a first-born child of primal astonishment, curiosity is a second-born. If astonishment is overdetermined, if perplexity mixes the overdeterminate and indeterminate, curiosity dominantly stresses the determinate. Often we think of wonder in this third modality as confronting problems. This is understandable—the "It is!" of first astonishment turns into the "What is?" (indeed into "What the hell is it?") of perplexity, turning now into the sober "What is it?" of curiosity. With this last form of the question, we ask about the determinate being there of beings, or the determinate forms or structures or processes. We move from ontological astonishment before being towards ontic regard concerning beings, their properties, patterns of developments, determinate formations, and so on. It is essential to the becoming of our mindfulness that we move into curiosity. The overdeterminate is saturated with determinations, not an indefiniteness empty of determinacy. The question "What is it?" turns towards the given intricacy of this, that and the other, and there can be something even reverent in this turning, for it too shares in our porosity to the astonishing givenness. We can marvel as these given intricacies, coming to admire, and even be in awe of such immanent richness. Curiosity releases the self-surpassing energy of our questing to know in the mode of determinate questions bearing on this richness.

Our curiosity can be engrossed in this richness, and this can be a kind of loving fascination. In it can also lurk a will to possession. Curiosity is not in essence a will to determination that masks a will to power. But there can be an engrossment that shows, as we say, an "unhealthy" curiosity. The phrase suggests that there are forms of inappropriate curiosity. It is not curiosity itself that always determines what is appropriate. For there can be a desire to know whose quest is questionably intrusive. Intrusiveness shows the lack of reverence before the intimate. Intrusiveness shows no respect for the properly secret. Intrusiveness knows no measure that would recall it to the ontological humility of our patience of being. Intrusiveness refuses all distance between us and the mystery of being. Intrusiveness collapses being's granted otherness into its own self-determining measure, into itself as the measure that determines what is other.

Where curiosity about determinacy is more than will to power, it retains a certain fidelity to the basic ontological porosity of our being that is communicated in the primal astonishment. This does not always happen. Quite to the contrary, our determinate questioning is tempted to think of itself as the measure of the "too-muchness," now seen as capable of being defined in terms commensurate with our own power to determine. If now there is determination, it is determination proportionate to our power to determine. The porosity is overtaken, overridden by a will to determinacy that loses touch with the *passio essendi*, while the endeavour to be and to know hides a will to power: a will to overtake emergent form, a refusal to be overtaken. In this refusal the seed of the death of wonder is sown—even in the seeming full flowering of the bloom of wonder itself.

On this point discriminations are necessary. Curiosity is, in the main, expressed in our power to pose determinate questions to determinate problems, each in principle soluble. A problem is not a problem if it is insoluble. The possibility of a solution is built into the definition of problem as such.[6] I illustrate by returning to Aristotle. Early in his

6. Gabriel Marcel has important things to say on problem and mystery in, for instance, *The Mystery of Being*, 2 vols. (London: Harvill, 1951). On the issue of solubility Kant shows his usual cautious equivocity when he writes: "Astonishment (*Verwunderung*] is the obstruction of progress in our thought, on account of a hindrance—a kind of fear—that it hits upon something for which it finds no rule." This obstruction does not entirely meet Kant's approval, yet he is not entirely dismayed, and his epistemological house-keeping allows him to balance the books of cognition. He finds a silver lining in the cloud of astonishment. For astonishment might lead to "a hope of finding one [a rule] in the future, if it is an actual event [e.g., a nebula]." Were this to happen, Kant is willing to extend some approval, and "therefore, astonishment is actually quite pleasant." Admiration is given Kant's scientific *imprimatur*, but the miraculous earns no *nihil obstat*. Thus: "Admiration is an always continuing astonishment which remains even when we know the cause (rule) for it. Actual miracles which allow us no hope—strike down the mind. My cognition is *eo ipso* forbidden to me, all pondering comes to an end. But objects of admiration elevate the mind." Kant's *Lectures on Metaphysics* in *The Cambridge Edition of the Works of Immanuel Kant*, trans. Karl Ameriks and Steve Naragon (Cambridge: Cambridge University Press, 2001) 371. Though touched by wonder at the starry skies above and the moral law within, Kant does not impress one as gifted with much of the primal ontological astonishment in his philosophizing; though as I indicated in note 5 above, there is a kind of irritated perplexity that will not fit into the mould of scientific solubility. I thank Garth Green for bringing the above passage in Kant to my attention.

Metaphysics, in announcing our desire to know, he acknowledges its likeness to myth and the marvelling going with myth. This is closer to the first sense of wonder as astonishment. That the myth or religious story is invoked is much to the point, since primal astonishment and religious piety are close relatives in the family of ontological reverence. When in the *Metaphysics* (983a13ff.) Aristotle returns to the issue, as illustration he gives an example from *geometry*. Here we begin with a question, and this is a wonder and a curiosity, but the problem is determinate, and we proceed by intelligent means to solve the problem. We take determinate intelligible steps (in geometry say, we construct, or extend a line here), and the solution becomes perspicuous to our univocal intelligence. When we solve the problem the wonder dissolves. This is ingredient in the logic of a certain kind of curiosity. But if we think the first and third modalities of wondering are the same, we seriously err, and cover over that is most challenging in the first modality—that it cannot be dispelled, though it can be deepened.

I might add this recourse to geometry is recurrent through the philosophical tradition, especially in modernity with thinkers like Descartes, Spinoza, Kant, Husserl. Even Vico is not free of the attitude that connects wonder with determining curiosity: "Wonder is the daughter of ignorance; and the greater the object of wonder, the more the wonder grows" (*New Science*, 184). Then: "Curiosity—that inborn property of man, daughter of ignorance and mother of knowledge—when wonder wakens our minds, has the habit, wherever it sees some extraordinary phenomenon of nature, a comet for example, a sundog, or a midday star, of asking straightaway what it means" (*New Science*, 189). Answering the question of curiosity is, one takes it, an overcoming of the ignorance of the wonderful, which now, converted into knowledge, is no longer wonderful. Vico's remarks here are not discerning enough about the differences between astonishment, perplexity and curiosity.

A certain *will to univocity* is at work in all this, and one might trace a line of continuity from ancient Aristotle to more contemporary positivisms on this score. The underlying view is something like this: to be is to be intelligible, and to be intelligible is to be determinate; what cannot be made determinate, cannot be said to be intelligible, and indeed cannot properly be said to be. This creates huge questions, not only about the indeterminate and the overdeterminate, but also the self-determining. Early in modernity Pascal was a genius of *l'esprit de géométrie*, but he saw the absolute necessity for another modality of mindfulness for what cannot be determined in that univocal modality—*l'esprit de finesse*. Without *l'esprit de finesse*, Pascal might well have been a geometrical positivist instead of a tortured soul, racked by often unremitting perplexity, relieved by intermittent, unpredictable gleams of the astonishing. Finesse is a mindfulness porous to ontological nuance—dealing with everything of the overdeterminate, the indeterminate and the self-determining—and yet not merely self-determining, nor merely indeterminate. Finesse is a mode of mindful participation in the overdeterminate. Finessed mindfulness comes out of attention, secretly mothered by astonishment, fostered by perplexity in excess of the problematic, for it finds the finitely problematic itself problematic—a problem it cannot solve through itself alone. Something in excess of finite determinability keeps coming back to haunt it.

L'esprit de géométrie might want in principle to univocalize all concepts, and perhaps every matter in question; but to affirm *l'esprit de finesse* is not to deny the space of intelligibility. Finesse, like first astonishment, does not negate that space; rather there would be no space of determinate intelligibility without them. Yet they communicate of what, as intelligible, is trans-determinate: hyper-intelligible, if you like. This might look like a surd if you think intelligibility is exhausted by what is subject to univocalization; but this surd is not at all absurd, if approached from the direction of the opening up of, our being opened up to, the hyper-intelligible. Finesse, like first astonishment, is not unintelligible but, as hyper-intelligible has more to do with what I call the *intimate universal*. You cannot univocalize it, but this does not mean it is absurd. To be is not exhausted by being determinate (or self-determining either); being intelligible is also not exhausted by determinate intelligibility. With finesse as companion, there is no need for philosophy to celebrate either the irrational or become hostile to reason. Reasonableness in the mode of finessed mindfulness can be the safeguard of astonishment and perplexity—keeping itself open in porosity to religion and art. Indeed it draws attention to the opening in *scientific thinking itself* to the marvellous intelligibility of finite processes, a marvel putting us in mind of what exceeds finitization. Wonder beyond the barbarism of reflection makes a call on a renewal of *l'esprit de finesse*.

Since solubility is built into the notion of the problematic, if we lack a sense of wonder before and beyond curiosity, we risk deeming any question that does not admit of solubility as beyond all consideration. If it is not a problem in this soluble sense, no longer need it engage our curiosity—until it is turned into the proper form of a soluble problem. Without finesse, this attitude can lead to a tyrannical determination of the questionable—a questionable determination of the questionable, since it means we will only accept as a problem what we deem capable of solubility on the terms we determine or dictate. All the great mysteries of life will be banished, as not properly problems. The extremes of positivism give doleful witness to this form of stupidity. All the great questions are redefined as pseudo-problems; but though banished they do not vanish—they keep coming back again and again in the first two modalities of wonder, the astonishing and the perplexity. We can, of course, try to white them out in a certain curious construction of the image of the world. The elephant is in the room but no one is allowed to say there is an elephant in the room. The elephant is the monster of the transproblematic mystery of being. It is also the transproblematic mystery of our own being. For we can live the existential pretence that we are not questions to ourselves, running from the porosity and patience, the very endowments that allow our very running on all false ways taken for the direction of truth. This is the counterfeiting of the true in the name of univocal truth. There is a defect of basic truthfulness about all this.

The univocalizing of being that this determinative curiosity tries to effect, I should say, is not untrue to things. There are forms of univocal truth, there are precisely formulated questions which are to be answered in clear and distinct answers. They answer to what is fixed as determinate in the unfolding of happening. The happening of becoming takes on more or less fixed forms and our determinate curiosity can be slaked by knowing what these are. There are many forms of univocity here, answering different kinds of curiosity—for instance, mathematical, scientific, and common sense. Nevertheless, the determination of curiosity comes out of something preceding and exceeding determination. What we determine as univocal is itself emergent from a milieu of being that is more than any and every univocal fixation. In this emergence what precedes determinate univocity and what exceeds it is not a mere indeterminacy. Rather the determinate gives evidence of being sourced in the overdeterminate. If we are only curious about the more-or-less fixed forms that come to form, we lose sight of the more than determinate forming and source of forming. Attention to this is not outside of wonder, but it is not wonder in the modality of determinative curiosity.

My point, then, is not to deny a kind of truth to the univocal determinacy settled by curiosity and settling it; it is to argue against the kind of scientism that follows when we claim this is wonder and that this exhausts the processes of being and knowing. With this scientism the outlook takes hold that the univocalizing approach is the one and only approach. This is a contraction of the plurivocity promised in the other modalities of wonder. Determinate curiosity has its place within the embrace of the more original sense of wonder, and while it occludes it, it cannot itself even function, much less prosper, if it does not dip back again and again into the primal modality of the originating astonishment. Scientism is itself a kind of sclerosis of this astonishment, even as it seems to be a hypertrophy of determinative curiosity. Science is also born from it, though its form takes shape in terms of the second-born wonder of curiosity. The mother of mindfulness in astonishment has to be honored, and the original porosity seeded again and again with determinate figures of intelligibility—though none of the latter exhaust the fecundity of the former.

When we ride on the surge of curiosity and forget that these figures have been shaped by our univocalization, curiosity colludes in the loss of the sense of ontological mystery at its most deep. There is a contraction of wonder relative to all the hyperboles of being. There is a loss of marvel before the idiocy of being—this is now a mere surd, taken for granted as an otherwise absurd thereness, not an intimate otherness given as a marvel granted, rousing ontological delight and appreciation. There is the dimming of the marvel of aesthetic happening—the world is stripped of the qualitative textures of sensuous communication and becomes a bare neutral thereness, less than the glorious, saturated creation it is. There is a mutation in the erotics of selving and the urges of an *eros turranos* induce a turn of the desire to know and its love of the true in the direction of will to power. Finally, there is an occlusion of the agapeics of community where the overdeterminacy of ethical and religious love have an absolute intimacy and into which the intrusiveness of determinative curiosity can only be a desecration. Knowing becomes raping.

Consider how the eros of knowing can mutate into an *eros turannos*, generating shapes of curiosity that are carriers of a restless will to power. Why do I want to know? Because I want to enter into the secret of what is other. Why enter? Not because I love the secret but because possessing it, I advance myself, I overtake its intimacy, I take over. There is a love of the secret that is truly a love; there is a curiosity that wants to know the secret for its own purposes and not for love of the secret. Reverence for the other is not among those purposes. Likewise, thinking, qualified by such curiosity, is in the mode of a grasping. Reaching out to grip what is there, it seeks to get a handle on it to turn it this way or that, as I so determine it to turn. This move from the indeterminacy (really overdeterminacy) of the secret to determinative cognition is governed by the rule of self-determination. I am not really curious about the other—in my desire to grasp it, grip it, handle it, it is my own power that is at issue and that is pursued. Grasping is a grabbing. It is not a caressing or being caressed. It is not a conceiving wherein I must await impregnation,[7] wherein I must become porous to what is sent to me, and the embrace of love in conception is not grasping. Conceiving, in this fecundating sense, is a truer metaphor for knowing in allowing reference to the porosity, to the *passio essendi*, and the seed of knowing sown and gestating in the soul. Curiosity in the determinative modality does not await the arrival of love but hurries things along according to its own more itchy desire. This is the loveless love that, as the medievals deemed it so, masks concupiscence.[8]

---

7. It is true that "conceiving" as *concipere* coming from *con-capere*, implies "to take in," "to hold," "to take to oneself." But while the hold, the taking in is there in conceiving, as something generative, there is more of the pathos in conceiving, more of the receiving in being impregnated, and becoming fruitful. There is something of the *poros*; and the *co-natus* is not just self-active but entails a being "born with."

8. *Concupiscentia oculis* is not quite a quaint, monkish concern. Think only of "the look" of Sartre. It is connected with beholding and "beholding from." Consider here different ways of looking and beholding. First, Spinoza: his first biographer, Colerus, tells us that for recreation, for fun, Spinoza would throw flies into the web of a spider and when the fly met its fate, its death, Spinoza would laugh with glee. Schopenhauer saw something dishonorable in this. Schopenhauer was an atheist, but he borders on the apopleptic at Spinoza's attribution of divinity to devouring nature. Big fish eat little fish, Spinoza and Schopenhauer would agree, but Schopenhauer could not stomach Spinoza's glee (see Arthur Schopenhauer, *Parerga and Paralipomena*, trans. E. F. J. Payne [Oxford: Oxford University Press, 1974], vol. 1, 73). In Spinoza's looking the *conatus* is in the ascendant. Second, Augustine: there is a touching remembrance in his *Confessions* when perchance he finds himself watching as a lizard or a spider catches flies (*Confessions*, 10,

Such a univocalizing curiosity alters our feel for the ethos of given being. Loss of ontological mystery is coupled with loss of aesthetic marvelling, and if there is an ontological mood it tends in the long run in the direction of disgust. Paradoxically we strip the world of its aesthetic marvel, but disgust is itself a living repudiation of such stripped down thereness. An ontological karma comes home to us: the valuelessness of being we sought to crystallize does not, in the end, strike us as valueless but as repulsive. Disgust at the blank surd comes home to us, as our curiosity is more and more inclined to be a revolted and revolting interference with the intimacy of being. Secretly we might want to bring the world alive again but now this is expressed just exactly in the mode of the repulsive and the disgusting.[9] Repulsion seems to bring us alive once again, and the barbarism of reflection becomes coupled with the now voyeuristic carelessness to, now sadistic interference with, the intimacy of being. This intimacy is to be guarded with reverent finesse but we now revolt against this. A kind of aesthetic sadism, devoid of wonder, may seem fascinated by the flesh but erotic hatred has usurped the place of love. For the flesh of the world is only first neutered to make it all the easier to quell all compunction in advance of the assault now unleashed on its worthless and revolting thereness. The barbarism of reflection is not only our going mad and wasting our substance, it is our ransacking the substance of the flesh of all things.

---

35). He is caught up in looking, but his response is quite different from Spinoza's: it is not the glee of ascendant *conatus*; there is patience deeper than concupiscence, and this too can be present in transforming *how one beholds*. In Augustine's recounting, a praise of God's mysterious work displaces, indeed replaces, the concupiscence of curiosity and of the eyes. One might say that first astonishment comes again with this praise.

9. Consider the distance charted by this change: Kant said that the disgusting was the one feeling art should not represent; now some parts of the fashionable avant-garde seem to find nothing more thrilling than the disgusting. It used to be *de gustibus non est disputandum*, now it seems *de disgustibus non est disputandum*. The disgusting is to be tasted with gusto. A sign of this: Umberto Eco edited a book titled *On Beauty* (London: Secker & Warburg, 2004), and it received demure attention, while by contrast his *On Ugliness* (New York: Rizzoli, 2007) was a loud success. Would Kant have been disgusted at the ironical gesture of Piero Manzoni, who a few years before his death in 1963 canned ninety samples of his own excrement and called these works *Merda d'Artista*? Cans (I mean works) have been bought for large prices by the Museum of Modern Art in New York, the Pompidou in Paris, and the Tate Gallery in London. Since their first creation, about half of the works (I mean cans) have exploded, due, it seems, to a fault in the design. The issue of excrement comes up when, down below, I discuss posthumous wonder.

When this happens in our metaphysical attunement to being, the wonder given in astonishment becomes little more than confrontation with the world as a worthless whole. It is there but it has no value in itself, for itself. It arouses no marvelling or admiration or celebration. It is there and that is all—valueless process, ongoing from nothing to nothing, and between nothing and nothing. What then could be there in the worthless whole that could call forth our festive affirmation of its "to be" as glorious or good. These latter seem merely subjective appraisals or projections. And given our diminished porosity to first astonishment, ultimately there seems no reason why we should project thus and not otherwise. It is more likely to be otherwise, since human beings cannot help but value or deem worthy. What is not worthy to be deemed worthy cannot be maintained as merely neutral, for as unworthy to be deemed worthy it is already unworthy and hence its purported neutrality must tilt us over to the side calling forth our antagonism or negation. Otherness as such then becomes estranged thereness over against one. We know ourselves as strangers in this strangeness.

Strangely enough, the experience of strangeness, at an extreme, can be the prelude to some return to wonder in the modalities of perplexity and astonishment. What we find lacking wonder brings forth in us another wonder. A world lacking wonder perplexes us; but perplexing us, we might begin to wonder at it again. For its emptiness, or hostility, or repulsiveness, or even its mere thereness seem to us *uncanny*, and so we are brought to the border of *looking again*, and beholding the strangeness of being as intruding again into mindfulness. It intrudes—beyond our intrusiveness—and not violently either, but simply by the silent constancy of its continuing to be there. If we let ourselves be stilled by that silent constancy, we are opened again either to perplexity or astonishment. We see the neutral in a light that is not neutral and that cannot be neutral. In fact, there is a light that should not be there at all, if being itself does not light itself. Even as supposedly neutral, this light is there, and so there is more than the neutral, turn we or not towards granting it to be so. It is so. Even the darkness of the world when understood as darkness is understood as darkness, and so is illuminated by a light that should not be there at all, if darkness is the ultimate.

I stress that the particular understanding of curiosity now under review turns the teleology of wonder into a movement from the indeterminate to the determinate, and thence from determination to determination, all the way to the totality of determinations which are held to exhaust the whole. This kind of curiosity turns against the indeterminate, for this purportedly cannot be grasped, for only the determinate is thus graspable. Really behind this grasping there is a secret hatred of the overdeterminate: all perplexity troubled by the "too-muchness" is deemed an oppressive equivocity and as such no longer to be abided. There is no abiding with the mystery of given being. There is to be nothing abiding about the mystery of given being. If we conceive the teleology of knowing thus, and claim that this is the one and only path to the end of true knowing, the end result must be the evacuation of spiritual seriousness not only in art, and religion but also in philosophy. For there is no room now for *thaumazein* in the modality of agapeic astonishment or in the modality of erotic perplexity. Great artworks, like religious reverence or awe, may offer us striking occasions of originating wonder—ontological admiration, appreciation of being. If such wonder is entirely impelled out of its initial hiddenness by determinative curiosity, the porosity is no longer kept open in philosophical mindfulness. Philosophy, lacking the initiative of originating wonder, must itself atrophy, its ontological astonishment or perplexity substituted for by the virtuosity of technical cleverness or the scholasticism of second-hand commentary. It becomes treasonous to the wiser patience of first astonishment.

POSTHUMOUS WONDER—RESTORING THE WORLD

Where does curiosity end? It has no end. Wonder ends if determinative curiosity is taken to exhaust all the ways of wondering. Nevertheless, curiosity does not have the last word, as it does not have the first. Now to end, I want to consider the resurrection of astonishment and perplexity in the "too-muchness" of being as given, as communicated, and in whose mystery we always participate. We cannot make this mystery an object of curiosity simply just because the power to objectify lives from a participation in it which is non-objective. This participation calls on the promise of a new porosity, equally transobjective as transsubjective, in a mindfulness exorcized of the bewitchments of the barbarism of reflection. I want to suggest a second wonder, both as perplexity and astonishment, in the hyperbolic dimension of the overdeterminate, posthumous to every form of determinate curiosity.

We will not cease from extending our curiosity; the accumulation of more and more determinate cognition—this will continue. That accumulation certainly can serve our technical determination of things, and a more self-regarding humanist self-determination. But if there is something before and beyond determination and self-determination, namely, the overdeterminacy of given being, there is the ever fresh source of the originating and the outliving wonder. It gives birth, it will be re-born. Let determinative science perfect itself, there is still something recessed, taken for granted, reserved. My point now is not hostility to determinate science and its progress but memory of this overdeterminacy making determination possible, and recessed in the self-determination that regards itself alone. When we canonize the primacy of self-determination we sign the death warrant of surprise, and so also of wonder as astonishment; but this death warrant can never be fully executed, since it would be the death of self-determination also. Wonder lives on between life and death. If this seems like an enfeebled condition, it can be seen otherwise. It can be seen as what I will call a *posthumous wonder.*

What I mean is related to the idea of posthumous mindfulness, an idea I have proposed in a kind of metaphysical thought experiment.[10] It goes something like this: Suppose, having died, one were to awaken anew to life, thinking as if one had outlived one's given interim of life, beholding given life from beyond the normal division of life and death. Suppose we were to think from out of the future when we will be dead, about what is worthy of affirmation here and now. How would one behold now the interim of life, as if from beyond death, looking on life as if posthumously, wondering what in life is worthy of ultimate love—affirmed as worthy in the good of its "to be"? Such a posthumous mindfulness could not be defined in terms of instrumental reason. One is to behold now what is worthy to be affirmed for itself, not what is useful for this or that utilitarian project. Beyond the instrumental relation of means and end, posthumous honesty would serve nothing but praise of the worth of being, of truly worthy being. Suppose one were restored to the world in the following sense. Imagine what it would be like if, having died, one were restored to life a hundred years hence? What would one love to behold again, behold with a kind of love? What would one mourn to see utterly destroyed? What are the nameless things we now love which we

10. See, for instance, William Desmond, *Philosophy and its Others: Ways of Being and Mind* (Albany: SUNY Press, 1990) 278–81.

would delight to greet again? What are the intimates of being we would love to greet us? Would not our wonder about such intimates be much more than curiosity? Would it not be love? The thought of posthumous mind jolts us *back into life now*, and makes us wonder about such intimates that *now* bless being, without which life would be metaphysically bankrupt. Posthumous mindfulness makes us wonder about what we so love now that its loss or desecration would grieve us to the roots on our restoration to life.

Curiosity killed the cat, information made him fat, the old rhyme says, but as the old saw does not say, one of the curious things about curiosity is that it keeps on being *re-born*, even when it is a killer. There is in all curiosity the eros of transcending which is more than *penia* or lack, and it always finds itself, unbidden, the beneficiary of unpremeditated *poros*, the porosity fluid to the infusion of divine festivity beyond death. "Dionysus is Hades," Heraclitus says; but Hades is also Dionysus. Life passes into death, death passes into life. In the porosity of being, passing into being and passing out of being themselves pass into each other, and passing out is passage now that also passes on the gift of being. This is not reflection; this is happening; and reflection can become mindful of what is happening—beyond the barbarism of reflection.

Reflection *per se* is not barbaric. Reflection is an indirect relation to the things themselves, indirection inevitable with the mediating powers of minding. What is barbaric is reflection which has become the prisoner of its own mediating abstractions—the abstractions do not serve the true things themselves but themselves alone. Thought just thinks itself, not what is other to thought thinking itself. It revolves in an autistic circle of its own self-reflection, not truly reflecting the "too-muchness" of given being in its otherness. Reflective thought thus at home with itself produces an alienation from what is other to thought. The barbarism is the hall of mirrors of thoughts that reflect only themselves. The human being is sunk in its own self-reflecting subjectivity—not porous to the given world in its wonder, not out in the between. Retracting the milieu of the between, the intermedium of being, into its own self-mediation, it finds nothing but itself. Thinking circles around itself and while it seems most at home with itself, its self-circling is a kind of self-imprisonment. The barbarism of reflection is a mirror of this—a mirror that escapes the determination of self-mediating thought in enacting the karma of its estrangement. This enactment is a loss of porosity, loss of the *passio essendi*, a malformation of the *conatus essendi*, now no longer a *co-natus*, a coming to birth with, but an energy of self-being that circles around itself and betrays the ecstasis that natively puts minding outside of itself in the between. The opening of mindfulness beyond reflection does not deny reflection; it entails a fresh ontological-metaphysical patience, a piety posthumous to consummated curiosity.

If there is wonder after and beyond the barbarism of reflection, as there is wonder before, it cannot be defined just in terms of determinate cognition and its indeterminate extension. The latter might follow, if the vector of endless curiosity was all, curiosity that is in the business of putting wonder in its place, killing it in the act of overcoming it. Posthumous wonder outlives curiosity in the hyperbolic dimensions of the overdeterminate. None of the hyperboles of being are treated truthfully in the barbarism of reflection, indeed they hardly arises as troubling perplexities. But none ever ceases to be astonishing and perplexing: not the idiocy of being, the "that it is at all"; not the aesthetics of happening, the glory of creation in its beauty and sublimity; not the erotics of selving, source of human greatness and wretchedness; not finally, the agapeics of communication, promising the ennoblement and sanctification of human life, and the blessing with good of all of the "to be." All these hyperboles again and again call us to the wakefulness of astonishment and perplexity beyond the curiosity otherwise so well served by determinate cognition. Posthumous wonder is not the enemy of determinate curiosity, but conducive to a kind of unknowing knowing of something metaphysically ultimate in the idiotics of being, the aesthetics of happening, the erotics of selving and the agapeics of community.

It is important to grant that there can be a *second hyperbolic perplexity* as well as astonishment. Posthumous wonder might have to pass through a kind of purgatory. I illustrate by recalling a curious claim of the poet Yeats: "Descartes, Locke, and Newton took away the world and gave us its excrement instead. Berkeley restored the world."[11] The claim about Descartes, Locke, and Newton seems outrageous: What is this excrement? Part of the point is the implication that they evacuate the world of the richness of its qualitative textures, given before the work of modern science and its abstractions, and consigned thereafter to sec-

11. W. B. Yeats, *Explorations*, selected by Mrs. W. B. Yeats (London: Macmillan, 1962) 325.

ondary status at best. Such abstractions would be constructions of the curiosity that kills. (Wordsworth: "We murder to dissect.")

But how could Berkeley restore the world? Berkeley, in his own curious way, claims that to be is to perceive or be perceived. Does not this dissolve rather than restore the world? Nevertheless, Berkeley is insistent that he merely affirms, by contrast with the concepts of scientific "reflection" and "metaphysical" constructions, this given world as it is. My interest is not the further interpretation of Berkeley but this question: What would it mean to restore the world? It would mean granting that we have lost the world—I would say, lost the sense of the wonder of it. Restoring the world would ask restoring the sense of this wonder. It is not that what is there is lost, for it is still there. The matter concerns *how it is between us and it*. (This "between" is relevant to the interpretation of Berkeley's curious claim about being and perceiving or being perceived.) We have lost the relation between us, or substituted one that lacks wonder for love of the world. This is what it would mean to give excrement. The last word would be given to the offal, the surd as the mess. There would be nothing of the good of the "to be" in this mess, this cess. Instance: *ecological excrement*—the foul, fetid dump rampant technical hubris has often, alas, made of nature.

Restoration of the world would have to pass through the purgatory of this second hyperbolic perplexity. The excrement is within us as well as without. I mean here also to recall our modern, post-modern pride in not "naively" reaching above the Cave to the sun of light, but in digging the hole deeper, under the underground, all with bravado for the dirty truth. We reduce even the ambiguous chiaroscuro of daily life to excrement down under. We move in the shadow of Schopenhauer: down below, there is something other to the principle of sufficient reason, the dark origin; there is nothing of intrinsic worth in being as given. This takes us away from the *thaumazein* of minding we find in ancient *theoria*. Without this sense of celebrating contemplation, all reason becomes easily instrumental—servant of a secret tyrannical other: will to power, *eros turannos, thanatos*. I do not say that first wonder or posthumous wonder ask us to deny a sense of the surd in the idiocy of being—a surd in being other to our determinative and self-determining reason. The issue with the second perplexity is whether this surd is absurd, or mysterious with a meaning hard to decipher. Is the mysterious surd just excrement? Do we go under the underground into the shit—as if that were the "truth."

Heraclitus again: "Dionysus is also Hades." The second perplexity rather must wonder: "What the hell is this?" Are we staring at Medusa—the Gorgon's head that gave Nietzsche terror? But if this is the "truth," we petrify. Nietzsche wanted to protect us from that petrification, to console us with art—we have art to save us from "truth," he says. As an artist Yeats, by contrast, sought restoration not protection. He admired Nietzsche but held Vico in high esteem. Facing the horror the second perplexity puts the question: What would restore, who could restore the world in its wonder? What would we behold from the green earth, to what would we be beholden? Who will offer the posthumous wonder, beyond the barbarism of reflection? Would one not need a god to bring life out of hell, and beyond Hades bring us to life again?

Because the posthumous wonder cannot be the product of our self-determination, we cannot determine its *advent*—its coming again and again, its *corso e ricorso*, to speak with Vico This coming again is not a matter of reviving our capacity for wonder, as if it were some power we could turn on at will. Rather we are capacitated by wonder, and its coming anew is inseparable from us knowing a certain incapacitation vis-à-vis both determinate and self-determining power. Being stricken anew by wonder is more a grace than our work. We can do nothing to earn it, though we can divest ourselves of false fixations of determinate finitude, and untrue self-inflating forms of self-determination. Mostly we are clogged on ourselves—we are full of ourselves, too full. We have to let the emptying of this clogging—get ourselves out of the way of the flow, the overflow. We cannot just will this, but we have to be willing. We can make ourselves ready, but the ripening does not come through ourselves alone.

Unclogging the primal porosity is not unlike becoming as a child again. Of course, many acknowledge something of the child in astonishment, but this can be double-edged. The childlike might be seen as *only childish*. One thinks of August Comte's law of the three stages. First, religion for the childhood of the human race—here we are closer to wonder, and hence childish credulity. Then, metaphysics and its abstractions for the adolescence of humanity—abstractions distance us from the first immediacy of astonishment, but they are not yet true thought. Finally, positive science for the maturity of the human race—now disciplined, regulated curiosity knows positively of reality in all its detail. The teleology of knowing is mapped onto to an inexorable move from the indeterminate to the determinative. My question returns: What then of the overdeterminate? This (Comtean) teleology consecrates the amnesia of the reverent child and henceforth there can be no true ground for any reverence, or source of originating wonder. Human astonishment goes off into a fugue state. It ceases to know itself any more, except as a lack or deficiency to be overcome by more determinate, positive cognition. Paradoxically, it then courts a new childishness, an infantilization of the human spirit, at the apex of its "positive" knowledge, because of this its fugue state.

By contrast with Comte's three stages, one might recourse to Vico's *corso e recorso* of the three ages of gods, heroes and men. Between the triads of Vico and Comte we find surface similarity but the animating spiritual impulse is quite other. The age of the gods is an age of original wonder, the age of heroes is intermediary between gods and mortals, the age of men brings the age of reflection. While at the extremity of the last, the barbarism of reflection takes holds, nevertheless, in the *corso e recorso* of time the gyre turns again and again. There is always a once more. I am not quite subscribing to Vico's cyclical view of history; rather I am suggesting that in every "once more" it is the overdeterminacy of the mystery of being that abides, and we too must abide it, and abide with it, for we abide in it. It never is overcome, for it is over and beyond all our overcoming. It surpasses all our surpassings, and indeed is the endowing source that enables our own self-surpassing. The abiding of the overdeterminacy is not to be described in the becoming of any cycle, or in any linear development from an indefinite beginning to a determinate or self-determining end. It possibilizes the promise of every becoming or cycle but is more than every determination or self-determination in each becoming or cycle. Vico's ideal eternal history suggests the mystery of providence, and this is perhaps too determinate a way of putting it. The persistence of piety in Vico is a sign suggesting this abiding mystery: for him in the end there is to be no vanishing of piety, though the barbarism of reflection ends without piety. The barbarism of reflection ends in disaster just because of the squandering of piety and the originating wonder that seeds all our seeking and surpassing, though never exhausted by any. If the barbarism of reflection were the absolute end, Vico's clear message would be that this would be the destruction of the human being.

After Hades: resurrected astonishment in the modality of posthumous wonder; wonder that has died to the instrumentalizing of all being following from the absolutizing of determinative curiosity; wonder promising release beyond the self-determining that circles around itself alone; wonder reborn to the hyperbolic overdeterminacy of being, in its idiotic, aesthetic, erotic, and agapeic communications. *Corso e ricorso*: it will come again. (Instance: Vico coursing again in James Joyce: it comes again, the origin in the end—and now with comic wonder—ringing the true with laughter—beyond horror the hurrah of a divine guffaw—this is the festive laughter of posthumous mind.) For the artist can cooperate in the restoring wonder: the glory of the sensuous, the sounding that resonates in our porosity of soul, the word that utters intimate secrets. The religious person can participate in restored wonder: sacred amazement at the unbelievable goodness of God. The philosopher can think in the services of this restoration: wonder at the mystery of thinking, at the thereness of the "to be" at all, and the splendid enigma of the good of the "to be." The artist might have to spend a season in hell, to be readied for looking beyond horror. The religious person might have to be released from the heaviness of moralizing earnestness and accept the holy lightness of good beyond moral good and evil. The philosopher might have to descend into the noplace of the porosity, from which minding, in its endowed poverty, can begin again its ascent to thought, in the dark night of the sacred.

Contributors

Elisabeth Blum teaches philosophy at Loyola University Maryland in Baltimore. She published about Avicenna's concept of matter and other topics of medieval and Renaissance philosophy. Recently she translated Giordano Bruno's *Spaccio de la bestia trionfante* into German (forthcoming). A book of fiction, *Von Dickköpfen, Vagabunden und Engeln*, was published in 2003.

Paul Richard Blum is T. J. Higgins Chair of Philosophy at Loyola University Maryland in Baltimore. Most recently he has published *Philosophieren in der Renaissance* (2004). He edited the volume *Philosophers of the Renaissance* (forthcoming).

Gábor Boros is head of the Department of Modern and Contemporary Philosophy at Eötvös University in Budapest, Hungary. He is mainly interested in topics of seventeenth- and eighteenth-century philosophy as well as ethics and the philosophy of emotions. His books include *Heilsfragment. Rekonstruktion der Spinozanischen Ethik* (1990) and *René Descartes' Werdegang. Der allgütige Gott und die wertfreie Natur* (2001). Most recently, with Herman De Dijn and Martin Moors, he coedited *The Concept of Love in 17th and 18th Century Philosophy* (2008).

Roland Breeur teaches modern philosophy at the Katholieke Universiteit Leuven (Belgium). His major interests are modern French philosophy (Descartes to Fénelon) and contemporary phenomenological thought.

Jianhong Chen is currently associate professor of philosophy at Nankai University, Tianjin, and is particularly interested in political philosophy and theories of religion. He has published articles and books on Leo Strauss, Carl Schmitt, Thomas Hobbes, and Plato.

William Desmond is professor of philosophy at the Institute of Philosophy at the Katholieke Universiteit Leuven, and David Cook Visiting Chair at Villanova University. He is author of many books, including *Being and the Between* (winner of the Priz Cardinal Mercier, and the J. N. Findlay Award), *Ethics and the Between* (2001), and most recently *God and the Between* (2008). He was also past president of the Hegel Society of America (1990–1992), of the Metaphysical Society of America (1994–1995), and of the American Catholic Philosophical Association (2008–2009). His work spans metaphysics, ethics, aesthetics, and the philosophy of religion.

Patrick R. Frierson is an associate professor of philosophy at Whitman College in Walla Walla, Washington. He is the author of *Freedom and Anthropology in Kant's Moral Philosophy* (2003) and numerous articles on Kant, Adam Smith, and Descartes. He teaches a variety of courses on topics in ethics, modern philosophy, Kant, and nineteenth-century philosophy.

Michael Funk Deckard is assistant professor of philosophy at Lenoir-Rhyne University (North Carolina). His research and teaching include aesthetics, early modern philosophy, ethics, the relationship of science and religion, and hermeneutics.

Dorottya Kaposi completed her PhD in 2007 at the University of Paris–Sorbonne (Paris IV) and at ELTE University (Budapest). Her dissertation treated the Cartesian doctrine of the will and the problem of individuality. She is a teaching fellow at ELTE University in Budapest and teaches modern philosophy and ethics. In 2006, Kaposi was the first secretary of the thirty-first international conference of the Association des Sociétés de Philosophie de Langue Française (ASPLF) in Budapest.

Péter Losonczi teaches philosophy and religious studies at the University of West Hungary (Szombathely). His fields of research are early modern and contemporary philosophy of religion, history of philosophy, interreligious dialogue, and the so-called postsecular context. Some publications include *Reflecting Diversity: Historical and Thematical Perspectives in the Jewish and Christian Tradition* (coedited with Géze Xeravits, 2008), *Religio Academici: Essays on Scepticism, Religion and the Pursuit of*

Knowledge (coedited with András Szigeti and Miklós Vassányi) and the forthcoming *From Political Theory to Political Theology* (coedited with Aakash Singh).

Tamás Pavlovits is associate professor in the Department of Philosophy at the University of Szeged, Hungary. He specializes in seventeenth-century French philosophy, ethics and moral philosophy, philosophy of religion, and metaphysics. In 2003 he defended his PhD dissertation titled "The Force of Reason in Pascal" at the University of Paris–Sorbonne (Paris IV). In 1999 he translated into Hungarian and edited Pascal's short writings and correspondence. He has coedited two collections of articles with Chantal Jaquet: *Les significations du „corps" dans la philosophie classique* (2004), and *Les facultés de l'âme à l'âge classique* (2006). His book titled *Le rationalisme de Pascal* was published in 2007.

Roberta Sabbath teaches at the University of Nevada, Las Vegas. She focuses on Tanakh-inspired literary traditions, including in her current monograph-in-progress, *Re-Membering the Body: The Jewish Sense of Wonder*. She is the editor of *Sacred Tropes: Tanakh, New Testament, and Qur'an as Literature and Culture* (2009).

Baldine Saint Girons is professor of philosophy at the University of Paris X–Nanterre. She has recently published *L'acte esthétique* (2008), *Les Marges de la nuit: Pour une autre histoire de la peinture* (2006), *Le Sublime de l'Antiquité à nos jours* (2005), and *Les Monstres du sublime: Hugo, le genie et la montagne* (2005). Her specializations are philosophy of the seventeenth and eighteenth centuries, aesthetics, and the philosophy of art.

Veronika Szántó has studied philosophy and biology. Her fields of interest include Western philosophy of the seventeenth and eighteenth centuries, the history and philosophy of life sciences, environmental thought, and bioethics. She is a junior researcher at the Institute for Philosophical Research at the Hungarian Academy of Sciences, Budapest, and she is completing her PhD dissertation on the relationship between radical politics and natural philosophy in seventeenth-century England.

Miklós Vassányi is assistant professor in philosophy at Károli Reformed University (Faculty of Arts, Department of General Humanities) in Budapest. His dissertation is called "Anima mundi: The Rise of the World Soul Theory in Late German Enlightenment and Early Romanticism." His specialization is in the history of early modern and of modern philosophy (Locke, Diderot, Schelling) and in the philosophy of religion (physico-theology, religious toleration).

Koen Vermeir is a fellow at the Radcliffe Institute for Advanced Study at Harvard University, and he is a fellow of the Fund of Scientific Research (Flanders), associated with the Katholieke Universiteit Leuven. His research has focused on seventeenth- and eighteenth-century natural philosophy, wondrous technologies, and magical thought. He has published articles in diverse books and journals, including the *British Journal for the History of Science*, *Intellectual History Review*, and *Studies in History and Philosophy of Science*, and has edited four books in the history of philosophy and aesthetics.

Index of Authors

Abdala the Saracen, 6
Abramson, Kate, 220, 223
Adorno, Theodor, 140
Agrippa, Henry Cornelius of Nettesheim, xviii, 15–24, 52, 54
Akiva ben Joseph, Rabbi, 269
Albert the Great, 10, 16
Alembert, Jean Le Rond d', 83, 95, 248
Alsted, Johann Heinrich, 51
Allaci, Leone 234–35
Allison, Henry E., 210, 276
Alter, Robert, 278
Altmann, Alexander, 274, 279, 281
Aquinas, St. Thomas, 40, 42, 139–40, 264
Aristotle, xxiii, 15–16, 40–41, 53, 66, 108–12, 119–20, 124, 130–31, 135, 171, 200, 226, 241, 263, 316, 319, 332–33
Arnauld, Antoine, 158, 173
Asclepius, 6
Aubrey, John, 128
St. Augustine, xvii, 12, 139–40, 158, 162, 168, 337–38
Avicenna, 349

Bacon, Francis, 52, 59, 63, 65, 67, 127, 148, 238
Bacon, Roger, 10, 16
Baier, Annette, 211–12, 220
Bakos, Gergely, 196
Balzac, Honoré de, 235
Barker, Matthew, xviii, 74
Barnouw, Jeffrey, 138
Baumgarten, Alexander Gottlieb, 274, 276
Beeckman, Isaac, 67
Bellini, Giovanni, 246
Benedict, Ruth, 250
Benjamin, Walter, 280
Bergson, Henri, 172
Berkeley, George, xii, xiii, xxi, 189–209, 343–44
Berman, David, 189
Blake, William, 132
Blumenberg, Hans, xvii, xix, 139–41, 161–62, 174, 177
Boileau-Despréaux, Nicolas, 235
Bouhours, Dominique, 235
Boyle, Robert, vi, xiii, 57, 59, 76, 150
Bradatan, Costica, 194–96
Brahe, Tycho, 59
Bramhall, John, 129
Breuer, Edward, 279
Bryson, Michael, 154
Brooke, John Hedley, 152
Bruno, Giordano, xviii, 24–31, 36, 38, 59, 60, 120, 349
Brown, William P., 192, 194
Buckley, Michael J., 191
Buffon, Georges-Louis Leclerc Comte de, xviii, 81–83, 86, 94, 97–98, 102–3
Burke, Edmund, xxii, 210, 233, 248–59, 274–76, 321
Burns, William, 58
Burton, Robert, 234

Campanella, Tommaso, xviii, 36–39
Campbell, Mary Baine, xvi, 45

Cardan, Jerome, 234
Casaubon, Meric, 234
St. Catherine of Siena, 14
Charleton, Walter, xviii, 74
Chernus, Ira, 268, 269
Chesterton, G. K., 173
Cicero, 8, 111, 226
Cohn-Sherbok, Dan, 268, 269
Clark, Stephen R. L., 194
Clark, Stuart, 61
Cochin, Charles-Nicolas, 243
Coetzee, J. M., 218
Colerus, Johannes, 337
Collins, Jeffrey R., 137
Comenius, John Amos, 149
Comte, Auguste, xxiii, 345–46
Copenhaver, Brian, 60
Copernicus, Nicolas, xv, 24, 60, 85, 120
Copleston, Frederick, xii–xiii
Corneille, Pierre, 235
Crousaz, Jean-Pierre de, 236
Curley, E. M., 183, 184
Cusanus, Nicolaus, 124

Daniel, 268
Dante, Alighieri, 245
Darwin, Charles, xv, 95, 153, 160
Daston, Lorraine, xvi, xviii, xxi, 45, 53, 62–63, 65–66, 70, 211
Davidson, Donald, 220
Decorte, Jos, 195, 207
De Dijn, Herman, v, 212, 220, 349
Dee, John, 54
Delon, Michel, 253
Del Rio, Martin, 55
Derham, William, xviii, 76–77, 79, 85–86, 88–89
Desargues, Gerard, 121
Descartes, René, xii–xiii, xv, xix–xx, xxii, 65, 67–69, 75, 107–18, 122–23, 129, 157–62, 174–86, 219, 226–27, 232, 254–55, 333, 343, 349–50
Dostoevsky, Fyodor, 326–27, 330
Diderot, Denis, 83, 94–96, 352
Dionysius [Pseudo], 12
Dryden, John, 217–18
Dodds, E. R., 249
Dury, John, 149–50

Eagleton, Terry, 215
Eco, Umberto, 338
Empson, William, 153–55
Epicurus, 75, 118, 171
Erasmus, Desiderius, 19–20, 230, 234
Ezekiel, 269

Fabricius, Johann Albert, xviii, 78–79
Fénélon, François, xviii, 75, 91–92, 235
Ficino, Marsilio, xviii, 1–5, 14, 39, 40–41, 60
Finlay, Christopher J., 210, 213, 218–20
Fisher, Philip, xvi
Fontenelle, Bernard le Boyer de, 298
Foucault, Michel, 206–7
Francis, St., 14
Frasca-Spada, Marina, 228
Freud, Sigmund, xxii, 226, 256–58
Freund, Richard A., 268
Fumaroli, Marc, 234

Galilei, Galileo, 51, 57, 162, 197
Gaukroger, Stephen, xiii, xviii
Genghis Khan, 38
Gerard, Alexander, 243
Gervase of Tilbury, 45, 50
Gillespie, Neil C., 144, 150
Gouhier, Henri, 158

Habermas, Jürgen, xv

Harrison, Peter, 58
Hartlib, Samuel, 149–50
Heidegger, Martin, xix, 127, 140–42
Heitner, Robert R., 280–81
Henry, John, 61
Hepburn, R. W., 256
Heraclitus, 327, 342, 345
Herder, Johann Gottfried, xviii, 81, 84
Heschel, Abraham Joshua, 260
Hieronymus, 12
Hill, Christopher, 149
Hobbes, Thomas, xiii, xix, 127–42, 174, 177, 178, 180, 227, 349
Homer, xxii, 130, 249–52, 277
Huizinga, Johan, xvii,
Hume, David, xi–xii, xxi–xxii, 137–38, 143, 159, 210–232
Hurlbutt, Robert H., 143
Husserl, Edmund, 159, 333
Hutchison, Keith, 61
Huxley, Thomas Henry, 160

Isaiah, 193
Israel, Jonathan, xiii

James, Susan, 219
Jerusalem, Johann Friedrich Wilhelm, 39
Jessop, T. E., 204–5
Johnson, Samuel, 255–56
St. John, 4
Joyce, James, 347

Kant, Immanuel, xxii, 74, 76–77, 80–81, 90, 99–102, 159, 186, 233, 235, 256, 274, 279, 284–308, 316, 325–26, 332–33, 338, 350
Kaplan, Aryeh, 269
Kassell, Lauren, 58, 61
Kearney, H. F., 149
Keats, John, 323
Kenny, Neil, xvi
Kepler, Johannes, 59
Kircher, Athanasius, ix, 48, 50, 68–69
Kolakowski, Leszek, xvi, xx
Korsgaard, Christine, 296
Kristeller, Paul Oskar, 32
Kuhn, Thomas, 70

Lacoste, Jean-Yves, 190
Lamport, F. J., 282
La Tour, George de, ix, 239–41
Le Brun, Pierre, ix, 64
Leeuwenhoeck, Antonie, 79
Leibniz, Gottfried, xii, 77, 79, 91–92, 97, 273–76, 280, 350
Lenowitz, Harris, 267, 269, 271–72
Lesser, Friedrich Christian, xviii, 79
Lessing, Gotthold Ephraim, 274, 280–82
Lewis, C. S., 146, 153, 154
Lindberg, David C., xiii
Locke, John, xii, 74, 94, 127, 343, 352
Longinus, 233–35, 242, 274–76, 278
Lowth, Robert, 274–75, 278
Luria, Isaac, 269–70

Machiavelli, Niccolo, 24
Maimonides, Moses, 263–4, 270
Malebranche, Nicholas, xi, xiii, xx, 157–73, 201, 207, 219, 226
Marcel, Gabriel, 332
Marr, Alexander, xvi, 54, 152
Martinich, A. P., 128, 130
Maupertuis, Pierre Louis Moreau de, xviii, 80–81, 86, 88, 94, 96–97
McGinn, Colin, 160
Medici, Cosimo II De, 51
Meier, Jean-Paul, 273, 276
Mendelssohn, Moses, xxii, 267, 273–83
Méré, Chevalier, 122–23
Mersenne, Marin, 115, 162

Merton, Robert, 57
Michelangelo, ix, 243–47, 258
Milton, John, xx, 143–56, 217–18
Mohammed, 38
Montaigne, Michel de, 230
Montesquieu, Charles Secondat de, xxii, 238, 243–48
More, Henry, 234–35
More, Thomas 19
Moreau, Denis, 161, 172–73
Moses, 11, 12, 266–68, 275

Nadler, Stephen, 158–60
Newton, Isaac, 59, 96, 303, 343
Nietzsche, Friedrich, xxiii, 161, 326–27, 345
Nieuwentyt, Bernard, 75–76, 83, 87–88, 96, 103–4
Norton, David Fate, 230

Paley, William, xviii, 84
Paracelsus, 59
Parmenides, 324
Park, Katharine, xvi, xxi, 45, 53, 62–63, 65–66, 70, 211
Pascal, Blaise, xiii, xix, 119–26, 230, 327–28, 333–34, 351
Patritius, Franciscus, 35
Patrizi, Francesco, xviii, 31–36, 49–50
St. Paul, 12, 74
Pico della Mirandola, Giovanni, xviii, 5–15, 54, 60
Piles, Roger de 236
Piranesi, Giovanni Battista, ix, 256–58
Plato, 1, 3, 13, 33, 40, 102, 119–20, 130, 171, 226, 228, 234, 237, 247, 316, 320, 324–26, 349
Pluche, Abbé, xviii, 70, 72–73, 77, 78
Pompilius, Numa, 38
Pomponazzi, Pietro, xviii, 14–15, 18
Pope, Alexander 249
Porphyry, 9
Porta, Giambattista della, 55–56, 59
Pythagoras, 13, 38

Rapin, René, 235
Ray, John, xviii, 75, 85, 144, 150–52
Régis, Pierre Sylvain, 158
Renault, Laurence, xvi, 109
Reuchlin, John, 20
Rodis-Lewis, Geneviève, 176
Rogers, John, 144, 154, 156
Rousseau, Jean-Jacques, 76, 83

Saint-Pierre, Jacques-Henri Bernardin de, xviii, 83–84, 103
Sartre, Jean-Paul, 337
Scaliger, Jules César, 234
Schelling, F. W. J., 258, 352
Schleiermacher, Friedrich, 190
Schliesser, Eric, 211, 213
Schmaltz, Tad M., 158
Schmitt, Carl, 140, 180, 349
Scholem, Gershom, 263, 268–71
Schopenhauer, Arthur, xxiii, 326, 337, 344
Schott, Gaspar, 51
Shaftesbury, Third Earl of, xxii, 237–41
Shakespeare, William, 43, 277, 280,
Smith, Adam, 211–12, 214, 216–18, 227, 231–32, 350
Sophocles, 277
Spinoza, Baruch, xii, xiii, xvi, xxi, 120, 124, 138, 174–86, 226, 274, 279, 333, 337–38, 349
Sprat, Thomas, 150–51
Stephen, Sir Leslie, 130
Stevin, Simon, 66
Strabo, 33
Strauss, Leo, 138, 349
Suárez, Francisco, xviii, xix, 40–42
Suetonius, 247

Taylor, Charles, xv, xviii, xxii
Thomas, Keith, 61
Thorndike, Lynn, 59
Thucydides, 128
Tillyard, E. M. W., 156
Tirosh-Rothschild, Hava, 265, 272
Toland, John, 94, 96–97, 209
Townsend, Dabney, 213
Trescho, Sebastian Friedrich, xviii, 81, 102
Trismegistus, Hermes, 4, 6
Troyes, Chrétien de, 262

Urmson, J. O., 200–201

Vallée, Gerard, 281
Vico, Giambattista, xxii–xxiii, 237, 241–43, 278, 309–11, 327, 330, 333, 345–47
Virgil, 234, 280
Voltaire, 91, 235, 248

Wallis, John, 129
Ward, Seth, 129
Weber, Max, xviii, 56–57
Webster, Charles, 59, 150
Weiskel, Thomas, 250–51
Westfall, Richard, 150–51
William of Paris, 10
Wolff, Christian, xviii, 77, 97–98, 102–3, 274, 276
Worden, Blair, 156
Wordsworth, William, 277, 314–15, 322, 344
Wren, Christopher, 150, 151

Yack, Bernard, 140
Yassif, Eli, 266
Yates, Frances, 59–61
Yeats, William Butler, 343, 345

Zoroaster, ix, 4, 53
Zvi, Sabatai, 267, 270–72, 283

Index of Subjects

Alchemy, 58, 62
Apology (apologetics), 76, 189–209
Astrology, 16, 36, 55, 62, 285

Bible (Scripture), 14–15, 41, 58, 62, 82, 131–33, 136–37, 153, 156, 191–97, 206, 209, 265, 267, 282. *See* also Torah
Body, 4, 26, 48, 76, 81, 96, 113, 133, 146, 158–66, 172–73, 175, 212, 235, 239, 244, 246, 249–51, 255, 268

Cause (causation) xi, 1–2, 32, 41–42, 45–47, 66, 69, 80, 87, 90, 97, 99–101, 131–32, 135, 144, 146, 153, 158–59, 181, 213, 219–22, 249, 253, 291, 293, 301, 332
Christianity (Christian), 1, 9–15, 19–24, 29, 36, 38, 73–74, 76, 78, 127, 137, 139–41, 149, 189–90, 197–98, 200, 208–9, 237, 241, 260, 262–64, 266, 282
Cosmos, 15–6, 24, 70, 148–49, 156, 195, 208, 269
Cosmology, 2, 4–5, 85–87, 156
Curiosity (*Curiositas*), xii, xvii, xx, xxii, xxiii, 42, 48, 51, 55, 67, 110, 124, 127, 133–42, 161, 168–71, 211–12, 215–17, 220, 224, 227–32, 248, 254, 276, 311–13, 316, 318, 321, 323, 327, 329–47

Disenchantment, xviii, xix, xxii, 56–65, 70

Empiricism, xii, 3, 42, 56, 212
Enchantment, xviii, xix, 56–63, 65, 70
Enthusiasm, 59, 68, 128, 233, 234–43, 252, 255, 259
Evil, 10, 20, 30, 74–75, 90–94, 132, 271, 280, 347

Fear, 40–41, 51, 57, 67, 70, 77, 126–27, 137, 149, 164, 174, 181, 220, 225–26, 251–52, 254, 256, 261, 307, 332
Fideistic (fideism), 14, 19
Freedom, 103, 116, 165–66, 177, 179, 186, 238, 247, 289, 291–94, 301–2, 350

God (godhead), xi, xiii, 3–7, 9, 11–12, 18, 20–22, 26, 28–30, 34, 37–39, 46–47, 49, 51–52, 54, 67, 72–76, 78–82, 87, 89, 91–93, 98–103, 115–16, 123, 131–37, 143, 147–52, 155–56, 158–68, 170, 172, 177, 190–98, 200–208, 235, 239–41, 250, 261, 269, 270, 274–76, 279–82, 289, 291, 310, 329, 347

Hermetism, 60
Hylozoism, 94–97

Idolatry, xx, 23, 55, 156

Imagination, 49, 61, 68, 111, 133, 167, 170, 173, 179, 180–85, 198, 201, 212–14, 217–20, 223, 225–28, 235, 238, 250, 256–57, 261, 267–68, 306
Infinite (infinitude), 24–27, 34, 60, 77, 85, 87, 89–90, 103, 115, 120–26, 147, 167–69, 206, 216–17, 219, 225, 281–82

Kabbalah, 9, 11, 12, 20, 55, 263, 269, 270, 271–12, 283

Liber mundi, 194–97, 202, 203
Love, 7–8, 11, 134–35, 163–67, 174–75, 178, 179, 181–83, 195, 212, 214, 215, 220, 223–32, 240, 277, 315, 317, 319, 320, 328, 336, 337–38, 341–42

Magic, xviii, 9–11, 13–23, 30, 36, 38–39, 43–45, 49, 51–63, 69
Mathematics, 16, 44, 58, 61–62, 121, 157, 228
Matter, 4, 80–81, 94–96, 99–101, 144, 146, 153, 160
Mechanistic, 39, 192, 193, 196, 198, 203
Metaphysics, xxi, 39, 79, 158, 161, 191, 197–98, 207, 241, 285, 287–89, 291, 294, 301, 302, 316, 325, 345
Miracle(s), 1–5, 11, 14–15, 17, 20–24, 32, 36–39, 42, 46, 49, 52, 54, 120, 132, 136–38, 141–42, 177, 180, 205–6, 208, 279, 312, 332
Motion, 75, 80, 88, 95, 96, 97, 103, 133, 135, 149, 160, 164–65, 167, 204, 321

Naturalization, xvi, 57, 63

Occasionalism, 158–61
Occult, 9, 10, 13, 15–17, 45, 52–53, 61, 253
Original sin, xx, 12, 162–63, 165, 171

Passions, xix, 67, 114, 118, 163–64, 170–78, 183, 186, 198, 211–15, 219–32, 248, 255
Physico-theology, 72–104
Politics, 57, 59, 129, 138, 154, 215, 218, 230–31, 237
Preternatural, 52, 70
Prophecy, 37, 49, 54, 267
Providence, 20, 84, 87, 92, 201–5, 208, 240, 310, 346

Rationalism, xii, 18–19, 56, 60, 235, 274, 327
Revelation, 78, 86, 102, 140–41, 195, 202, 235, 279

Secular/Secularization, xiii, xv, xvii, xxi, 9, 14, 19, 24, 57, 140, 174, 186, 262–63, 266, 273
Soul, xxi, 2–5, 13, 15–16, 34, 41, 67, 82, 102–3, 111, 114–18, 139, 149, 160–67, 170–72, 198, 203, 212–13, 216–17, 221–22, 225–28, 232, 241, 247–48, 255, 276–77, 280, 289, 311, 316, 327, 334, 337, 347
State, 129
Sublime, xxii, 10, 107, 168, 236, 241–43, 247, 248, 252–58, 273–76, 278, 281–82, 286, 301–8, 319
Substance, 5, 29, 52–3, 75, 82, 88, 90, 93, 96–7, 198–200, 202, 215, 289
Supernatural, 4, 10, 16, 17, 19–20, 24, 36–39, 46, 47, 52, 54, 55,

73, 89, 102–4, 113, 115, 133, 138, 159, 208, 269, 312

Torah, 262–33, 266, 268, 275, 279
Transcendence (transcendent), xx, 3, 27, 89, 95, 101, 170, 208, 261, 284